Soil Physical Conditions and Plant Roots

Authors

Jan Gliński, D.Sc., Ph.D.

and

Jerzy Lipiec, D.Sc., Ph.D.

Institute of Agrophysics
Polish Academy of Sciences
Lublin, Poland

CRC Press
Taylor & Francis Group
Boca Raton London New York

CRC Press is an imprint of the
Taylor & Francis Group, an **informa** business

First published 1990 by CRC Press
Taylor & Francis Group
6000 Broken Sound Parkway NW, Suite 300
Boca Raton, FL 33487-2742

Reissued 2018 by CRC Press

Library of Congress Cataloging in Publication Data

Gliński, Jan.
Soil physical conditions and plant roots/authors, Jan Gliński and Jerzy Lipiec
p. cm.
Includes bibliographies and index.
ISBN 0-8493-6498-1
1. Plant-soil relationships. 2. Crops and soils. 3. Roots (Botany) 4. Soil physics. I. Lipiec, Jerzy, 1946- . II. Title.
S596.7.G57 1990
631.4'3 — dc20 89-7280

A Library of Congress record exists under LC control number: 89007280

Publisher's Note
The publisher has gone to great lengths to ensure the quality of this reprint but points out that some imperfections in the original copies may be apparent.

Disclaimer
The publisher has made every effort to trace copyright holders and welcomes correspondence from those they have been unable to contact.

ISBN 13: 978-1-315-89760-8 (hbk)
ISBN 13: 978-1-351-07670-8 (ebk)

Visit the Taylor & Francis Web site at http://www.taylorandfrancis.com and the
CRC Press Web site at http://www.crcpress.com

FOREWORD

For a long time, information on the relations between physical properties of soil and plant roots has been accumulating. Recently, the increase in research activity and the application of modern methods of root studying have also brought about significant changes in views on various phenomena. This led to the necessity to summarize modern views on this subject in order to suggest reasonable directions for further meaningful research.

This monograph attempts to explain the importance of physical properties of soil by showing how they affect root growth and functions; and on the other hand, how roots themselves change their environment. Emphasis is placed on the interactive effects of soil physical factors. The need for this monograph has now become more urgent because of an increasing use of heavy machinery traffic on the soil and narrow crop rotation in modern agriculture, resulting in drastic changes in soil physical characteristics that in turn affect root growth and functions. Basically, the present monograph includes the entire field of relations between soil physical properties and plant roots which have been intensively investigated recently. An attempt has been made to analyze the possibilities of the root system's modification by both soil and plant management.

The great number of publications has hindered an all-inclusive integration. We apologize to all scientists whose valuable contributions have not been included or inadequately treated in this monograph.

This monograph reflects our collaboration with many scientists, with whom we discussed the problems, and who broadened our opinions and encouraged us to write this book. We particularly wish to thank Professor H. U. Taylor for his helpful suggestions. We are indebted to all the authors who have sent us reprints of their papers.

PREFACE

Plant production is closely related to soil environment, influencing both root growth and function. Root penetration and distribution in soil depend on its chemical, physical, and biological properties. Root-soil relations are very complex and dynamic in the rhizosphere. Root environments are rapidly changing especially in respect to soil physical conditions such as mechanical properties, moisture, aeration, and temperature.

Soil and plants are very complicated objects from the point of view of their composition and functions. The relationships between them are much more complicated.

Soil-plant system relationships have been analyzed in a number of valuable reviews.[40,134,188,277,683,842,1023,1024,1065,1180,1190] These books, except those by Arkin and Taylor[40] and Russell,[1024] devote little attention to the physical aspects of the problem.

On the other hand, recent years have brought an increasing number of papers concerned with the effect of physical properties of soil on the growth and function of root systems. These, however, are scattered among a number of publications, and to our knowledge, there is no book presenting such relations comprehensively.

Data on the interrelations between root systems and soil physical characteristics constitute an indispensable element in the modeling of root growth and processes in natural environment, and especially in the prediction of water and nutrient uptake by plant roots.

The need for designing such models arises from the fact that investigations of root systems are very time- and labor-consuming.

Among papers on soil-root relationships, most are concerned with problems of soil aeration, water, and compaction in relation to root growth and development. These, of course, are the main soil factors influencing plants.

For these reasons, writing such a specialized monograph on soil physical conditions and root-system relations would seem fully justified.

The book is addressed to research workers and advanced students in soil and plant sciences and may also be of interest to agronomists, ecologists, and related specialists.

Jan Gliński
Jerzy Lipiec

THE AUTHORS

Jan Gliński, D.Sc., Ph.D., is Professor of Agricultural Sciences, Pedologist, and is a Director of the Institute of Agrophysics of the Polish Academy of Sciences in Lublin.

Earlier he worked at the Soil Science Department of the Agricultural University in Lublin, where he carried out investigations on soil chemistry and especially on trace elements in soils and plants and instrumental methods of their determination.

In the Institute of Agrophysics, Professor Gliński leads the group that works on electrical and aeration properties of soils. He has published more than 15 papers on this subject alone from a total of 150 of his scientific publications. His last book with W. Stepniewski, *Soil Aeration and Its Role for Plants,* was published by CRC Press in 1985.

As a member of the International Society of Soil Science, he participated in the Congresses of this Society in Madison (1960), Bucharest (1964), Moscow (1974), Edmonton (1978), and Hamburg (1986) where he was elected Vice-Chairman of Commission I (Soil Physics) of the Society.

Jerzy Lipiec, D.Sc., Ph.D., is Associate Professor of Agricultural Sciences, Pedologist, and is an employee of the Institute of Agrophysics of the Polish Academy of Sciences in Lublin and a Head of the Laboratory of Basic Problems of Soil Improvement in this Institute. Earlier he studied agronomy at the Agricultural University in Lublin, where he earned his B.Sc. in 1969. He earned his Ph.D. in 1976 and D.Sc. in 1984.

The main subject of his research is the study of soil improvement methods and the relation between soil physical properties and crop growth. He has authored about 35 papers on those problems.

DEDICATION

This book is dedicated to Professor Bohdan Dobrzanski (1909—1987), founder and long-term director of the Institute of Agrophysics of the Polish Academy of Sciences in Lublin.

TABLE OF CONTENTS

Chapter 1
SOIL PHYSICAL FACTORS INFLUENCING ROOT GROWTH 1
1.0 Introduction 1
1.1 Texture 1
1.2 Surface Area 2
1.3 Structure 5
1.4 Pore Space 7
1.5 Density and Bulk Density 8
1.6 Water 8
1.7 Temperature 13
1.8 Aeration 15
1.9 Consistency, Compression, and Compaction. 21
1.10 Interrelation of Soil Physical Characteristics. 29

Chapter 2
ROOTS AND THEIR FUNCTIONS 33
2.0 Morphological and Anatomical Characteristics of Root Systems 33
2.0.1 Root Morphology 33
2.0.2 Root Anatomy 34
2.0.3 Root Growth 35
2.0.4 Mechanical Properties of Roots 39
2.1 Root Metabolism 40
2.2 Root Exudation 40
2.3 Sorption and Transport of Water and Nutrients by Roots 41
2.3.1 Sorption and Transport of Water 41
2.3.1.1 Absorptive Capacity of Roots 41
2.3.1.1.1 Age of Roots 41
2.3.1.1.2 Water Uptake Rate and Root Permeability (Hydraulic Conductance) 42
2.3.1.2 Water Absorption 46
2.3.1.2.1 Osmotic (Active) Absorption of Water 46
2.3.1.2.2 Passive Absorption of Water 47
2.3.1.3 Water Movement in Roots 47
2.3.1.3.1 Radial Water Movement 47
2.3.1.3.2 Longitudinal (Axial) Water Movement 49
2.3.1.4 Translocation of Water in the Soil by Roots 49
2.3.2 Sorption and Transport of Nutrients 49
2.4 Relations Between Roots and Above-Ground Parts 53

Chapter 3
SOIL-ROOT INTERFACE 57
3.0 Introduction 57
3.1 Physical Aspects 57
3.1.1 Root-Soil Contact Area 57
3.1.2 Root-Soil Water Relations 58
3.2 Physicochemical and Chemical Aspects 58
3.3 Biological Aspects 60

Chapter 4

INFLUENCE OF SOIL PHYSICAL FACTORS ON ROOT GROWTH AND FUNCTIONS

4.0 Texture 65
4.1 Structure 66
4.2 Mechanical Impedance 75
4.2.1 Artificial Media 75
4.2.2 Soil Conditions 79
4.2.2.1 Root Growth 79
4.2.2.2 Morphology of Roots 82
4.2.2.2.1 Root Branching 82
4.2.2.2.2 Root Hairs 83
4.2.2.2.3 Root Distribution 84
4.2.2.3 Root Anatomy 84
4.2.2.4 Root Functions 85
4.2.2.4.1 Water Uptake 85
4.2.2.4.2 Nutrient Uptake and Chemical Composition of Roots 87
4.2.2.4.3 Photosynthate Requirements and Oxygen Supply and Uptake 89
4.2.2.4.4 Symbiotic Fungi, Nodulation, Biological Activity, and Root Diseases 90
4.2.2.5 Root Growth and Penetrometer Pressure 91
4.2.3 Modeling of Root Growth in Relation to Mechanical Impedance 96
4.3 Water 97
4.3.1 Root Growth 97
4.3.2 Root Morphology 97
4.3.3 Root Porosity 97
4.3.4 Root Functions 98
4.3.5 Drought Tolerance 100
4.3.6 Irrigation 104
4.3.7 Waterlogging 107
4.3.8 Drainage 108
4.3.9 Root Diseases 108
4.3.10 Modeling of Water Uptake by Roots 109
4.4 Oxygen 112
4.4.1 Diffusion of Oxygen To and Within Plant Roots 112
4 4.1.1 Radial Diffusion 112
4.4.1.2 Longitudinal Diffusion 119
4.4.2 Root Respiration and Factors Affecting It 122
4.4.2.1 External Factors 123
4.4.2.1.1 Oxygen 123
4.4.2.1.2 Carbon Dioxide 127
4.4.2.1.3 NPK Fertilization 127
4.4.2.1.4 Ethylene 127
4.4.2.1.5 Temperature 127
4.4.2.1.6 Soil Mechanical Impedance 129
4.4.2.1.7 Soil Salinity 129
4.4.2.1.8 Redox Potential and Soil Acidity 129
4.4.2.2 Internal Factors 130
4.4.2.3 Respiration of Root Nodules 133
4.4.3 Metabolic Activity and Energy Charge of Roots Under Anoxia 133
4.4.3.1 Carbon Dioxide Production 133
4.4.3.2 Ethanol Synthesis and Loss from Roots 133
4.4.3.3 Ethylene 136

4.4.3.4 Other Anoxic Metabolites 137
4.4.3.5 Microbial Colonization and Activity 138
4.4.3.6 Energy Charge 138
4.4.4 Oxygen Deficit and Root Growth and Water Uptake 138
4.4.4.1 Root Growth Rate 138
4.4.4.2 Root Growth Force and Pressure 141
4.4.4.3 Water Uptake 141
4.4.5 Nutrient Content and Uptake by Roots Under Oxygen Stress 142
4.4.6 Adaptation of Plant Roots to Atmospheric Stress in the Soil 143
4.4.6.1 Anatomical Adaptation 144
4.4.6.2 Morphological Adaptation 150
4.4.6.3 Shallow Rooting 151
4.4.6.4 Metabolic and Physiological Adaptation 151
4.5 Temperature 152
4.5.1 Root Growth 152
4.5.2 Water Uptake 155
4.5.3 Nutrient Uptake 156
4.5.3.1 Nitrogen 157
4.5.3.2 Potassium 157
4.5.3.3 Phosphorus 157
4.5.3.4 Magnesium and Calcium 157
4.5.4 Chemical Composition of Roots 158
4.5.5 Root Nodulation and Nitrogen Fixation 158
4.5.6 Root Rot Diseases 160
4.5.7 Root Zone Heating 160
4.5.8 High and Low Temperatures 160
4.6 Light Penetration Through Soil 161
4.7 Interactive Effects of Soil Physical Properties on Roots 161
4.7.1 Soil Mechanical Impedance and Texture 161
4.7.2 Soil Mechanical Impedance, Moisture, and Aeration 164
4.7.3 Soil Mechanical Impedance, Pore Size Distribution, and Aggregate Size Distribution 167
4.7.4 Soil Mechanical Impedance, Temperature, and Moisture 170
4.7.5 Modeling of Interactive Effects 171

Chapter 5
INFLUENCE OF ROOT SYSTEMS ON SOIL CONDITIONS 173
5.0 Introduction 173
5.1 Structural Stability 173
5.2 Bulk Density and Soil Strength 174
5.3 Porosity 176
5.4 Mechanical and Mineralogical Composition 176
5.5 Water and Ions 178
5.6 Biological Activity 178

Chapter 6
MODIFICATION OF ROOT DEVELOPMENT 179
6.0 Introduction 179
6.1 Conventional and Reduced Tillage 179
6.2 Deep Loosening 181
6.3 Tillage and Root Diseases 184

6.4 Soil Management ... 186
6.5 Sowing Date ... 186
6.6 Soil Moisture Regulatory Practices ... 187
6.7 Mulching ... 187
6.8 Chemical Treatments and Plant Hormones ... 187
6.9 Plant Breeding and Selection ... 189
6.10 Fertilizer Application ... 190
6.11 Optimum Root Size ... 190

Chapter 7
METHODS FOR STUDYING ROOTS ... 193
7.0 Introduction ... 193
7.1 Root Quantity and Distribution ... 193
7.1.1 Root Observation Techniques ... 193
7.1.2 Nuclear Magnetic Resonance Imaging (NMR) ... 195
7.1.3 Radiography ... 195
7.1.4 Excavation Techniques ... 195
7.1.5 Root Counting Techniques ... 195
7.1.6 Prediction of Root Growth Based on the Relationship Within the Root System ... 196
7.2 Root Water Potential ... 196
7.3 Root Cation Exchange Capacity (RCEC) ... 196
7.4 Water Use by Roots ... 196
7.5 Root Processes and Growth ... 197
7.6 Separation of Roots from the Soil ... 197

Chapter 8
A LOOK AHEAD ... 199

REFERENCES ... 201

INDEX ... 247

Chapter 1

SOIL PHYSICAL FACTORS INFLUENCING ROOT GROWTH

1.0 INTRODUCTION

Soil is defined as the unconsolidated natural material on the surface of the earth that has been subjected to various environmental factors. The main function of soil as a medium for plant growth is to provide mechanical support for the plant and to cover plant requirements concerning nutrients, water, heat, and air.

From the physical point of view, soil is a three-phase system: solid, liquid, and gas (Figures 1 and 2). About 50% of the total volume of the soil commonly, consists of its solid fabric-mineral matter derived from the parent rocks, with small quantities of relatively stable organic matter localized mainly in the surface layers of the soil. This solid fabric is penetrated by irregularly shaped pores, or voids, filled either with air or with water.

The main aspect of soil physics for plant productivity is to preserve suitable proportions between solid, liquid, and gaseous phases. These proportions are found to be optimal only in some soils and when climatic conditions and human activity are proper. An example of changes of solid, liquid, and gas proportions in loess soil under human activity is shown in Figure 3.

1.1 TEXTURE

The solid phase of soil consists of discrete units, called elementary (primary) particles or granules. These particles may vary widely in size, shape, and composition.

The spectrum of particle sizes is divided into ranges, called soil size fractions or soil separates on the basis of their equivalent diameters. There are several classification systems in use which differ only slightly. The one recognized by the International Society of Soil Science (ISSS) consists of the following distinctions:

Skeleton particles	ϕ >2,000 μm
Stones	ϕ >20,000 μm
Gravel	ϕ 20,000—2,000 μm
Fine earth	ϕ <2,000 μm
Coarse sand	ϕ 2,000—20 μm
Fine sand	ϕ 200—20 μm
Silt	ϕ 20—2 μm
Clay	ϕ <2 μm

The above soil size fractions are often divided into subgroups. The most significant of these is the fine or colloidal clay with equivalent diameters of 1 to 0.2 μm.

Because of their various degrees of fineness, chemical composition, and structural makeup, soil size fractions play a fundamental role in the formation of physical, physicochemical, and chemical properties of soils. Coarse fractions (sand particles) are of comparatively large size and have a small surface area, 0.001 to 0.01 m^2g^{-1}, as compared to those of silt or clay particles. They are mainly composed of quartz grains. For this reason, their chemical activity is almost negligible, but they play a role in physical soil properties through increasing the size of spaces between particles, thus facilitating movement of air and drainage water. The silt fraction has a larger surface area (about 0.05 $m^2\ g^{-1}$) and a faster weathering rate and release of soluble nutrients for plants than sand. Silt particles in soil show a tendency to stick together or to adhere to other particles, influencing crust formation, after wetting.

Most chemical and physical reactions in the soil are connected with the fine clay fraction

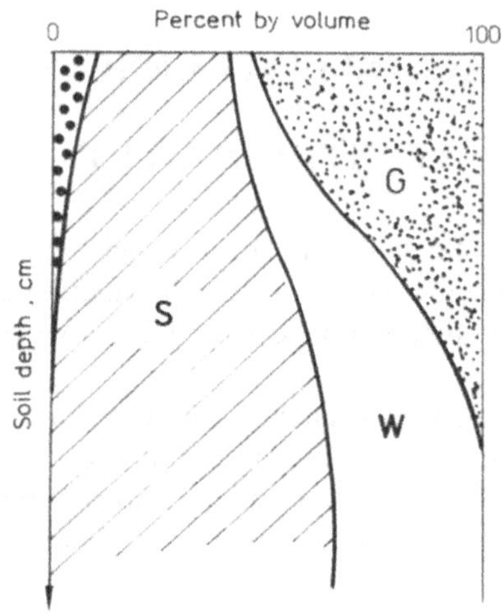

FIGURE 1. Three-phase soil system: S, solid mineral part of soil; O, organic matter; W, water; and A, air.

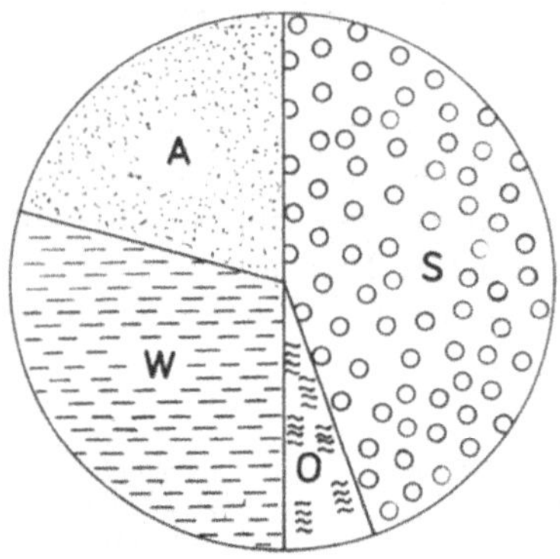

FIGURE 2. An example of the distribution of phases within soil profile: S, solid phase; W, liquid phase; and A, gaseous phase.

which has a surface area of about 800 $m^2 g^{-1}$. The clay fraction in most soils is composed of minerals that differ greatly in composition and properties as compared to sand and silt. They are of crystalline (clay minerals and crystallized oxides or hydroxides of silicium, aluminum, iron, and manganese) and noncrystalline (organic matter and oxydes of silicium, aluminum, and iron having different degrees of hydration) origin. Because of their colloidal nature and surface charge influencing high sorption capacity (Table 1), they can store water and certain available nutrients.

A soil is never composed of only one size fraction. Taking into account weight percentages of sand, silt, and clay, textural soil classes have been distinguished (Figure 4).

1.2 SURFACE AREA

Almost all soil properties are to a large extent dependent on the level of disintegration and

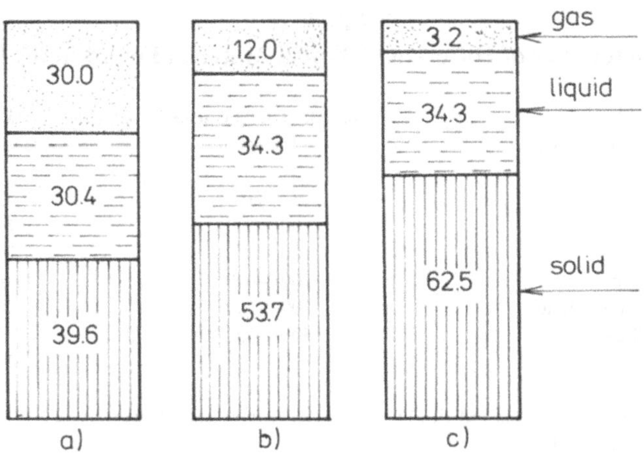

FIGURE 3. Changes in three-phase system of a loess soil. (a) After plowing; (b) after wheat harvest; (c) after compaction with tractor. (Modified from Turski, R., Domzal, H., Borowiec, J., Flis-Bujak, M., and Misztal, M., *Soil Science, Practicum for Students of Agronomy*, Agriculture University, Lublin, Poland, 1977, 34.)

structure of soil particles. The extent of the surface of a disperse soil system is expressed in terms of specific surface area, which signifies the square meters of surface per gram of disperse phase ($m^2\ g^{-1}$).

According to the method of determination used and the kind of soil material under examination, three surface areas in soil, external, total, and internal, can be distinguished. External surface area (Sa) is a surface which is determined with an apolar adsorbate, e.g., nitrogen, argon, and CCl_4. It is equal to the geometric surface calculated from the granulometric composition and to the surface of walls of microcapillaries of dispersed soil particles. Total surface area (Sp) is a surface measured with a polar absorbate, e.g., water vapor and ethylene glycol. Internal surface area is found from the difference between total and external surface area and is closely connected with the internal structure of some clay minerals.

Methods of determination of the surface area are based on the isotherm BET equation:

$$\frac{P}{V(P_o - P)} = \frac{1}{V_m C} + \frac{C - 1}{V_m C} \cdot \frac{P}{P_o} \tag{1}$$

where V is the total volume of gas adsorbed; V_m is the volume of gas adsorbed when the entire adsorbent surface is covered with a complete monomolecular layer; P is the pressure of the gas; P_o is the saturation pressure of the gas; and C is a constant involving the heat of adsorption in the first layer and the heat of condensation. The surface area is calculated from the following formula:

$$S = 0.269\sigma_m V_m \tag{2}$$

where σ_m is the surface occupied by a particle of adsorbate in a monolayer (for nitrogen, 16.2^{-20} m^2; and for water, 12.8×10^{-20} m^2).

Investigations carried out by Stawinski[1114] showed that soil surface area is not only a function of the disintegration of mineral components of soil and their composition, but also of the content

TABLE 1
Characteristics of Different Colloidal Material Found in the Soil

Colloidal particles	Specific surface area (m^2/g)	C.E.C. (meq/100 g)
Kaolinite	5—20	3—15
Illite	100—200	10—40
Mica	100—200	20—40
Montmorillonite	700—800	80—100
Sepiolite — atta/pulgite	100—200	20—30
Alumo — silicates	300—600	± 60
Humus		150—250

Data from Buckman, H. C. and Brady, N. S., *The Nature and Properties of Soils*, 7th ed., Macmillan, London, 1969.

and the quality of organic matter. It is a parameter which combines a few soil properties and decisively influences almost all of the physicochemical processes occurring on the surface of soil particles.

The numerical relationships between the total surface area for numerous Polish soils and other soil properties are as follows:

$$S = 1.12x + 4.14 \tag{3}$$

where x is the colloidal fraction content at a humus content of 1 to 2%;

$$S = 1.88x' + 6.54 \tag{4}$$

where x′ is the colloidal fraction content at a humus content of over 2%;

$$T = (S - 0.2) : 1.36 \tag{5}$$

where T is the total sorptive capacity of the soil;

$$W = 0.5285 + 1.37 \tag{6}$$

where W is water unavailable to plants;

$$S = 1.7\ (K_h : K_f)\ \%\ \text{O.M.} \tag{7}$$

where K_h denotes humic acids, K_f indicates fulvic acids, and O.M. is organic matter; and

$$(S_p - Sa)\ k = \%\ \text{Montmorillonite} \tag{8}$$

where k is a proportionality coefficient.

For the plow layers of mineral Polish soils, the surface area is in the range from 1 to 500 $m^2\ g^{-1}$, and its mean value for sandy soils is 35 $m^2 g^{-1}$, for silty soils is 72 $m^2 g^{-1}$, and for loamy and clay soils is 120 $m^2\ g^{-1}$.[315]

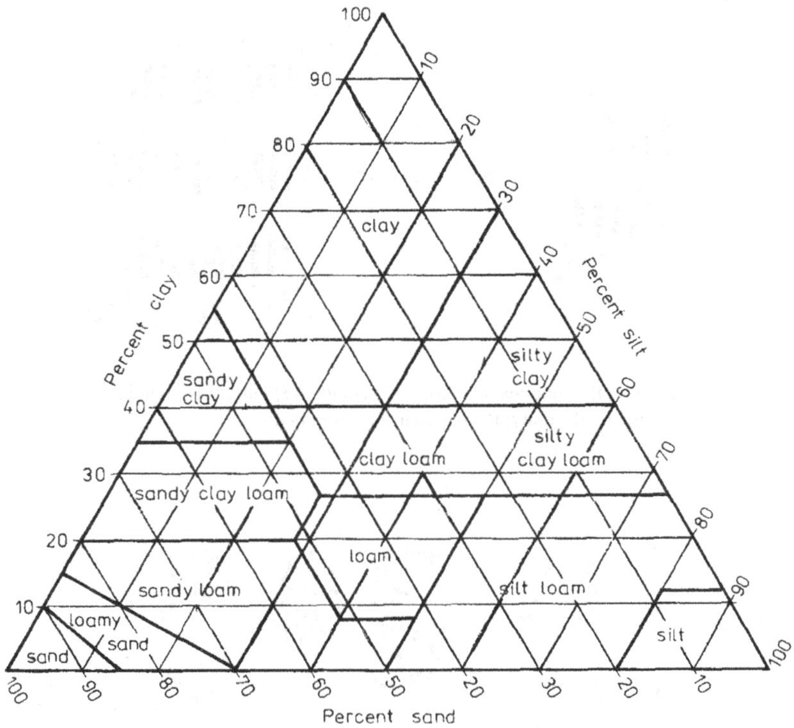

FIGURE 4. Textural soil classes.

1.3 STRUCTURE

Soil structure is the spatial arrangement of the elementary soil particles into compound particles, aggregates, or peds and the stability of these units. The main structural types are shown in Figure 5. Soil aggregates can be subdivided into micro- and macroaggregates. The former are called clay domains when they consist of clay particles (Figure 6). They have a diameter from 250 to 500 μm. Macroaggregates have a diameter up to 10 mm, but if larger, they are called clods. Forces which keep elementary particles together are

1. Intermolecular forces (Van der Waals forces and H bonds)
2. Capillary forces resulting from the existence of a meniscus
3. Chemical forces including the influence of the adsorbed counterions

During the formation of macroaggregates, the following phenomena exist:

1. Chemical stabilization or cementation
2. Drying out effect which gives rise to capillary forces
3. Bonds made of a colloidal substance (mineral and organic) and also microbial gums between coarse particles and/or microaggregates
4. Tension and pressure exerted mechanically on the soil so that locally sites having greater density will occur, which, afterwards, will form the clods when the soil is broken up again

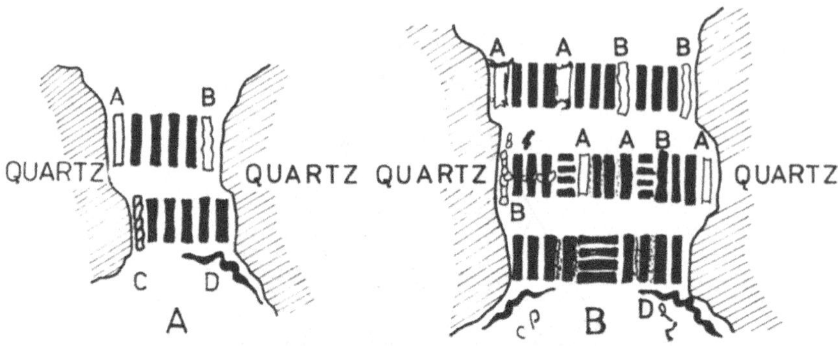

FIGURE 5. Various soil structures: (1) dusty; (2) cloddy; (3) nutty; (4) granular; (5) columnar; (6) blocky; (7) prismatic; (8) platy; (9) lamellar; (10) thin platy.

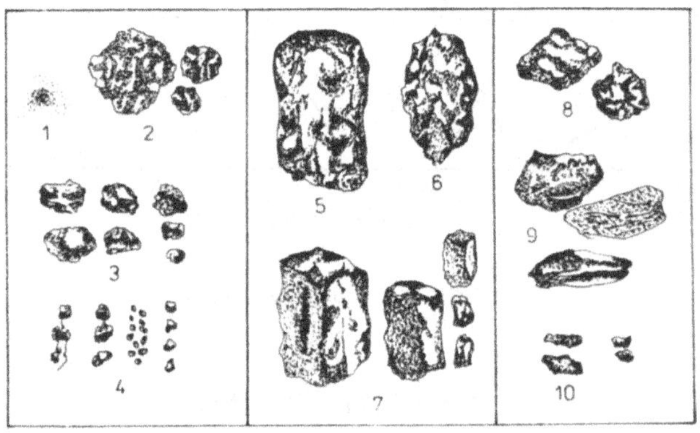

FIGURE 6. Schematic presentation of different possible linkages between clay domains, quartz particles, and organic matter in an aggregate: (A) weak linkage due to water meniscus; (B) strong linkage due to humus or sesquioxides; (C) moderately strong linkage due to a nonionized polymer; and (D) strong linkage due to a bitumen micel or ionized polymer. (Modified from Emerson, W. W., *J. Soil Sci.*, 5, 233, 1954.)

Soil aggregates are more or less stable, i.e., resistant to compression or to the action of water. They are in a stage of continual changes under the influence of wetting and drying, freezing and thawing, tillage, and biological activity.

Soil structure is an important property because it influences not only physical conditions (aeration, heat and water economy, and soil mechanical impedance), but also microbial activity, availability of plant nutrients, and decomposition of soil organic matter. To find a relationship between soil structure and soil fertility and crop yields, various soil structure indicators are used,[294] e.g., numerical aggregation index after drying (LWA_s), numerical aggregation index after wet sieving (LWA_m), water stability index (WW), water stability coefficient (K), soil structure index (W), coefficient of soil structure (W_s), index of soil dispersion (W_r) and index of optimum agricultural aggregation (W_a).

These indicators are obtained by measuring aggregate composition by means of sieving the air-dried soil samples using a set with meshes of, respectively, 7, 5, 3, 1, 0.5, and 0.25 mm in diameter and water stability of the individual aggregate fraction according to a modified Baksheev method.[316]

1.4 PORE SPACE

Between soil particles arranged singly or in aggregates, there is a system of pore space on which plant roots, microorganisms, and soil animals depend for the storage and movement of water and air. The spatial arrangement of the solid particles and voids is called soil fabric.

The volume percentage of the total bulk of soil not occupied by solid particles is commonly called porosity (E). It is defined as:

$$E = (V_l + V_g):V_t \tag{9}$$

where V_l is a volume of liquid, V_g is a volume of gas, and V_t is a total or bulk volume of a quantity of soil. The porosity of mineral soils is in the range of 0.28 to 0.75 $m^3\,m^{-3}$ and that of organic soils ranges from 0.55 to 0.94 m^3m^{-3}. Optimum porosity for plants is usually considered at 0.50 $m^3\,m^{-3}$. Soil porosity consists of gas-filled or air-filled porosity (E_g) and liquid-filled porosity (E_l), which is equal to the volume fraction of liquid (q). For optimum plant growth, the aeration porosity should be at least 0.10 to 0.12 m^3m^{-3} within 2 or 3 d after irrigation or heavy rainfall.[678]

Storage availability and transport of soil solution and soil air are not so much dependent on the porosity as on how the total pore space is partitioned, i.e., on the so-called pore size distribution. Soil pores differ in size and shape as a result of textural and structural arrangement. Based on the diameter at the narrowest point, pores may be classified as follows:

macropores >100 μm
mesopores 30—100 μm
micropores <30 μm

Distribution of pore space according to the size of a pore is usually measured from the curve relating water content to matric potential (compare to Figure 8) and directly with the use of a mercury porosimeter (compare to Figure 23). The quantity of pores between 0.2 and 60 μm is the major factor which determines the reserves of water which are available to plants. These pores hold water against gravity. The larger pores freely drain water under gravity and the fine pores retain water as inaccessible to plants. For root growth, which in the majority exceeds 60 μm in diameter, a sufficient number of continuous pores which they can enter freely is important. Since roots represent only a small fraction of the total soil volume (usually less than 5%), they find necessary conditions for an easy penetration, and thus the characteristics of pore space in soil are more important for water and air movement than for root penetration.

In soil undergoing change of volume as a result of compaction, swelling, or shrinking, its total volume is not constant and there are then advanatages in using the so-called void ratio (or void index) instead of porosity. It is defined as:

$$e = (V_l + V_g):V_s \tag{10}$$

where e is the void ratio and V_s is the volume of the solid. The relation between E and e is

TABLE 2
Examples of Porosity and Void Ratios for Some Soils

Soil	Porosity (E)	Void ratio (e)
Surface soil of wet clay	0.58	1.37
Surface soil of loamy texture	0.52	1.07
Subsoil of sandy texture	0.39	0.65
Sandy loam soil compacted by heavy traffic	0.28	0.39

Data from Marshall, T. J. and Holmes, J. W., in *Soil Physics*, Cambridge University Press, London, 1979, 10.

$$E = e\,(1 + e) \tag{11}$$

or

$$e = E\,(1 - E) \tag{12}$$

Values of e may vary in a wide range from 0.39 to 1.37 which is shown in Table 2.

1.5 DENSITY AND BULK DENSITY

The density of the solid phase of the soil is defined as a ratio of mass of solid phase of dry soil to its volume, expressed in Megagrams per cubic meter ($Mg\ m^{-3}$). This property varies for mineral matter between 2.60 and 2.85 $Mg\ m^{-3}$ (2.70 $Mg\ m^{-3}$ as average). For the soil organic matter, it is from 1.20 to 1.50 $Mg\ m^{-3}$ (on the average 1.40 $Mg\ m^{-3}$). Except for precise scientific work, 2.65 $Mg\ m^{-3}$ can be taken as the average density of the solid phase of mineral soils (when the organic matter content does not exceed 20%).

For the soil, the term of bulk density as the volume-weighed mean density of the three phases is often used. Average bulk densities found in surface soils are 1.60 $Mg\ m^3$ for clay and loamy soils and 0.25 $Mg\ m^{-3}$ for peat soils. Dry bulk density is needed for the conversion of gravimetric to volumetric water content. Wet bulk density together with dry bulk density is used to estimate the volume fraction of air-filled pore space.

1.6 WATER

Water content in soil is the amount of water lost from the soil upon drying to the content of weight at 105°C, expressed either as the weight of water per unit weight of dry soil or as the volume of water per unit bulk volume of soil. It is worthwhile to mention that soil water is not pure water, but it contains a wide range of solutes. The water in solution is characterized by a lower vapor pressure, freezing point, and chemical potential and a higher boiling point and osmotic pressure in comparison to pure water.

Water from precipitation or irrigation is stored in pore spaces between soil particles and aggregates, inside the aggregates, and on the surfaces and interlayers of clay minerals. Storage and retention of water by soil is a result of attractive forces between solid and liquid phases. These forces, named "matric forces", enable the soil to hold water against processes such as gravity, evaporation, uptake by plant roots etc. The terms matric potential (ψ_m), matric suction, and soil moisture tension or pressure are used.

TABLE 3
Matric Potential in Different Units Showing its Order of Magnitude over a Broad Range of Soil Water Conditions

Condition at the quoted potential	Matric potential			
	kPa or J kg^{-1}	m	bar	pF
Saturated or nearly so	-10^{-1}	-10^{-2}	-10^{-3}	0.0
Near field capacity	-10	-1	-10^{-1}	2.0
Near permanent wilting point	-1.5×10^3	-1.5×10^2	-1.5×10	4.2
Air dry at relative vapor pressure of 0.48	-10^5	-10^4	-10^3	6.0
Conversion from a matric potential of h meters	9.8 h	h	0.098 h	log 100 lhl

Data from Marshall, T. J. and Holmes, J. W., in *Soil Physics*, Cambridge University Press, London, 1979, 40.

Matric potential is the equivalent negative pressure in the soil water. It is equal to the equivalent pressure that must be applied to the soil water to bring it to hydraulic equilibrium through a porous permeable wall or membrane, with a pool of water of the same composition. The matric potential is expressed in several units in soil physics. The comparison of these units and their reference to different soil water conditions is given in Table 3.

There are three mechanisms for binding of water to the solid matrix:

1. Adhesion of water molecules to a solid surface by London-Van der Waals forces
2. Capillary binding of water
3. Osmotic binding of water in double layers

The adhesive forces, together with the cohesive forces between water molecules, form the basis for capillary binding of soil water. Water is retained by soil particles in a wide range of energy (Figure 7) when plant roots can absorb it in a rather narrow range (50 to 350 kPa).

The relation between matric potential changes and changes in water content (by weight or volume) is called moisture-retention curve or pF curve. It is a complex function, different for different soils as illustrated for three soils in Figure 8. It shows a hysteresis effect if the soils had been dry and were then wetted to the same fixed matric potentials (Figure 9). This phenomenon can be partially explained on the basis of irregularly shaped pores.

Soil water is one of the most dynamic soil components in which content can change rapidly. One very important factor which influences this dynamic is water movement. Water movement in soil is of vertical and horizontal character. It includes the infiltration of water, saturated and unsaturated, flow through the soil, including upward and horizontal movement and evaporation. The rate of movement is controlled by the size and continuity of the pores containing water and by its viscosity. Considering the movement of water within soils, it is necessary to make a distinction between the bulk movement of the soil solution and the movement of pure water as vapor.

The soil solution percolates downward through the noncapillary pore space by mass flow under the influence of gravity and moves through the capillary pore spaces and in films surrounding soil particles under the influence of surface tension or matric forces. Pure water diffuses as vapor through air-filled pore spaces along gradients of decreasing vapor pressure. It has been proven that the two phases of water liquid and vapor can move independently of one another, each along its own gradient in potential.

The infiltration rate is caused by gravitational forces, but there may be some horizontal movement caused by matric forces when water is applied to a point on the surface of a dry soil, as in an irrigation ditch, or is moved toward roots. The major factors affecting infiltration of

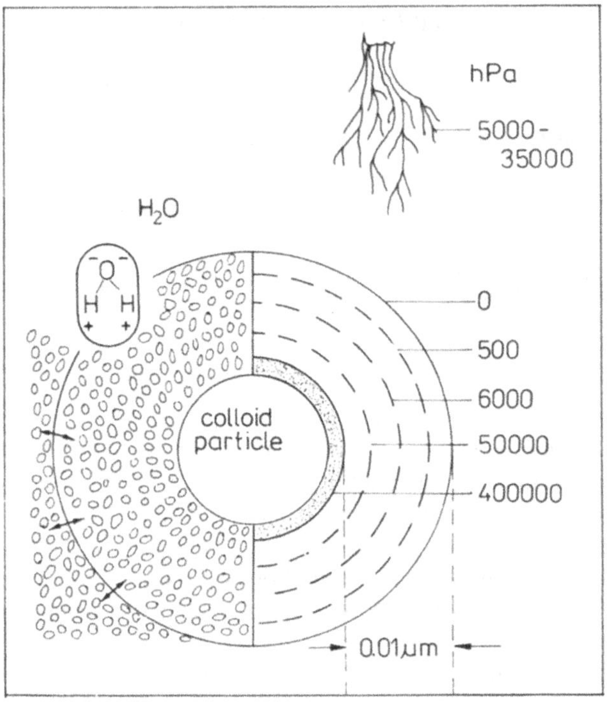

FIGURE 7. Water binding forces on soil colloid particle.

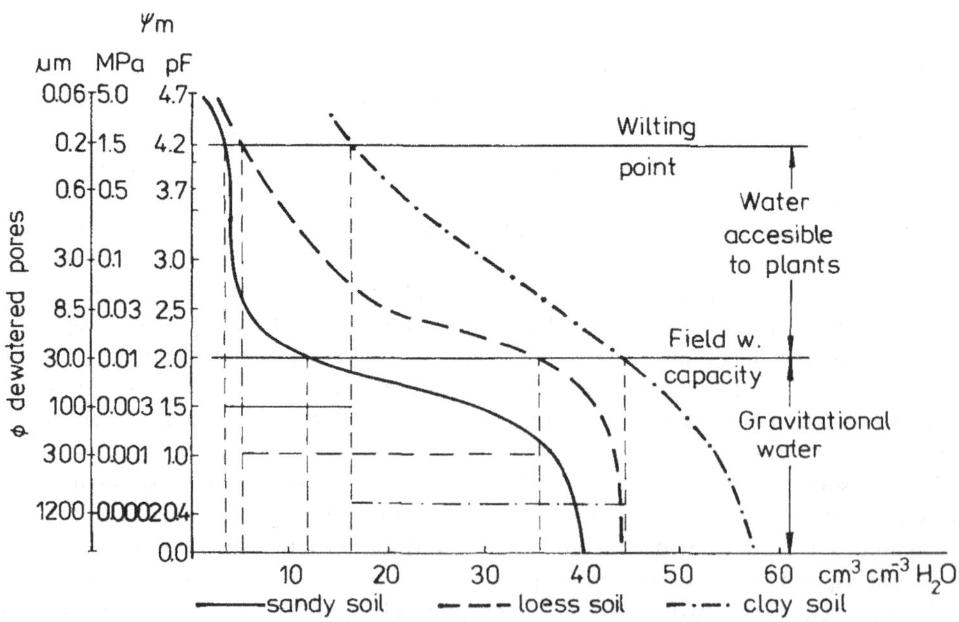

FIGURE 8. Moisture-retention curves of three soils and size distribution of pores as given by the moisture characteristics of these soils.

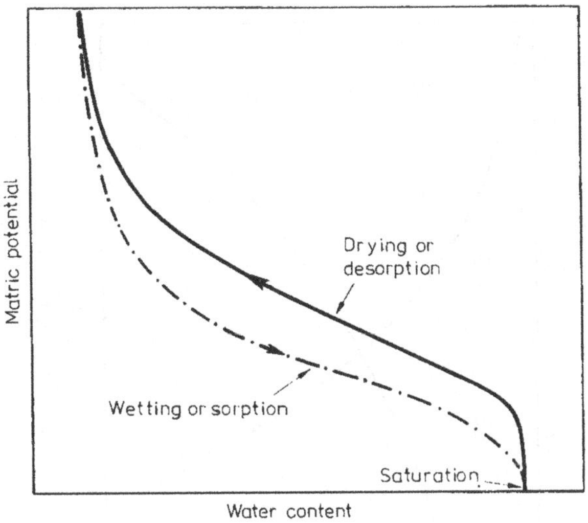

FIGURE 9. Hysteresis effect resulting in differences between matric potential at a given water content in wetting and drying cycles of soil.

water into soil are initial water content, surface permeability, internal characteristics of the soil affecting hydraulic conductivity, such as pore space, degree of swelling of soil colloids, and organic matter content, duration of rainfall or irrigation, and temperature of the soil and water.

Liquid water flow occurs in response to a hydraulic potential (in saturated soils) or a matric potential gradient (in unsaturated soils) and is not closely connected with the water content gradient. A measure of the ability of the soil to conduct the flow of water is called hydraulic conductivity (K). This term concerns both the so-called saturated flow in saturated soils with water and the unsaturated flow, or capillary conductivity, in unsaturated soils.

The values of hydraulic conductivity vary widely, ranging from 0.0025 cm h^{-1} in the least permeable soils to 25 cm h^{-1} in the most permeable soils. A hydraulic conductivity less than 0.25 cm h^{-1} characterizes poorly drained soils, and that greater than 25 cm h^{-1} characterizes soils which do not hold enough water for good plant growth.

The theory of movement of liquid water in soil is based on Darcy's law which states that the quantity of water passing a unit cross section of soil is proportional to the difference in hydraulic head:

$$V = Q/A = K\,(\psi_2 - \psi_1)/L_{12} \qquad (13)$$

where V is the rate of discharge per unit area, Q is the rate of discharge of water through a cross section area A, which is taken normal to the direction of flow; K is a proportionality constant; and ψ_1 and ψ_2 are the hydraulic potentials measured at two positions separated by a distance L_{12} along a straight line parallel to the direction of flow.

Darcy's law written in vector notation is

$$V = -K\,\nabla\,\psi \qquad (14)$$

where $\nabla\,\psi$ is a gradient of the potential.

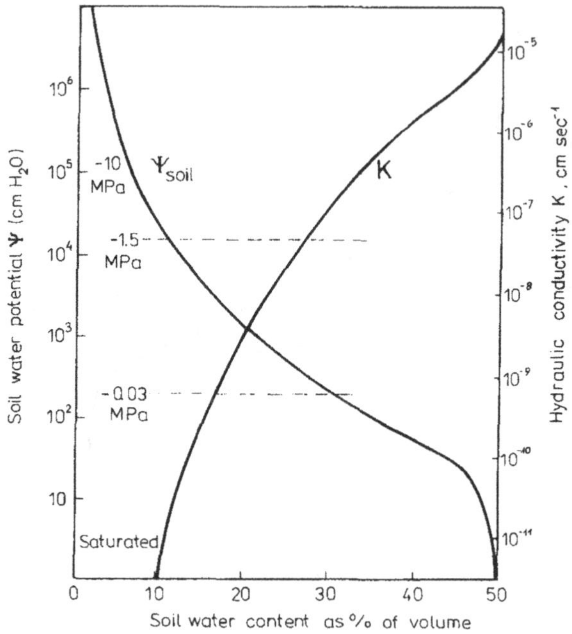

FIGURE 10. Diagram showing approximate decrease in hydraulic conductivity (K) and soil matric potential (ψ) with decreasing soil water content. (Data from Kramer, P. J., *Water Relations of Plants*, Academic Press, Orlando, FL, 1983, 78.)

The hydraulic conductivity of saturated soils is constant in time in a uniform soil and also in position. The hydraulic conductivity of unsaturated soils (of which the pore volume is only partially filled with water) is smaller than that of saturated soils.

When in unsaturated flow the hydraulic head or gravitational component plays a significant role in water movement, in unsaturated flow, the matric potential becomes more important. Darcy's law is applicable to unsaturated flow if the hydraulic conductivity is regarded as a function of water content. The relation between matric potential, water content, and hydraulic conductivity is shown in Figure 10.

The upward transport of water in a soil profile (capillary rise) occurs due to evaporation from the soil surface and/or transpiration by vegetation and following decrease of matric potential resulting in the upward movement of water against gravity. At the constant rate of evaporation, the rate of upward movement of water to the soil surface depends on the depth of the water table and the soil texture. The amount of water which soil can lose by evaporation after it has been wetted depends on the evaporation power of the air, and also on the rate of the downward movement of the water into the subsoil, and on the amount of water the surface soil can hold against drainage.

Evaporation is drastically reduced after the soil surface becomes dry because water movement through the surface layer as vapor is much slower than the movement as liquid. The rate of movement of water from soil to roots depends on the water potential gradient in soil and the difference between soil water potential and the potential at the root surface and on the hydraulic conductivity of the soil. Both of them decrease rapidly as the soil water content decreases (see Figure 10).

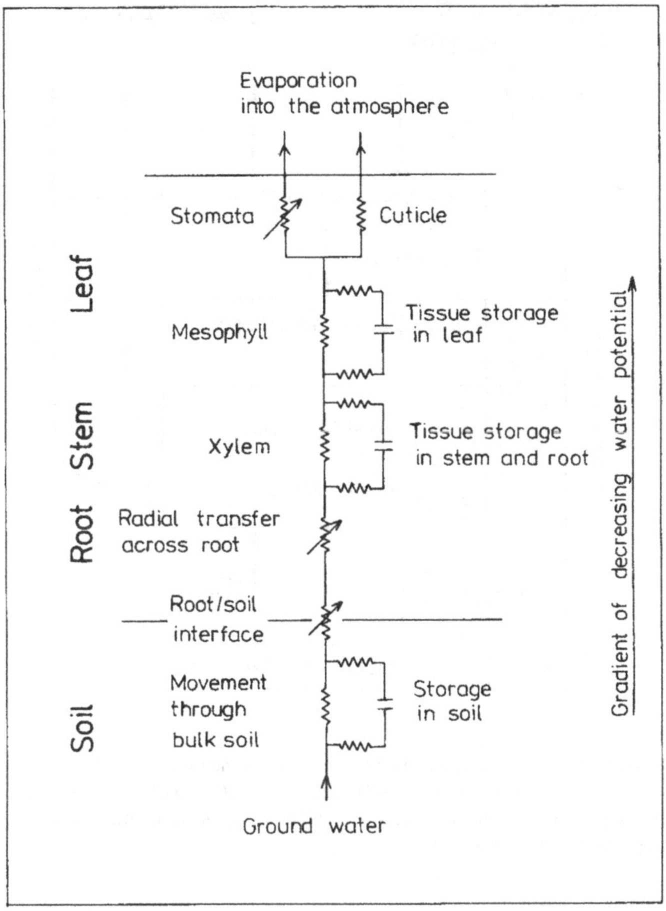

FIGURE 11. Diagrammatic representation of passage of water from the soil to the atmosphere through a plant. (Data from Russell, R. S., *Plant Root Systems: Their Functions and Interaction with the Soil*, McGraw-Hill, London, 1977, 94.)

Water movement through the soil-plant-atmosphere continuum is treated as a series of interrelated, interdependent processes analogous to the flow of electricity through the conducting system (Figure 11). Soil water movement is affected by temperature, both in the liquid phase and in the vapor phase, and by the effects of temperature on surface tension and on the vapor pressure of water, respectively. A general scheme of classification and movement of water in soil is presented in Figure 12.

1.7 TEMPERATURE

Soil derives its heat almost entirely from the sun and loses much of it by radiation back into the atmosphere. Examples of daytime heat balances are shown in Figure 13. Daytime heating is by short-wave radiation (mostly shorter than 2 μm) from the sun and the atmosphere. Nighttime cooling results from the loss of energy emitted from the surface of the soil as long-

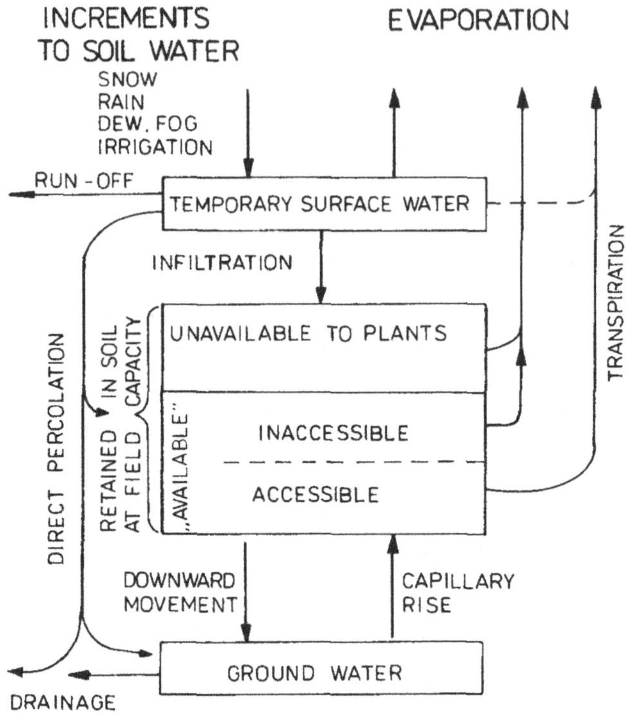

FIGURE 12. A general scheme of classification and movement of water in soil. (Data from Heinonen, R., *Soil Management and Crop Water Supply*, 4th ed., Department of Soil Science, Swedish University of Agriculture Science, Uppsala, Sweden, 1985, 4.)

wave radiation (mostly longer than 4 μm). The balance between incoming and outcoming energy at the soil surface also includes other items as shown in Equation is:

$$(1-\alpha)\,R_s = R_l + G + H + LE \qquad (15)$$

where R_s is the incoming short-wave radiation, a fraction a of which is reflected from the surface and so does not affect soil temperature; R_l is the long-wave radiation (outgoing minus incoming from the atmosphere); G and H represent heat flow into the ground and air, respectively; and LE is the latent heat consumed by the evaporation flux E, L being the specific latent heat of vaporization of water.

The amount of heat flow in soil influences soil temperature. It may vary because of:

1. Exchange of heat with the air — a combination of conductance and convection
2. Exchange of heat with the environment — radiation
3. Flow of heat within the soil — conduction
4. Chemical or physical processes connected with evaporation, wetting, and condensation

The following thermal properties and processes of soil are distinguished as follows:

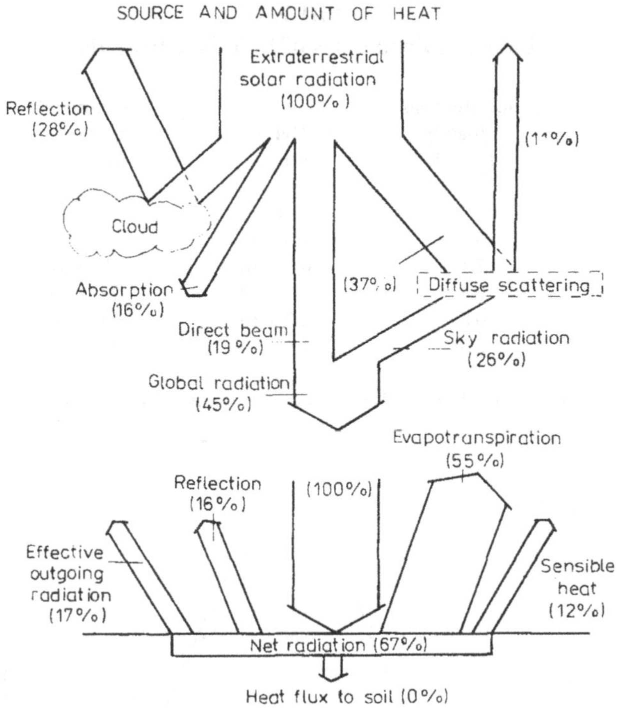

FIGURE 13. Daytime heat balance. (Data from Baver, L. P., Gardner, W. H., and Gardner, W. R., *Soil Physics*, 4th ed., John Wiley & Sons, New York, 1972, 255.)

1. Temperature: the intensity of heat in soil
2. Heat capacity: the amount of heat required to raise the temperature of a unit volume or mass of soil by 1°C
3. Thermal conductivity (conduction of heat): the ratio of the amount of heat flow per unit of area per unit of time to the temperature gradient (It is a measure of how much heat will be conducted through the soil under standard conditions. For a moist soil, the heat capacity depends on both the water content and the mineral and organic matter composition of the soil.)

Thermal properties of soil constituents are given in Table 4. Soil temperatures fluctuate due to the cyclic variations of the surface temperature. These variations are the diurnal (day-night) and annual (summer-winter) cycles. Examples of such variations are shown in Figures 14 and 15.

1.8 AERATION

Soil air fills that part of the soil volume that is not occupied by water. The air content in the soil (E_g), often called the air-filled porosity of the soil, is therefore equal to the difference between the total porosity of the soil (E_o) and its current moisture content by volume (θ):

$$E_g = E_o - \theta = E_o - \rho W \qquad (16)$$

where ρ is soil bulk density and W is moisture content by weight.

TABLE 4
Thermal Properties of Soil Constituents

Constituent	Specific heat capacity (kJ kg^{-1} K^{-1})	Density (kg m^{-3})	Volumetric heat capacity (Kj m^{-3} K^{-1})	Thermal conductivity (W m^{-1} K^{-1})
Air (20°C)	1	1.2	1.2	0.025
Water	4.2	1.0×10^3	4.2×10^3	0.6
Ice (0°C)	2.1	0.9×10^3	1.9×10^3	2.2
Quartz	0.8	2.7×10^3	2.0×10^3	8.8
Clay minerals	0.8	2.7×10^3	2.0×10^3	2.9
Soil organic matter	2.5	1.1×10^3	2.7×10^3	0.25

Data from Marshall, T. J. and Holmes, J. W., in *Soil Physics*, Cambridge University Press, London, 1979, 277.

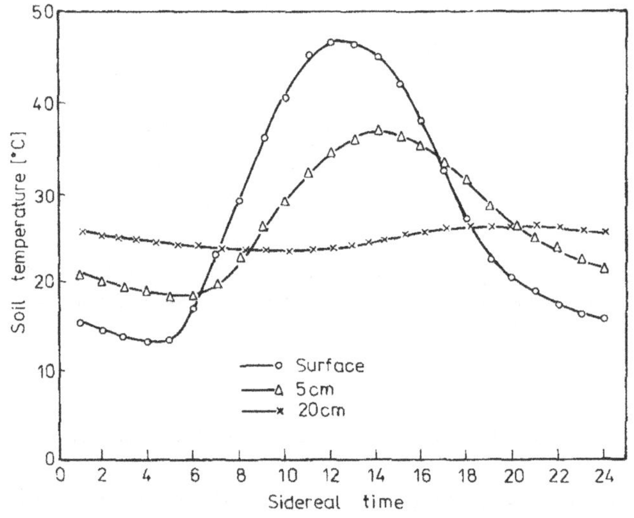

FIGURE 14. Diurnal variations in temperature with depth in a loamy soil. (Data from Baver, L. P., Gardner, W. H., and Gardner, W. R., *Soil Physics*, 4th ed., John Wiley & Sons, New York, 1972, 277.)

In the case of a nonswelling soil, the air content decreases in a linear manner with an increase in moisture content both by weight and by volume. In a swelling soil, the relationship becomes nonlinear. The content of air in a soil profile is subjected to dynamic changes depending on the changes in the total porosity and soil moisture content, although it does not usually exceed 0.4 m^3 m^{-3}.

The amount of air in the soil determines not only the reserve of oxygen present, but also, and primarily, the value of the oxygen diffusion coefficient, and it therefore affects the rate of gas exchange. In turn, the rate of gas exchange in the soil, together with the rate of soil respiration, determine composition of the soil air (Table 5).

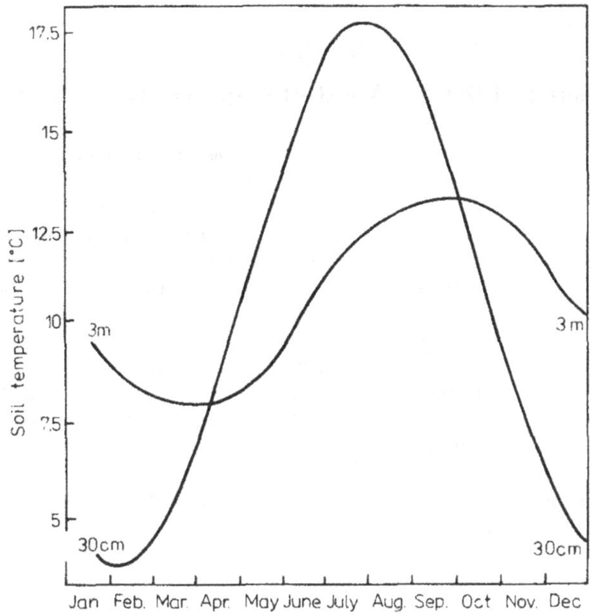

FIGURE 15. Annual variations of soil temperature with depth under short grass for an Oxford soil. (Data from Russell, R. S., *Plant Root Systems: Their Functions and Interaction with the Soil*, McGraw-Hill, London, 1977, 391.)

In comparison to the atmospheric air, soil air contains less oxygen (the quantity of which usually ranges from 0.1 to 0.2 $m^3 m^{-3}$, though it can even drop to zero) and more carbon dioxide (the content of which usually varies from 0.001 to 0.05 $m^3 m^{-3}$ and sometimes may even exceed 0.1 $m^3 m^{-3}$).

The content of argon and other inert gases remains practically unchanged, while the content of nitrogen varies within a range of several percent as a result of changes in the sum of oxygen and carbon dioxide. In exceptional cases, under anoxic conditions, with extremely high concentrations of carbon dioxide and in the presence of methane, the content of nitrogen may even drop to 0.4 or 0.3 $m^3 m^{-3}$.[367]

In certain cases, when soil oxygenation is insufficient, other gases, such as CH_4, H_2S, N_2O, C_2H_4, and H_2, may occur in the soil air. The presence of any of these, despite their low concentration, is an important indicator of the aeration status of the soil. Soil oxygen is primarily contained in the gaseous phase of the soil and is contained only to a small extent in the soil water. Its content in well-aerated cultivated soils usually does not fall below 0.15 $m^3 m^{-3}$.[447] Oxygen concentration in the soil usually decreases with depth in the soil profile. It also changes in the vegetation season with a tendency for a minimum to occur during summer because of higher temperatures. In poorly drained soils, oxygen concentration may be considerably lower, and under especially adverse conditions, it may even fall to zero. Factors increasing oxygen content in the soil air include all measures which loosen the soil.

The content of carbon dioxide in the soil varies dynamically in a manner opposite to that of oxygen content (Figure 16). Maximum concentrations of carbon dioxide in the soil air occur during periods of increased moisture content and soil temperature.

TABLE 5
Volumetric Composition of Dry and Moist Atmospheric Air of Different Temperatures

	Concentration ($m^3\ m^{-3}$)				
		Air saturated with water vapor at a temperature of			
Component	Dry air	0°C	10°C	20°C	30°C
N_2	0.7809	0.7762	0.7715	0.7629	0.7481
O_2	0.2095	0.2082	0.2082	0.2047	0.2007
CO_2	0.0003	0.0003	0.0003	0.0003	0.0003
Ar + other inert gases	0.0093	0.0092	0.0092	0.0091	0.0089
Water vapor	0.0000	0.0060	0.0120	0.0230	0.0420

Data from Gliński, J. and Stepniewski, W., in *Soil Aeration and Its Role for Plants*, CRC Press, Boca Raton, FL, 1985, 93.

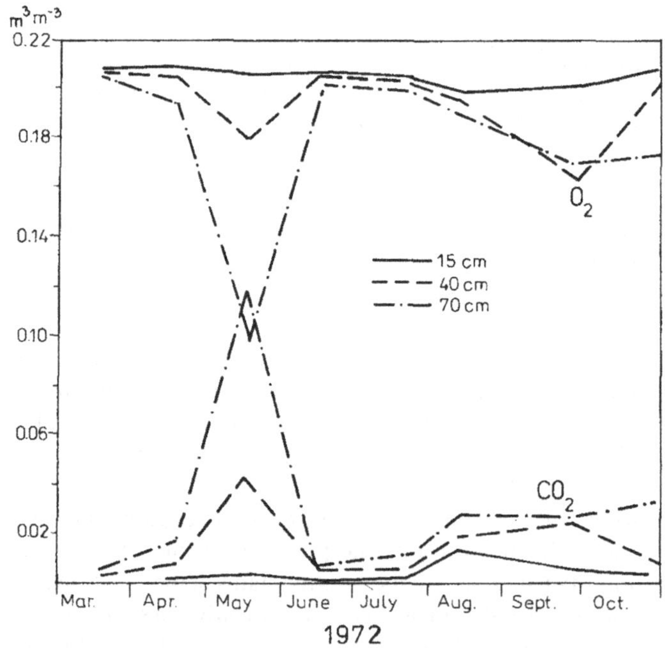

FIGURE 16. Oxygen and carbon dioxide content in the air of a loamy sand soil. (Data from Stepniewski, W., *Zesz. Probl. Postepow Nauk Roln.*, 197, 203, 1977.)

Concentration of nitrous oxide in soil is characterized by its great variability. It may reach an amount of 100 and sometimes 1000 and in individual samples, even 6500 $cm^3\ m^{-3}$. This concentration increases with the increase in the content of nitrates and of moisture levels.[447]

Ethylene concentration in soil does not reach high values, and even during soil incubation in closed containers, it does not exceed 40 $cm^3 m^{-3}$. Concentrations met under field conditions are generally from 0.01 to 10 $cm^3 m^{-3}$. The occurrence of concentrations higher than 1 $cm^3 m^{-3}$ is connected with periods of excessive moisture contents, application of slurries, and increased soil temperature, which are usually accompanied by a decrease in oxygen concentration. The maximum value encountered under field conditions was 18.4 $cm^3 m^{-3}$. The formation of ethylene in soils is also possible under oxic conditions.

Under anoxic conditions, methane appears in the soil, and in the presence of considerable quantities of organic matter in the soil, it may reach a concentration of several and even several tens of percent.[47] In these conditions, trace amounts of higher hydrocarbons, such as propane, ethane, and butane, are also found.

The exchange of considerable amounts of gases, particularly of oxygen and carbon dioxide, between the soil and the atmosphere (in the range of several tens of milligrams per hectare per year) takes place under the influence of both the pressure gradient (mass flow) and the concentration gradient (diffusion flow). These two kinds of flows may take place in the soil pores and also in the tissue of plants (internal exchange). The relative contribution of the above-mentioned mechanisms of gas exchange, as well as the two pathways, e.g., through the soil and through the plant, may vary within relatively broad limits. The values of the air permeability of soils usually fall within a range from 0.01 to 500×10^{-12} m^2 and depend directly on the gas porosity of the soil and on the square of the soil pore radii and indirectly on all the factors affecting these two properties, such as moisture, bulk density, or the specificity of a given soil unit. Factors that cause mass flow in the soil are variations in the soil temperature, atmospheric pressure, soil moisture (rain, irrigation, evaporation, and movement of the ground water table), and wind.

Another mechanism of gas exchange in the soil, of much greater significance than the mass flow, is concentration diffusion. This is a continuous process of gas transport in the direction of decreasing concentration. It is described by Fick's first law:

$$f_x = -D \frac{dC}{dx} \tag{17}$$

where f_x is the diffusion flow through a unit of surface area of the medium in which the diffusion takes place, in a unit of time, in an unaxial system; dC/dx is a gradient of concentration constituting the driving force of the flow; and D is a diffusion coefficient (the dimensions of which are the square of length per a unit of time); and also by another equation of general diffusion:[447]

$$\frac{\partial C}{\partial t} = \frac{D}{E_g} \nabla^2 C \pm q \tag{18}$$

where C is gas concentration in the air; D is a diffusion coefficient; and q is respiratory activity related to the soil volume unit.

In contrast to mass flow, the occurrence of diffusion flow is not induced by the presence of external factors, but only by the gradient of concentration which is a result of respiration processes.

In considering the process of gas diffusion in the soil, it is necessary to distinguish macrodiffusion (i.e., the diffusion in the whole soil profile disregarding the heterogeneity of soil in the microscale) and microdiffusion through water films surrounding plant roots. Principles and theory of gas diffusion in a porous body are widely discusssed by Gliński and Stepniewski,[447] Kowalik,[681] and Edling.[353]

Diffusive properties of a soil medium are usually characterized by means of a relative diffusion coefficient:

$$D/D_o = \gamma E_g^{\mu} \tag{19}$$

where D is a gas diffusion coefficient in soil and D_o is a diffusion coefficient of the same gas in

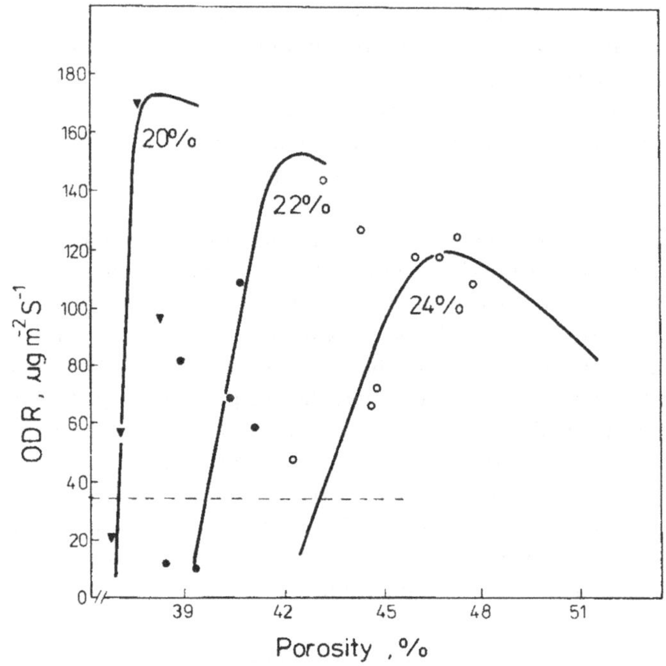

FIGURE 17. ODR changes under the influence of soil porosity and moisture content (% w/w) in a heavy marine silt soil. (Data from Boone, F. R., in *Trans. 7th Conf. ISTRO*, Uppsala, Sweden, 1979, 4:1.)

the atmospheric air, under the same conditions of pressure and temperature; E_g is air-filled porosity; and g and μ are empirical coefficients characterizing the material under investigation. The values of D/D_0 for air-dry soils are within 0.02 to 0.5 interval.

Microdiffusion consists in a possibility of oxygen uptake by plant roots that constitutes plant reaction to oxygen conditions in the soil. It is expressed by the so-called oxygen diffusion rate (ODR) that is determined on the basis of measurements of the diffusion current intensity, i.e., of the oxygen reduction current on the Pt electrode (being a model of a root absorbing oxygen) in a situation where the intensity of that current depends only on the rate of oxygen diffusion at the electrode surface from the surrounding medium. The ODR value is calculated from the equation:

$$\text{ODR}[\mu\text{g m}^{-2}\text{s}^{-1}] = \text{Mi/nFa} = 8.29\ 10^{-5}\ \text{i/A} \tag{20}$$

where M is the molecular weight of O_2 (32 g mol^{-1}); i is the current intensity in microamperes (μA); n = 4 (number of equivalents per a mole of oxygen); F is the Faraday (96,500 C $equiv.^{-1}$); and A is the area of the electrode square meters. The application of this method is limited to the matric potential up to 0.5 to 1 kPa when the whole electrode surface is covered by a moisture film enabling electrochemical reduction of oxygen.

ODR values in soils vary, most frequently, within a range from 0 to 200 μg m^{-2} s^{-1}. They decrease with an increase in soil moisture content and therefore increase with an increase in the matric potential and air-filled porosity of the soil (Figure 17). They also decrease with an increase in the soil compaction.[13]

Kirkham,[667] analyzing oxygen movement in the soil to a root zone during various tillage

TABLE 6
Sequence of Redox Processes in Soils

Succession of processes	Eh (mV) at the beginning of the process
Respiration: $O_2 \rightarrow H_2O$	>400
NO_3^- — reduction	+600 to +300
Formation of Mn^{2+}	+500 to +200
Production of Fe^{2+}	+400 to +180
S^{2-} — accumulation	+100 to −200
CH_4 formation	−150 to −280

Data from Ottow, J. G. G., in *Proc. Symp. Paddy Soils,* China, Nanjing, China, 1980, Institute Soil Science, Academica Sinica, Science Press, Beijing/Springer–Verlag, Berlin, 1981, 330.

practices, used an electrical analog. He found that the conductance of oxygen in soil was affected more by the width than by the depth of tillage.

The ODR is at present the best indicator of soil aeration and the potential oxygen availability for plant roots in the soil. The other characteristic of soil aeration status is redox potential (Eh). It is a very important indicator of soil reduction processes in water-logged soils and in conditions of oxygen exhaustion in the soil when other indicators such as ODR or O_2 content are insensitive and of little value. With the disappearance of oxygen from soil, a number of reduction processes (Table 6) which influence redox potential value take place.

The range of Eh changes in soil is from +800 to −400 mV. In normally moistened nonflooded soils, it is governed by oxygen concentration. Submergence of the soil causes its redox potential to decrease gradually.

According to Ponnamperuma et al.,[954] a redox potential of 300 mV (which is responsible for the reduction of oxidized forms of iron) was accepted as a boundary between reduced and oxygenated soils. So, an assumption was made that the time interval which is necessary for decreasing redox potential in flooded soils to 300 mV may serve as a characteristic of soil redox capacity in a given temperature. This characteristic, named "t_{300} indicator", was used by Giński and Stepniewska[446] for various Polish soils. They obtained t_{300} values in the broad range from 0.2 to 18 d at 18°C.

Characterizing soil aeration, it is necessary to mention soil respiration which is a result of the biological activity of soil living organisms. The soil respiration rate for different soils, measured on the basis of carbon dioxide evolution, is in the range from 0.1 to 28.4 dm^3 of CO_2 m^{-2} d^{-1} and from 1800 to 47,000 kg of CO_2 ha^{-1} $year^{-1}$.[447]

1.9 CONSISTENCY, COMPRESSION, AND COMPACTION

Soil consistency comprises the attributes of soil material that are expressed by the degree and kind of cohesion and adhesion or by the resistance to deformation or rupture. Every soil material has its consistency irrespective of whether the mass be large or small, aggregated or structureless, moist or dry. Soil consistency is also defined as the state of the soil, whether it is solid, plastic, or liquid. Depending on the degree of moisture which influences consistency the most, a soil may be saturated (pasty or soft), moist (friable or firm), or dry (soft or hard). The consistency of wet soil is characterized by its stickiness (the quality of adhesion to other objects) and plasticity (the ability to change shape continuously under the influence of an applied stress and to retain the impressed shape on removal of the stress). The consistency when soil is moist (moisture content midway between air dry and field capacity) is characterized by friability (the case of crumbling and crusting of soil). The consistency of soil material when dry is characterized by rigidity, brittleness, maximum resistance to pressure, more or less tendency to crush to a powder or to fragment with rather sharp edges, and inability of crushed material to cohere again when pressed together.

In some cases, cementation of soil to a brittle hard consistency caused by some cementing substances other than clay minerals, such as calcium carbonate, silica, and oxides or salts of iron or aluminum, may appear. Cementation is resistant enough to wetting. The following phenomena are attached to soil consistency:

1. Soil shrinkage (when soil is drying) or swelling (when dry soil is being wetted) connected with clay minerals of 2:1 type and organic matter content in soil, surface soil crusting, and sealing
2. Traction parameters (shear strength, tensile strength, friction, stiffness, pulverization, penetration resistance, compressive strength, rolling resistance, etc.) due to forces applied during traction affecting soil compaction and soil tillage which influence loosening and pulverization and also cutting, mixing, inversion, and sorting of the surface layers of the soil to improve physical conditions for crop production.

Soil consistency and all other properties connected with it are called mechanical properties of soil.

Compression may be defined as the change in the soil volume under an applied stress. It denotes the decreases in the voids ratio per increment of applied load or pressure. Compression increases with moisture content to a maximum and then decreases as the amount of the moisture becomes larger.

To examine the stress at the root/soil interface, the spherical cavity expansion theory of Vesić[1251] is used (see Chapter 5), with assumptions that the soil in the plastic zone (near the root surface) behaves as a compressible plastic solid, defined by the Mohr-Coulomb shear strength parameters, by effective cohesion and effective angle of shearing resistance, as well as by an average volumetric strain. Beyond the plastic zone, the soil is assumed to behave as a linearly deformable, isotropic solid (elastic zone) defined by Young's modulus of deformation and Poisson's ratio.

Soil compaction is defined as an increase in density and the closer packing of solid particles or reduction in porosity. Compaction can result from natural causes, including rainfall impact, soaking, and internal water tension. Artificial compaction occurs under the downward forces of machines and tractors which are usually of short duration in the case of moving vehicles and due to soil subsidence after tillage.

It is primarily the topsoil that is affected by vehicular traffic. When a tractor is driven on the soil surface, the influence of wheel pressures reaches through the whole topsoil, but, as a rule, rapidly decreases below it (Figure 18). Compaction influences in the topsoil depend mostly on the contact pressure between the wheels and the surface. The larger the total load on the wheel, the deeper the compaction effects will penetrate the soil. These effects are strongly affected by the soil water level.

Subsoil compaction occurs due to the permanent load of the topsoil. It is a common experience with many soils to find a compacted layer at the bottom of the plowed zone or a zone affected by other kinds of agricultural implements. These layers have been termed a plow sole, a subsoil sole, a disc harrow sole, and a traffic sole. All such soles reduce the permeability of the soil to water and restrict root elongation and proliferation.

A special case of soil compaction in the root bed occurs when zero tillage or direct drilling is practiced. In this system, the topsoil is not loosened by the primary tillage and therefore soil

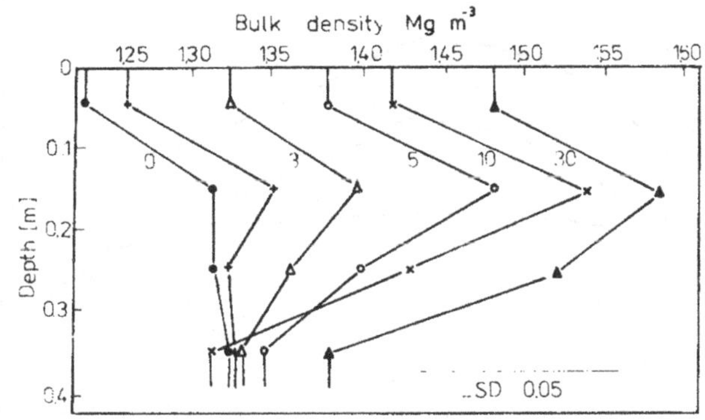

FIGURE 18. Variation of bulk density of loess soil (29 to 33% of clay) with depth as related to the number of wheel passes. (Modified from Canarache, A., Colibas, I., Colibas, M., Horobeanu, I., Patru, V., Simota, H., and Trandafirescu, T., *Soil Till. Res.*, 4, 199, 1984, With permission.)

compaction by traffic is allowed to add up from year to year. Moreover, compaction at harvest increases, when tire pressures are considerably higher than in the spring.

How a tractor tire influences soil compaction is shown in Figures 19 and 20. It is seen that the extent of soil compaction depends not only on the number of passes of the tractor, but also on the tractor weight and type of tires, tire pressure, speed, etc. Karczewski[640] showed that the bulk density increase below a rigid wheel was very much greater at a forward speed of 1 than at 12 km h^{-1}. The maximum intensity of compaction was 1.65 Mg cm^{-3} at 1 km h^{-1} compared with 1.49 Mg cm^{-3} at 12 km h^{-1}. The same level of soil compaction was reached at approximately twice the depth for slow speed than for high speed (Figure 21).

A good review of all the problems connected with soil compaction was done by Chancellor,[220] Eriksson et al.,[377] Barnes et al.,[91] Larson and Allmaras,[721] Larson et al.,[722] Soane et al.,[1094-1096] Kuiper and Kuiper,[694] and Håkansson et al.[506] Moreover, many new papers on the subject have appeared recently, but the principles and the views on the role of soil compaction for the soil and the plant have not changed drastically. Bulk density, porosity, and void ratio are measures of soil compaction.

For better comparison, the compaction levels from a crop production point of view, the so-called "degree of compactness", were designated.[377,505] Its measure is the actual dry bulk density of the soil as a percentage of the bulk density at the compacted, standard condition. The standard condition is the most dense condition of the saturated soil that could be obtained with a static pressure of 200 kPa. The compactness degrees that are obtained in the topsoil usually lie between 60 (implies a very mellow soil) and 100 (implies a strongly packed soil). The bulk density of the cultivated layer changes in the course of the year, depending on plowing and rain and frost action (Figure 22).

Compaction of soil in the field is almost always accompanied by soil deformation. During compaction, soil aggregates are destroyed more and the soil tends to become more homogenous. Compactibility increases very strongly with moisture content and it is accompanied by outflow of air to a certain amount (when air is entrapped).

During compaction, pores of large diameters are primarily destructed. Compaction also has large effects on pore size regularity. The number of large and nonregular pores decreases as soil density increases, whereas the amount of regular small pores initially increases (Figure 23). At

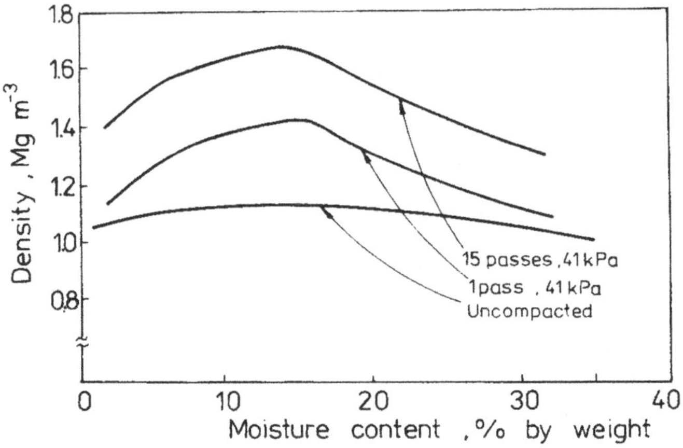

FIGURE 19. Maximum level of dry density in a sandy loam soil at varying moisture content under different treatments of tractor traffic. (Modified from Raghavan, G. S. V., McKyes, E., Chassé, M., and Mérineau, F., *J. Terramech.*, 14, 31, 1977a.)

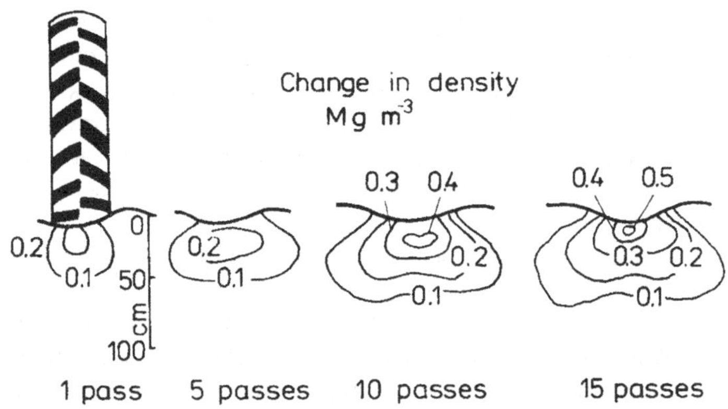

FIGURE 20. Distribution of change in soil dry density in a sandy loam soil under different treatments of tractor traffic. (Modified from Raghavan, G. S. V., McKyes, E., Chassé, M., and Mérineau, F., *Can. J. Plant Sci.*, 56, 505, 1976.)

large compaction, all pores change, except for the smallest ones. Compaction causes greater mechanical resistance or impedance to root extension, changes in soil aeration status (e.g., reduction in ODR[13]), and gas exchange between the soil and the atmosphere and the moisture characteristics of the soils (Figure 24).

The terms "mechanical impedance" or "resistance to penetration" or "cone index pressure"

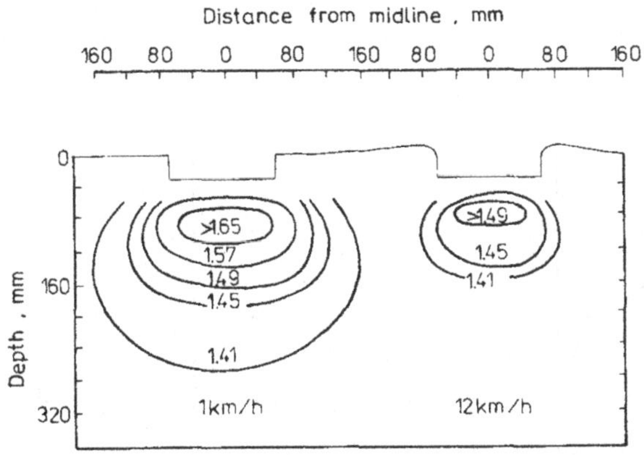

FIGURE 21. The effect of forward speed on the distribution of soil bulk density (Mg m^{-3}) below a tractor wheel. (Modified from Karczewski, K., *Zesz. Probl. Postepow Nauk Roln.*, 201, 69, 1978.)

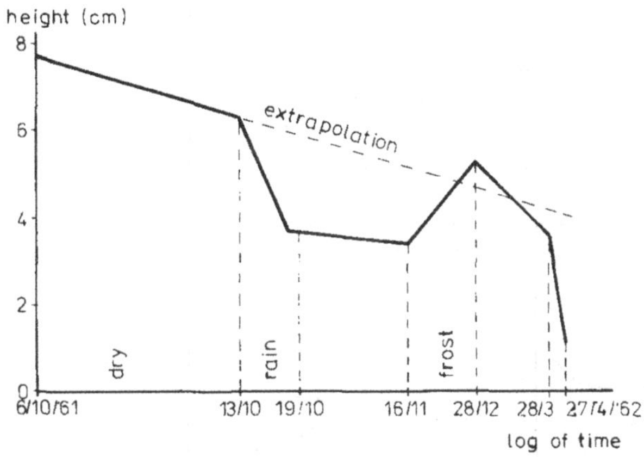

FIGURE 22. Mean surface height of clay soil in relation to log of time after plowing. (Modified from Koolen, A. J. and Kuipers, H., in *Agricultural Soil Mechanics*, Springer-Verlag, Berlin, 1983, 22.)

are integrated indicators of soil compaction, moisture content, texture, and type of clay minerals — in other words, indicators of soil strength under the conditions of the measurement. They refer to the growth response of the plant root as it is mechanically constrained by a volume of soil. It may be caused not only by the cohesive and massive soil layers, but also by surface soil crusts which can reduce or prohibit seedling emergence and root and shoot growth. Gupta and Larson[498] assumed that a penetration resistance of 2 MPa or greater can severely restrict root

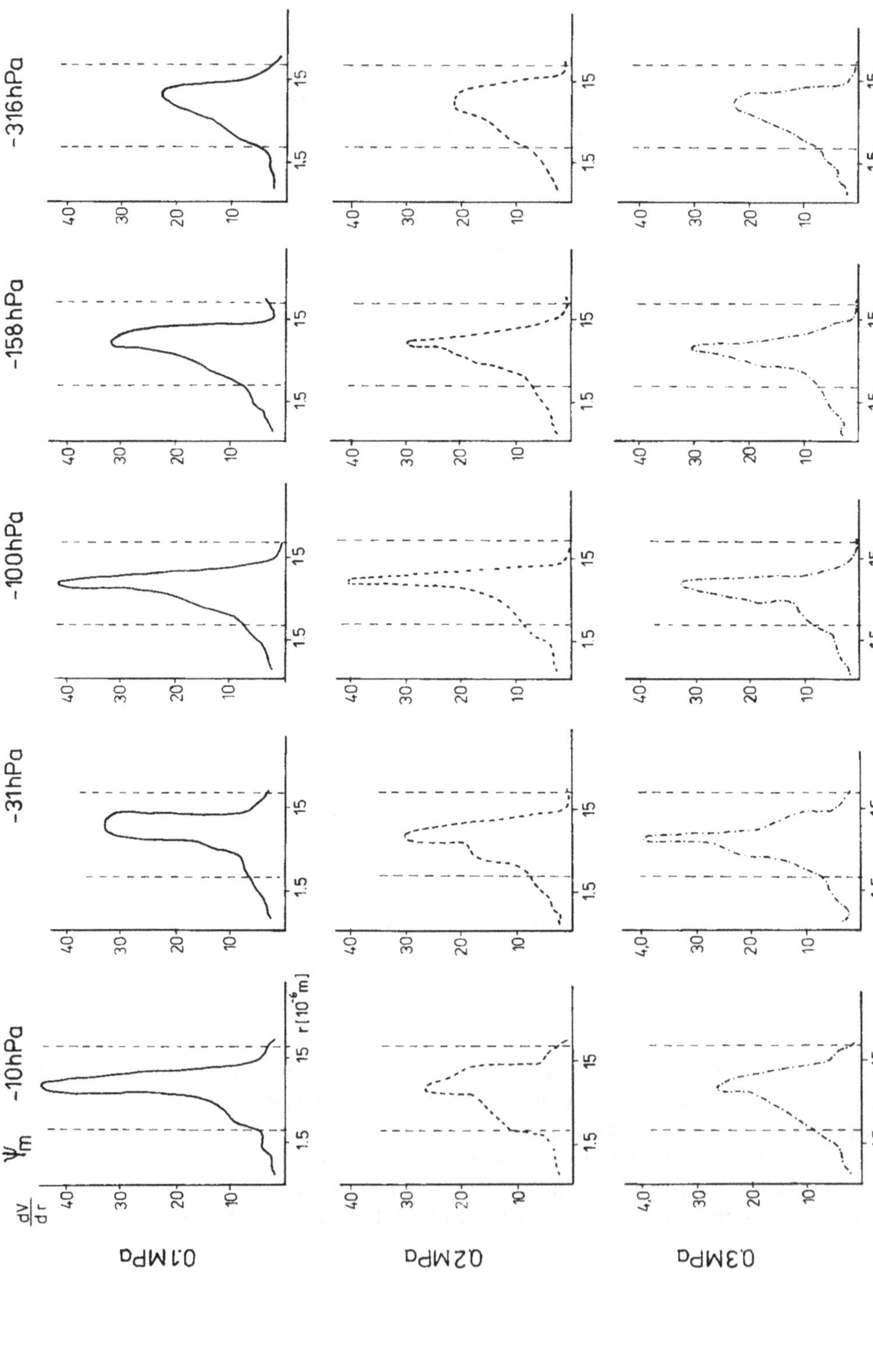
Ψm
dV/dr
r [10⁻⁶ m]
-10hPa
-31hPa
-100hPa
-158hPa
-316hPa
0.1MPa
0.2MPa
0.3MPa
1.5
15
10
20
30
40

FIGURE 23. Differential curves of pore size distribution for loess soil consolidated at various matric potential and different pressures (0.1 to 0.3 MPa); 1.5×10^{-6} m < r < 15×10^{-6} m — range of radius of pores containing water easily available for plants. (Modified from Konstankiewicz, K., *Zesz. Probl. Postepow Nauk Roln.*, 197, 79, 1977.)

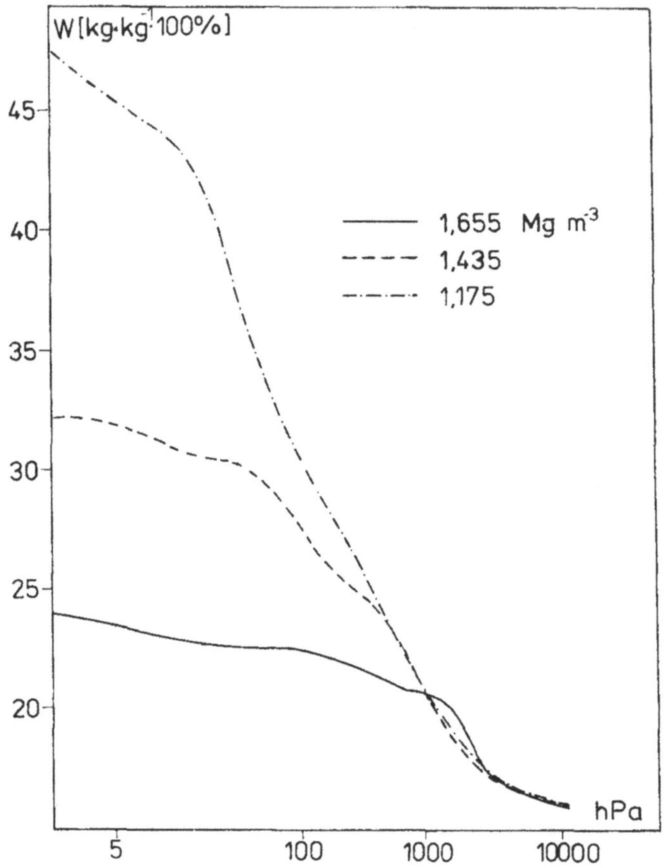

FIGURE 24. Water retention curves for loamy soil of different compaction level. (Modified from Walczak, R., *Zesz. Probl. Postepow Nauk Roln.*, 197, 11, 1977.)

penetration. Under wet conditions, soils with a high content of organic matter are more resistant to compaction than soils low in organic matter. During drying, the soil resistance to compaction increases most in soils low in organic matter.

The problem of natural soil regeneration is very closely connected with soil compaction. This problem has so far received relatively little attention. Bullock et al.[181] conducting experiments on the compaction of silty clay loam by wheelings, found that compaction decreased the macroporosity (pores >60 μm diameter) by over a half and that natural regeneration of porosity in the top 5-cm layer was achieved within 18 months, but the soil below remained compacted for a longer period of time. The regenerated structure consisted mainly of planar voids induced by physical processes unlike the unwheeled soil which was characterized by packing voids and biopores. On the other hand, Aura,[58] studying silty clay soils compacted with a tractor, found that the porosity of even severely compacted soils recovered almost completely from one spring to the next.

Håkansson[504] and Voorhees et al.[1260] reported that the effects of heavy soil compaction by vehicles with loaded axles persisted in the subsoil after a number of years in spite of annual

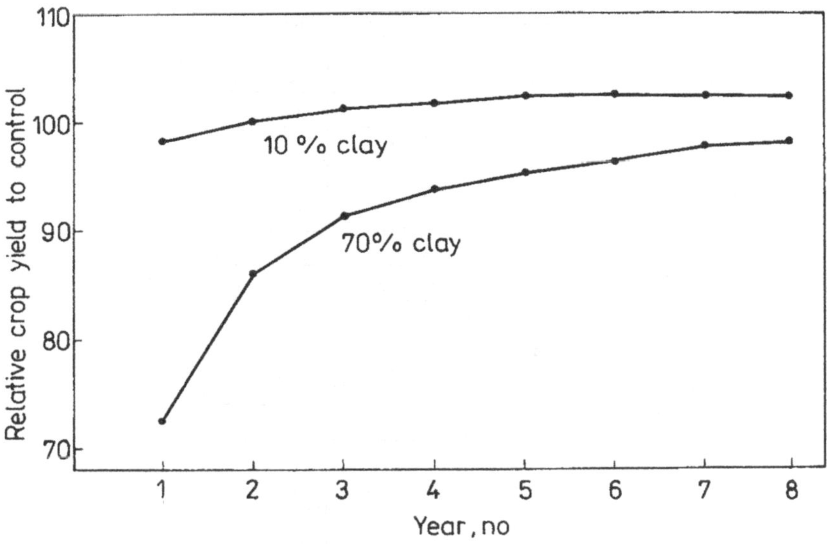

FIGURE 25. Expected crop yield from the heavily compacted soil as a function of clay contents and time. (Modified from Håkansson, I., *Soil Use Management,* 1 (4), 113, 1985.)

winter soil freezing to depths of 70 to 90 cm. This was reflected in the diminution of yields of various crops of which the extent was related to clay content of the soil (Figure 25)

Soil compaction is described by some numerical models, among which those by Raghavan et al.,[964,966] Gupta and Larson,[498] Smith,[1087] Bailey et al.,[66] and Larson et al.[723] are the latest. Smith's[1087] model is based on the prediction of soil specific volume changes arising from the changes in spherical stress caused by wheel loads and the tire/soil contact areas under consideration. The depth of soil to be modeled is divided into elemental layers and the spherical stress increase at the center of each layer, below the center of the wheel load, is estimated. The model may be used to compare the compaction caused by various types and arrangements of wheels.

Bailey et al.[66] proposed a three-parameter (three levels of moisture) multiplicative model of soil compaction. This model is capable of predicting soil density not only at high levels of soil compactness, but also at low (zero) stress levels with less error than the traditional model.

Gupta and Larson[498] developed a soil compaction model based on the compression relationship and the Boussinesq equation as modified by Soehne[1097] by introducing a concentration factor to describe different kinds and conditions of soils. Larson et al.[723] summarized various models to predict bulk density of soils. Raghavan et al.[964] elaborated an equation describing maximum density changes for clay soil after repeated passes of a tractor tire. Boone[151] discussed relationships between soil compaction and crop growth in the following aspects (1) when there is no statistically significant relation between soil compaction and yield, (2) when yield increases due to compaction, and (3) when yield decreases by compaction.

1.10 INTERRELATION OF SOIL PHYSICAL CHARACTERISTICS

In the earlier parts of this chapter, the main physical properties and processes in the soil have been described. These properties can be divided into two groups: (1) more stable properties, such as texture, structure, pore size, and bulk density, and more variable properties, such as water, air,

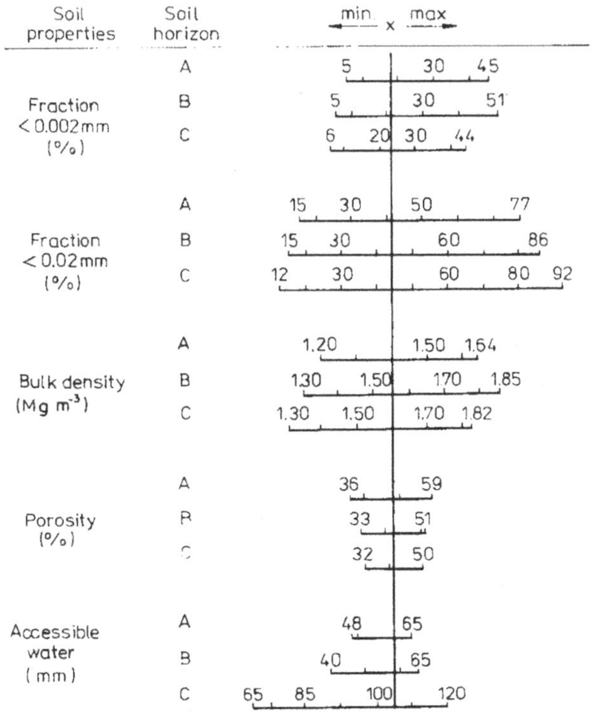

FIGURE 26. Model of optimization of soil properties for yields of winter wheat in the year 1976. (Modified from Slusarczyk, E., *Zesz. Probl. Postepow Nauk Roln.*, 220(I), 327, 1979.)

temperature, and consistency; and (2) properties directly related to plant growth (water, oxygen, temperature, and mechanical resistance) and properties indirectly related to plant growth (bulk density, aggregation, and pore size). Soil physical processes concern water and heat flux, gas diffusion, and soil compaction by means of heavy machines and soil loosening by tillage. Every soil physical property and process in question is modified by weather phenomena (rainfall, frost, solar radiation, and drought), erosion, and human manipulation (management practices which drastically change physical properties of upper soil layers). When, to these variations we add changes in the soil profile morphology (heterogeneity due to stratification resulting from geological and pedological processes) and soil situation in relief, we get an idea how complex the physical status of the soil is.

Soil physical properties and processes are more or less closely interconnected. A very good relationship exists between soil water, air, temperature, and mechanical resistance. They are all affected by texture, structure, bulk density, aggregation, pore size distribution, profile characteristics, etc. Efforts are made to find relationships (simple and multifactorial) between soil physical properties and processes mainly in the aspect of crop production. New methods, continuously improving measuring equipment, and computer techniques allow for better determination of these relationships for simulation and prediction of soil physical processes and crop yields.

Slusarczyk[1083] elaborated an empirical model of the optimal physical properties of soil for the requirements of field crops. This model is based mainly on the static physical features of the soil profile (texture, bulk density, porosity, and plant available water). To elaborate the model, high and very high yields (from 2 years) of various plants from field experimental localities throughout Poland were chosen. An example of such a model is given in Figure 26.

Canarache and Berindei[195] proposed a synthetic index called an "agrophysical index" to describe soil physical status. This index represents the arithmetic mean of the standarized values of several parameters (bulk density, macroporosity, mechanical resistance, hydraulic conductivity, infiltration rate, water stability of aggregates, and coefficient of dispersion).

A two-dimensional soil physical model was developed by Holzhei and Burkhardt[561] to study moisture and temperature interactions in the seed zone using a finite element formulation. Moisture and temperature conditions at various depths were obtained in response to changes in soil texture, soil structure, and tillage effects on the structure, furrow geometry, and meteorological conditions.

A relationship between pressure-pore size distribution-movement of water and air in soil-root resistance and root development were examined by Eriksson.[376] A mathematical model of the influence of the physical parameters of the solid phase (texture, humus content, and specific surface area) of soils on the water retention has been elaborated by Walczak.[1268] He also described problems of mass and energy transfer in the soil-plant-atmosphere system.[1269] Transport models for a soil-plant system based on numerical methods are well described by Campbell.[193]

Chapter 2

ROOTS AND THEIR FUNCTIONS

2.0 MORPHOLOGICAL AND ANATOMICAL CHARACTERISTICS OF ROOT SYSTEMS

2.0.1 ROOT MORPHOLOGY

Root systems are classified as tap (axe, primary, main, and seminal) or fibrous, depending upon the size and number of the individual roots as well as their origin. They arise directly from the stem. Additional roots grown from internodes near the base of the stem are called adventitious roots. Monocotyledons, especially grass, produce an abundant quantity of these roots, forming a fibrous root system. Roots arising from other roots are called laterals (secondary, tertiary, etc.). Various root systems are shown in Figure 1. According to their diameters, roots are classified as follows: very fine (<0.5 mm), fine (0.5 to 2 mm), small (2 to 5 mm), medium (5 to 10 mm), large (10 to 20 mm), and very large (>20 mm).[146]

Root systems of a individual tree (e.g., pine and tropical trees) may be sometimes increased by natural grafting to the adjacent trees, enabling the transfer of solutes between roots through connected fungal hyphae or by diffusion.[684] Quantitative data on the rooting depth and lateral spread of roots for many plant species have been tabulated by Taylor and Terrell.[1183] According to these data, maximum root depth may vary from 32 cm for *Poa sandbergii* Vasey and 40 cm for *Opuntia fragilis* (Nutt.) Haw. to 5330 cm for *Prosopis juliflora* (sw.) DC and 4500 cm for *Eucalyptus calophylla* R. Br. ex. Lindl., and lateral spread may vary from 10 cm for *Andropogon nutans* L. and *Medicago sativa* L. to 900 cm for *Yucca glauca Nutt.* and 1200 cm for *Redfieldia flexuosa* (Thurb.) Vasey.

Common root length densities for crop plants are up to 6 cm cm^{-3} in surface soil and values of 1 cm cm^{-3} are common at 0.5- to 1-m depth.[368,486,680] Data on root length, specific root length, and root diameters for various crops are summarized by de Willigen and van Noordwijk.[303]

One of the elements of root systems is root hairs. They originate as protuberances from epidermis cells. The root hairs appear a few millimeters behind the root apex, and in the main roots, the hair zone usually extends for several centimeters toward the base. The length of a root hair may vary from 80 to 1500 μm and the diameter varies only from 5 to 17 μm.[872] Root hair density (per millimeter of root length) for Gramineae plants varied from 19 to 98. Their numbers per 1 mm^2 of root surface are 50 to 100 and are 2.2 for loblolly pine and 5.2 for *Robinia peudoacacia*.[872]

Generally they are short-lived; frequently they persist only a few days and are easily decomposed by microorganisms. In some cases, however, they can live from a few weeks for wheat, to several months for sugar cane, to more than a year for several Compositae.[872] In some tree genera (*Cercis* and *Gleditsia*), root hairs become suberized and lignified and persist as nonfunctional brown hairs for years.[1149]

There is an opinion, although not uniform, that root hairs are capable of absorbing water and nutrients. Thay can also help in anchoring young roots in soil. The contribution of root hairs in the absorbing of water and nutrients is considered of little importance.[147,331,872,1024] However, many authors who studied phosphorus uptake by plants obtained evidence showing that root hairs enhance this process.[130,203,542,598,599,841] Root hairs contributed also in the uptake of potassium.[235,331] It appears from these studies that root hairs enhance the uptake of those nutrients which have relatively low mobility in soils. Werner and Kuhlman[1289] reported that root hairs of soybean and wheat when infected by *Rhizobium japonicum* accumulated many times more Fe, Co, and Ca than other parts of the root system. Danilova[285] indicated that root hairs renew the absorbing root surface behind the zone of elongation, where first mature xylem elements appear

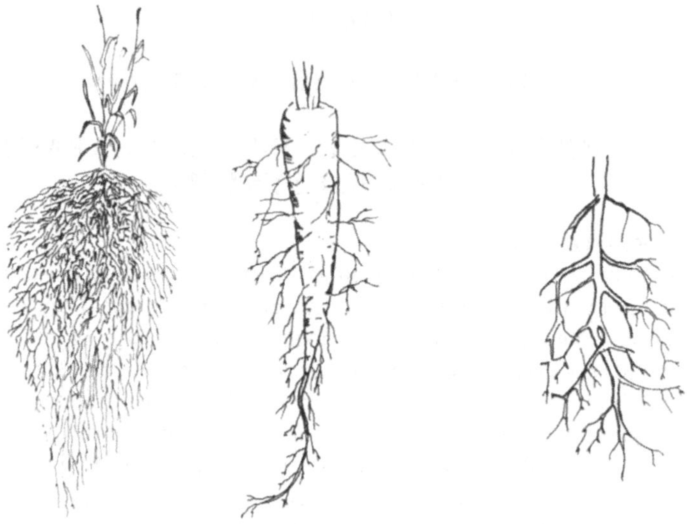

FIGURE 1. Root systems of various plants. (A) Fibrous root; (B) storage root; (C) taproot.

and translocation of minerals to above-ground parts begins. Root hairs are often neglected in root research, partly because of difficulties in their determination and because of the inadequate appreciation of their role.[716,977]

2.0.2 ROOT ANATOMY

The main constituent parts of the root are root cap, epidermis, cortex, vascular stele, and endodermis (Figure 2). The root cap is composed of loosely arranged cells. Not all roots have the root cap, e.g., short roots of a pine. It has no direct connection with the vascular system. Feldman[391] in his review concluded that the root cap constitutes a unique root tissue region as it perceives a variety of stimuli and translates them into a growth or a regulatory response. Its cells provide a slime cylinder, due to secretion of mucigel, in which the soft root can migrate even in hard soil. The root cap is also responsible for regulating the geotropic curvature of roots which is a result of inhibitor production.

The size of the root cap does not change significantly as growth proceeds. The sloughing off of the outer cells is compensated by a rise of the same rate of new cells. For numerous plants, such as maize,[89] bindweed,[946] and oats,[517] the cap is totally renewed every 6 to 9 d. Barlow and Hines[88] and Kadej[634] observed, with the scanning electron microscope, a cap regeneration in maize and peas. Root growth near the root apex is by division of cells and their rapid elongation and expansion.

Fine roots play an important role in biomass production of trees.[15,488,520,802] Although they may constitute less than 1% of the biomass of the mature tree, they account for as much as two–thirds of annual production. Fine roots may be replaced two or three times per year or even more often.[520,531,598]

The surface layer of the root is the epidermis. It is composed of relatively thin-walled,

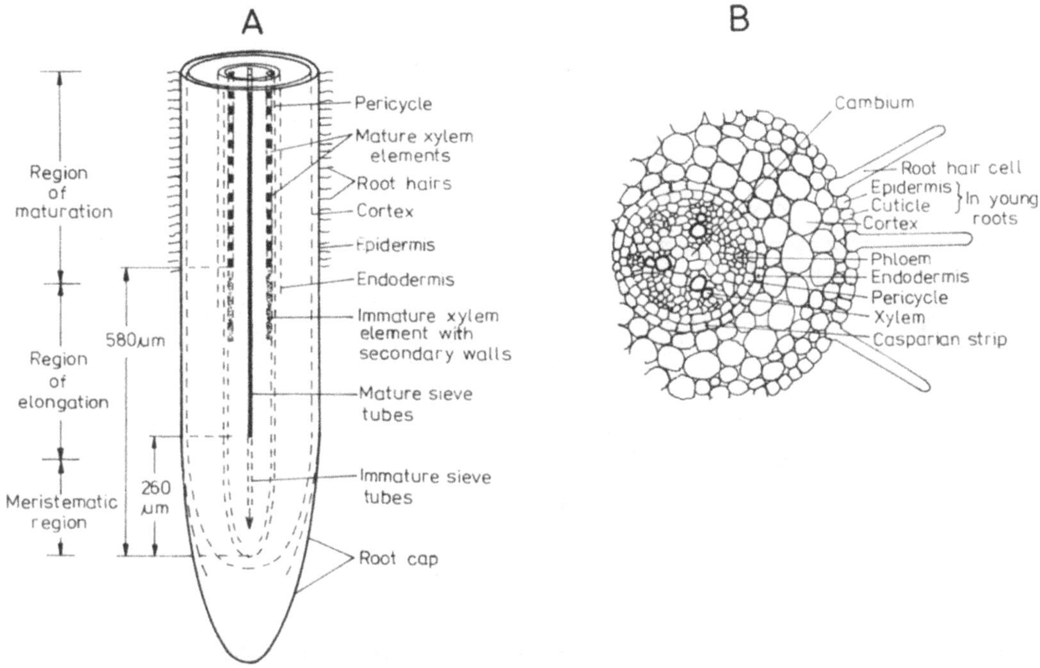

FIGURE 2. Diagrams showing root anatomy (A) longitudinal; cross sectional. (Modified from Kramer, P. J., *Water Relations of Plants,* Academic Press, London, 1983, 123.)

elongated cells, forming, in young roots, a dense external layer. The permeability of the epidermis depends on the degree of its suberization. One of the most characteristic features of the epidermis is the production of root hairs that are protuberances of epidermal cells. Both epidermis and root hair surface are covered with a hydrophobic lipid layer or cuticle which is not a serious barrier to water and solute entrance.

Outside the endodermis, there is the cortex. It is a layer of cells with intercellular spaces between them in the outer part which permit the longitudinal movement of gases. In some species the layer consisting of thick-walled cells with few intercellular spaces named the hypodermis is formed.

Vascular steles are composed of three main elements: the xylem which transports water and solutes to the shoots, the phloem through which metabolites pass downward, and the pericycle around them. The steles are surrounded by a sheath which is one cell thick and is called the endodermis. The first step of endodermis development is a suberin deposition forming Casparian strips (bands) which encircle the radial walls of the cell and may constitute more than one third of the total area. These strips and the suberized bands in other secretory tissues constitute impermeable barriers which isolate one region of extracellular space from another.

2.0.3 ROOT GROWTH

During primary growth of roots, one can distinguish four regions: the root cap, the meristematic region, the region of elongation, and the region of differentiation and maturation (see Figure 2). The initial stage of growth of mono- and dicotyledons is shown in Figure 3. In

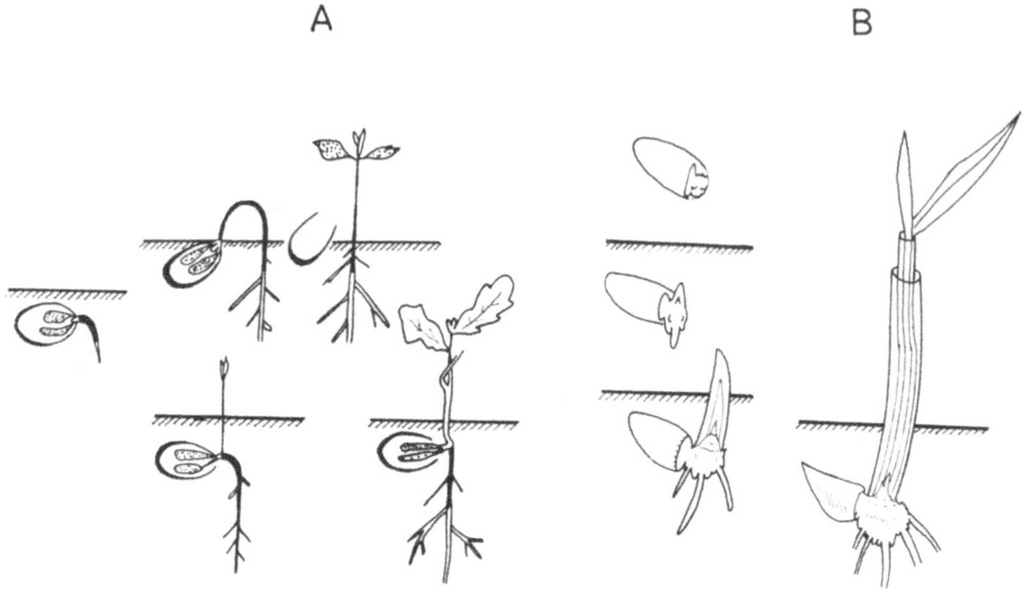

FIGURE 3. The initial stage of growth of (A) mono- and (B) di–cotyledons. (Data from Biologicheskaya Sovietskaya Enciklopedija, (Biological Encyclopedia), Moskva, Izd-vo 1986, 512 (in Russian).)

most of the monocotyledons, the root system develops from the apical meristem. It consists of small, densely arranged, thin-walled cells filled with cytoplasm. Close to the meristematic region, there is the region of elongation in which the cells are considerably elongated. In 1 d, each root tip may produce 18,000 new cells and extend up to 5 cm.[406] Within the next few millimeters, the cells become differentiated into tissues and form the region of maturation. Generally, older parts of the roots of the monocotyledons are of a cylindrical shape of an approximately uniform diameter. Secondary growth mainly concerning the dicotyledons consists in the increase in root diameter due to cell division tangential to the root axis (Figure 4).

Between the xylem and phloem, a cork cambium is formed. As cambial activity increases, the diameter of the stele, the cortex, and epidermis die and decay. The secondary growth is accompanied by lignification of the major part of the root system which restricts uptake of water and solutes. Suberin is wax synthesized by plant roots to control desiccation.

According to Trouse,[1207] roots of most crops can grow to a depth of 180 cm, spreading laterally more than 120 cm in less than a month when plants are grown in good aerial and soil environments. At the height of the vegetative period, as many as 14,000 new roots may emerge from a single plant on a single day. Dominant roots may grow 6 to 8 cm d^{-1} for many days; secondary roots grow slower and for fewer days.

The rate of root growth varies widely among the species and soil conditions, e.g., grasses, 12 mm d^{-1}; pine roots, 2.5 mm d^{-1}; maize, 60 mm d^{-1}.[684] The growth rate also depends on the type of roots within a given plant, e.g., in the case of small-grained cereals, the extension rate of the tap root is 20 mm d^{-1}, while in primary laterals it is 1 mm d^{-1}.[1024] Roots differ widely in their longevity. Some fine lateral roots of trees live only a week or two, whereas other roots live as long as the plant.[684]

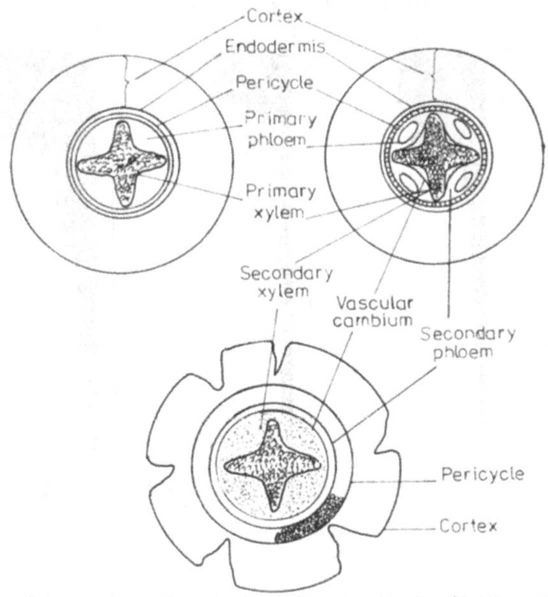

FIGURE 4. Diagram of the secondary growth of a root. (Modified from Kramer, P. J., *Water Relations of Plants*, Academic Press, London, 1983, 130.)

Geometrically, the root system is a set of filaments, the length of which is increased by the elongation and branching of its components. A simulation model of the growth of the root system was described by Lungley.[759] The size of the root system usually increases rapidly during the period of vegetative growth, then ceases or even decreases in weight during filling. A part of the root system often dies before the plant has matured.[1024] The total root system of the sugar beet can attain a total length of 1000 km ha^{-1} d^{-1} to a depth of 2 m.[12] During the vegetative stage of growth of graminaceous species, the relations between the total number, length, surface area, and volume of the root members remained approximately constant.[381,382,500,501] For the above-mentioned relations, a model of the extension and branching of a seminal root of barley and its use in studying relations between root dimensions have been proposed by Hackett and Rose.[502] This model may be useful to interpret measurements of rooting density obtained from fragments of a root taken from the soil. Root distribution in soil depends on plant and soil properties, but generally root quantity decreases sharply with depth (Figure 5) and distance from the plant row (Figure 6).

Madsen[785] studied extensively the distribution of spring barley roots in nearly 200 profiles of Danish soils of different texture and different climatic conditions. He found that root density decreased in a logarithmic sense with increasing soil depth. In pure sandy soils with less than 15% clay + silt (0 to 63 μm), the root density of barley decreases to one tenth for every 20 cm of soil depth, in soil with 15 to 35 % clay + silt for every 30 cm of soil depth, and in soils with more than 35 % clay + silt for every 40 to 50 cm of soil depth. In addition, root zone capacity maps for spring-sown cereals and winter-sown cereals have been constructed.[786]

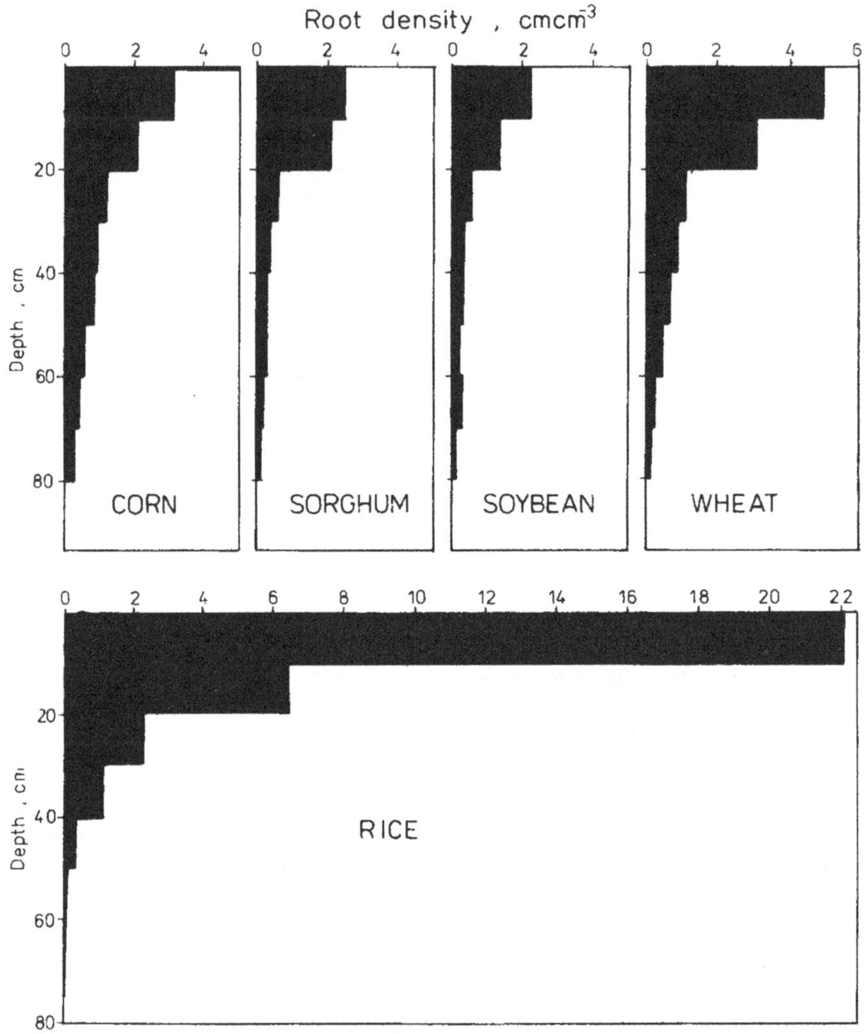

FIGURE 5. Root length density of five crops of flowering stage. (Modified from Hasegawa, S., Parao, F. T., and Yoshida, S., Special Seminar, March 21, 1979, Plant Physiology Department, Tsukuba.)

The rooting density changes with depth can be described by means of the empirical equations used by Rowse and Barnes[1017] (Equation 1) and by Gerwitz and Page[436] (Equations 2 and 3):

$$L_{vi} = (L^{1/2}/m)\exp[-Z_i/(mL^{1/2})] \tag{1}$$

$$P_i = 100[1 - \exp(-fZ_i)] \tag{2}$$

or

$$L_{vi} = L\langle[1 - \exp(-fZ_i)] - [1 - \exp(-fZ_{i-1})]\rangle/(Z_i - Z_{i-1}) \tag{3}$$

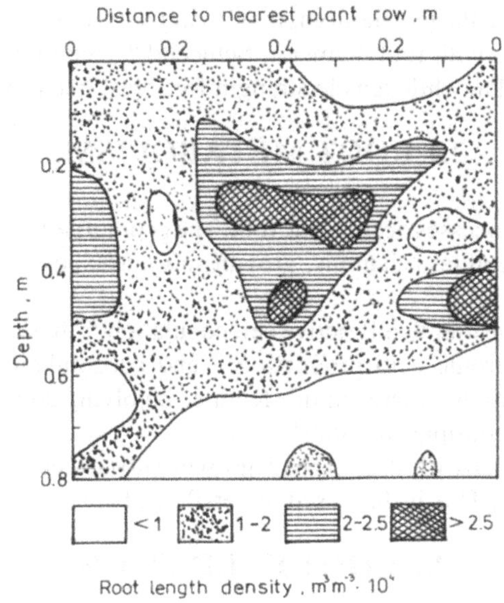

FIGURE 6. An example of soybean root-length density at 91 d, grown in loamy soil. (Modified from Arya, L. M., Blake, G. R., and Farrell, D. A., *Soil Sci. Soc. Am. Proc.*, 39, 437, 1975.)

where L_{vi} = root length density (L_v mm root mm^{-3} soil) for every soil layer (i); L = total root length per a unit length of surface (mm root mm^{-2} soil surface); Z = depth (mm); Z_i = depth (mm) to the midpoint of layer i or to the lower boundary of layer i; P_i = percentage of L at depth of Z_i (mm); $m = Z/(2.303\ L^{1/2})$; and $f = 2.303/Z$.

Modeling of root growth is very complex due to many variables and interactions between them affecting root growth. Lungley[759] proposed a numerical model in order to simulate growth and branching of the root system. The model allows calculation of the total length of roots or distances between them.

Recently, a mathematical simulation model of the three-dimensional root system architecture has been presented.[912] It is a numerical computer model with discrete time steps. The basic rules for axes growth and branching are a translation of the knowledge of the root system morphogenesis. From the three-dimensional branched pattern, it is easy to compute various outputs which can be compared to the observed data.

Based on the experiments and results already published, Pellerin and Tardieu[933] elaborated a concept of the general model for maize root system. Because of the variability of main root parameters, it seems, however, difficult to predict root distribution.

2.0.4 MECHANICAL PROPERTIES OF ROOTS

Mechanical properties of roots are characterized by tensile strength, root pressure, buckling stress, and elasticity. These features, except elasticity, are discussed in Chapter 4, Section III, in connection with mechanical impedance of soil.

Whiteley and Dexter[1296] characterized well root elasticity of 16-crop species using Young's modulus and two parameters, α and β, which account for nonlinearity. Values of the effective Young moduli ranged between 5 and 100 MPa for fully turgid roots. The parameter α ranged from 3.2 to 1.4 and β ranged from 1.1 to 0.7. Environmental factors such as soil water potential and nutrition had significant influences on the elastic properties of roots. Abe and Iwamato,[2] using new method of testing the tensile strength of living roots of *Cryptomeria japonica,* found the increase of this feature with root diameter. Schiechtl[1047] reported that tensile strength for grass roots was 3 to 10 MPa, while tensile strength for forest tree roots was 10 to 70 MPa.

2.1 ROOT METABOLISM

Numerous metabolic processes occur in the root. They comprise the synthesis of amino acids, nucleic acids, proteins, peptides, vitamins, growth substances, dyes and alcaloides, auxines, carotenoides, and porphyrins.[1152]

The root cap contains the mitotically active cell population.[769] It is an area of high respiratory activity connected with uronic acid production[895] and accumulation of plant growth substances.[1196] The root cap also has many unique features involving polysaccharide metabolism[87] as well as perception of geotropic stimuli.[631]

Under physical conditions of stress, plant growth substances are found in roots which regulate root development. This is discussed in detail in Chapter 4.2 and 4.4.

2.2 ROOT EXUDATION

A root in the soil releases organic matter which varies from simple organic molecules to cells and tissues that are sloughed in the process of growth. The main compounds released from roots are the following: exudates, secretions, mucillages, mucigel, and lysates.[1011]

Exudates are compounds of low molecular weight which leak from the cells into either the intercellular spaces and then to the soil via the cell junctions or directly through the epidermal cell walls into the soil. They are not metabolically mediated. A wide range of substances has been found in the exudates. There are soluble carbohydrates, organic acids, amino acids, and growth regulators (e.g., cytokinins and abscisic acid). The role of exudates is significant in the turnover of organic matter in the rhizosphere, in soil aggregate formation, and in nutrient and water absorption.

Secretions are compounds of low and high molecular weights, mucillages released as a result of metabolic processes. Mucillages being a readily available substrate for microbes are poorly defined materials (a) originating in the root cap and secreted by Golgi bodies. Golgi apparatus function is correlated with cellular development and the dictyosomes of the quiescent center characterized by low levels of activity indicated by their small size and absence of secretions.[242,843] The response of the Golgi apparatus to the cap removal is presently unknown, (b) hydrolysates of polysaccharides of the primary cell wall between the epidermal cells and sloughed root cap cells, (c) secreted by the epidermal cells including those secreted by root hairs, and (d) produced by bacterial degradation of the outer primary cell walls of old, dead epidermal cells.

The mucillages, according to Brams,[170] are hydrophyllic polyglucuromate and polygalacturonate polymers across which ions and uncharged solutes should be able to pass freely. Samtsevich[1036] describes a colorless gel-like excretion of a diameter of 1.8 to 4.7 μm and a length of 2.2 to 14 times greater than the root caps that produced them on the root caps of most agricultural crops in soil.

According to Morel et al.,[850] root mucillages exhibit a double granular and fibrillar structure. They are composed mainly of high-molecular-weight polysaccharides, $>15 \times 10^6$ Da, containing fructose, galactose, glucose, aralinose, xylose, mannose units, and especially a notable proportion of uronic acids (galacturonic acid). These compounds are supposed to display some properties of the pectin gels as well as to have hydrophyllic characteristics as cation exchangers. More detailed characteristics of mucillages are given by Oades.[893] Mucigel is the gelatinous material at the surface of the root. It is the root-soil microbial complex including plant mucillages, bacterial cells, and the colloidal, mineral, and organic matter from the soil.

Lysates are the compounds released from autolysis from older epidermal cells. The amount and kind of exudates are affected by the type of the root system, age and stage of development of plants, and by mechanical injuries.[507] Except for plant factors, environmental factors such as temperature and light via shoots, pH, anaerobiosis, mechanical forces, soil-borne microorganisms, and nutrients influence root exudation. In general roots can release 50 to 100 mg of organic matter per gram of root per day.[406]

2.3 SORPTION AND TRANSPORT OF WATER AND NUTRIENTS BY ROOTS

Factors affecting the absorption of water and nutrients by roots can be divided into two groups:[684] (1) those affecting the driving force, which is a gradient in water potential from soil into roots, and (2) those affecting the resistance to water movement through the soil and roots. The efficiency of the root system as an absorbing surface depends both on its anatomy and the extent, age, total surface, and permeability of the surface.

2.3.1 SORPTION AND TRANSPORT OF WATER

2.3.1.1 Absorptive Capacity of Roots

The absorptive capacity of the root system depends on its permeability or conductance as follows:[402]

$$L_R = L_P \times A_R \tag{4}$$

where L_R is the absorptive capacity or root system conductance in $cm^3\ s^{-1}\ MPa^{-1}$; L_P is the permeability or hydraulic conductance in $cm^3\ cm^{-2}\ s^{-1}\ MPa^{-1}$; and A_R is the root surface in square centimeters. Tachibana and Ohta[1157] found that the capacity for water and calcium uptake by cucumber plants was more closely related to the total surface area of the root system than to the root weight. The water uptake by roots is influenced by their distance from a stem base only in the case of very deep roots. Newman and Kramer,[875] summarizing the results on the subject, suggest that if roots extend about a meter or more from the stem base, the resistance to flow along the xylem can be large enough to cause appreciable decrease in water uptake rate with increasing distance.

Taylor and Klepper,[1179] in a rhizotron experiment conducted with cotton, showed that water uptake is proportional to rooting density and to the water potential difference between bulk soil and plant xylem.

2.3.1.1.1 Age of Roots

The range of water absorptive capacity of roots is related to the young roots. The main part of the root responsible for the absorption of water and minerals is situated a few centimeters behind the root tip (Figure 7), with the maximum range at around 20 cm.[1079] The essential role in absorption processes of this part is attributed to root hairs which increase the root surface being in contact with the soil.

Less is known about the function of the older suberized root as an absorbing surface. It is probable that a large part of the absorption of water and solute by some perennial plants occurs through roots that have undergone secondary growth and are covered with layers of suberized tissue.[684] Water absorptive capacity does not increase directly with the increase in length or area of roots because as the young roots appear, older roots mature and thus become less permeable.

Kramer and Bullock,[685] examining suberized roots of loblolly pine (*Pinus taeda*), found a significant increase of water uptake when root diameter increased. The problem of the importance of particular characteristics (root volume and root surface area) in water uptake is controversial. In some cases (i.e., rice), root aging and suberization and also low transpiration

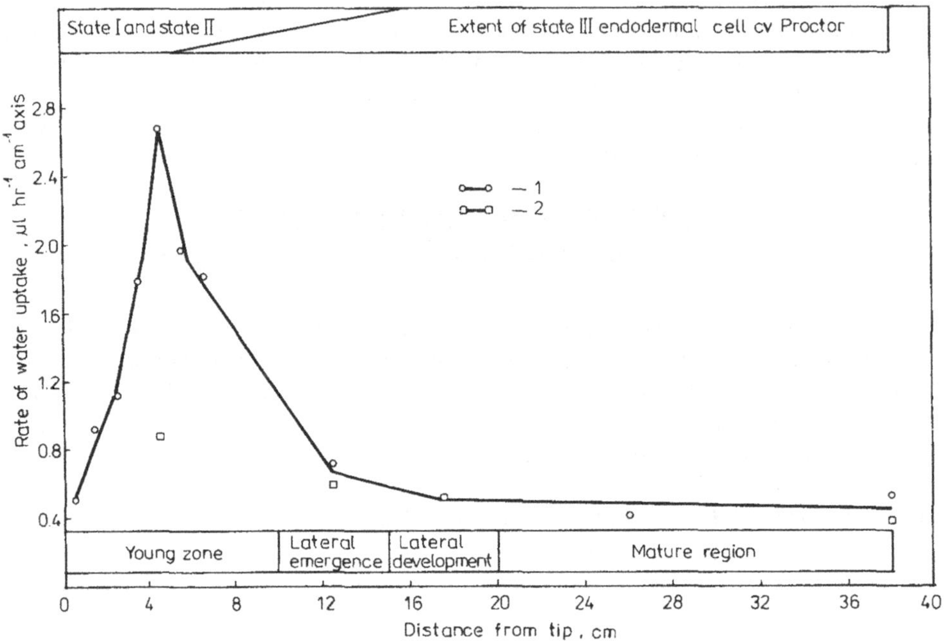

FIGURE 7. The rate of water uptake by intact segments along the length of the seminal axes of barley. (1) 1-cm segments with shoot; (2) 5-cm segments with shoot. (Modified from Sanderson, J., *J. Exp. Bot.*, 34, 240, 1983.)

requirements result in a negative correlation between root density and water uptake by roots (Figure 8).

There is no clear evidence to state that water uptake occurs mainly in younger parts of the root system. It is generally accepted that water uptake by roots is slowed just behind the tip due to poor xylem differentiation. Further regions are characterized by maximum water uptake, rates of which decrease again further from the tip due to suberization.

2.3.1.1.2 Water Uptake Rate and Root Permeability (Hydraulic Conductance)

An important feature of roots is their capability of water uptake. This feature can be expressed by a specific water uptake rate. The data on this characteristic for various crop plants of temperate and subtropical environments have been tabulated by Hamblin[509] and Taylor and Terrell.[1183] They varied from 0.00007 $cm^2\ d^{-1}$ for *Dactylis glomerata* L. to 1.98 $cm^2\ d^{-1}$ for soybean.

Specific water uptake rates and integrated root water uptake depend on soil water distribution in soil profile. In the case of better water status in deeper layers, the highest water uptakes are near the roots in these layers.[359,487] However, in the opposite situation, specific water uptake can be greatest in the shallower soil layers with higher root length densities.[590] The examples of differences in water uptake rates by two crops as a function of soil depth are shown in Figure 9.

Hydraulic conductance of roots is closely related to the plant species (Table 1), age and stage of development, and environmental conditions, especially soil and root water potentials influencing the resistance to water flow. It is believed that permeability of root hairs is much

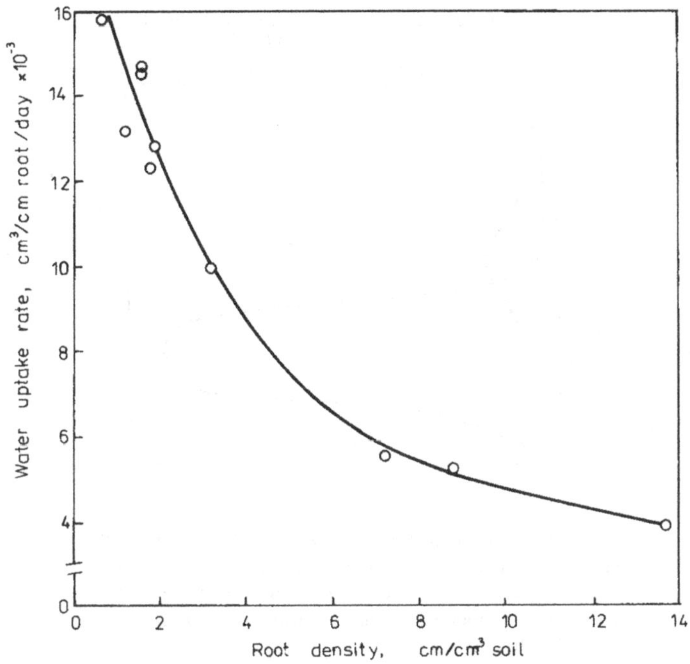

FIGURE 8. Maximum water uptake rate as a function of rice root density. (Data from Hasegawa, S. and Yoshida, S., *Soil Sci. Plant Nutr.*, 28, 191, 1982.)

greater than that of suberized roots. Increased suberization has been observed in dry soil by Leshem[735] and Lyr and Hoffman.[773] According to Newman,[872] suberization of root cells does not necessarily result in reduction of hydraulic conductance, since the deposit ion is often discontinuous and penetrated by plasmodesmata. Older suberized roots have been observed to contribute significantly to water supply in plants subjected to moisture stress.

Severe water deficit and high concentration of salt can result in decreasing root permeability due to dehydration of root cells and increasing suberization of the roots and, thus, increasing root resistance to water flow. The relation between root permeability and water uptake by three plants is shown in Figure 10. This relation has been also confirmed for sunflower by Hernandez and Orioli.[546]

Soil and plant resistance to water flow has been suggested as a significant factor limiting water uptake by roots. The relative importance of soil and plant resistance is discussed by many authors. In some papers it was suggested that the resistance is mainly localized in plants.[358,545,828,869,870,975,1179,1180] There are also various opinions about the location of the resistance to flow of water in roots[96] or in shoots.[106] Other studies[585,628] indicate that soil resistance can be of greater importance, particularly in soils of low hydraulic conductivity and where root density is low. It is suggested that at hydraulic conductivities of soil of about 10^{-8} to 10^{-9} m d^{-1}, the water uptake by roots is limited by soil resistance. Because these conductivities may occur in the range of available water, it can be more appropriate to use the hydraulic conductivity than the wilting point to indicate the limitation of water uptake.[585]

To characterize the actual transport processes, Fiscus[400] and Kaufman and Fiscus[647] proposed the term differential resistance. This term includes the hydraulic conductance (resistance coefficient) and diffusive and active uptake components.

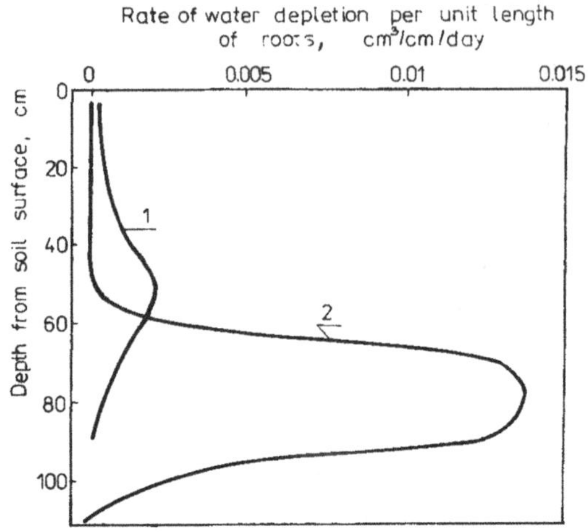

FIGURE 9. Water depletion rates expressed in terms of the length of root at various depths of soil profile. (1) Rain-fed rice; (2) rain-fed cowpea. (Modified from Angus, J. F., Hasegawa, S., Hsiao, T. C. D., Liboon, S. P., and Zandstra, H. G., *J. Agric. Sci. Cambridge*, 101, 669, 1983.)

TABLE 1
Hydraulic Conductance (K) of Roots

Species	Hydraulic conductance ($m\ s^{-1}\ MPa^{-1}$)	Ref.
Beans	$0.8—6.1 \times 10^{-7}$	398
Soybeans	$2.2—5.6 \times 10^{-7}$	398
Sunflower	1.8×10^{-7}	160
Soybeans	$0.8—2.3 \times 10^{-7}$	866
Sunflower	$0.8—2.2 \times 10^{-7}$	866
Maize	1.2×10^{-7}	866
Cotton	1.0×10^{-7}	1179
Barley	$0.3—4.3 \times 10^{-8}$	1126
Wheat	$1.6—5.5 \times 10^{-8}$	1006
Maize	$0.9—4.8 \times 10^{-8}$	1006
Fescue	1.5×10^{-8}	183
Onion	$4.3—9.0 \times 10^{-8}$	33
Clover	1.5×10^{-8}	183
Tomato	6.1×10^{-8}	871
Pine (suberized)	7.55×10^{-8}	1040
Pine (unsuberized)	1.95×10^{-7}	1040

Klepper[670] discussed resistances in the soil-root-xylem pathway and the low-resistance interconnections among various parts of the root system by xylem elements ("T-junction" and "junction resistance"). On this basis she explained the lack of significant deliveries of water from deeper wet soil layers to drier top layers by the root system.

So et al.[1092] measured *in situ* and simultaneously the flux of water into the roots of maize and the water potentials at the root-soil interface and in the xylem of a root growing in soil. They obtained a linear relation between the root resistance to water flow and the water potential of the

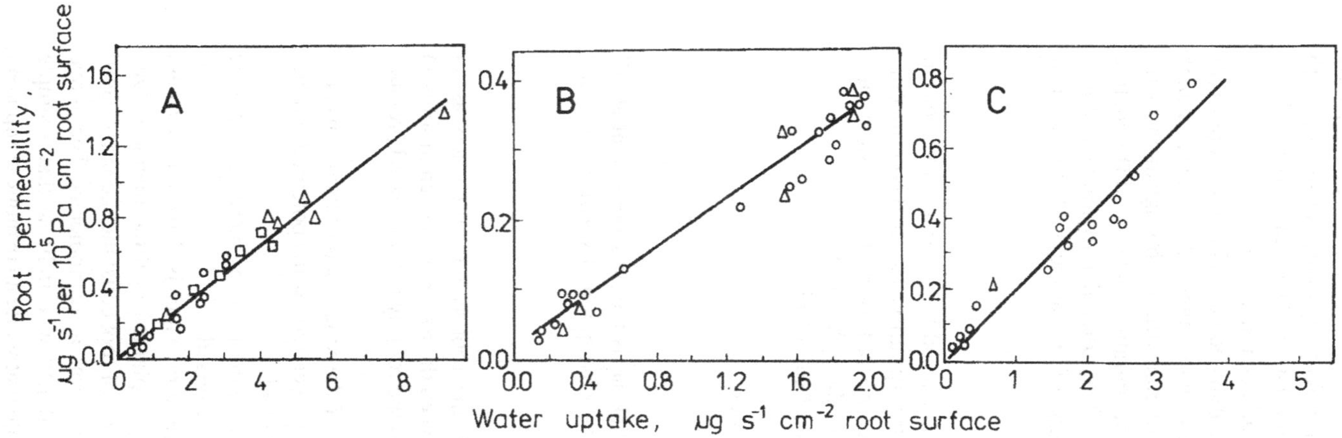

FIGURE 10. Water uptake and root permeability of 5- to 8-week-old (A) sunflower; (B) barley; and (C) maize. (Data from Aston, M. J. and Lawlor, D. W., *J. Exp. Bot.*, 30, 169, 1979.)

root tissue. The rhizosphere resistance was of the same order of magnitude as the root resistance when the soil moisture potential of the rhizosphere was from about −0.1 to −0.2 MPa.

Diurnal fluctuations in root resistance to water flow in the root system have been reported.[96,401,918,1077] How the transport coefficients change diurnally is known, but the question of mechanisms of the changes remains unanswered.[401] They can be related to the rhythms in root growth or carbohydrates or growth regulator transport to the root.[182,1210] The data obtained by Bunce[182] indicate that root/shoot interactions occur which affect apparent root resistance to water flow and complicate interpretation of the whole plant data on leaf water potential and transpiration in terms of the dependence on the root hydraulic characteristics.

As the rate of transpiration increases, the root water potential decreases, but not proportionally, due to capacitance effects and probable changes in the root resistance. Also the resistance to water movement in roots declines with the increasing rate of transpiration through them. Daily fluctuations of water absorption by roots in relation to the transpiration rate of 11 different tree species were measured by Braun and Marte.[171] Sands et al.[1040] reported that hydraulic conductance of the loblolly pine seedling roots was independent of the amount of mycorrhizal associations.

2.3.1.2 Water Absorption

Movement of water from the soil into the root system is caused by osmotic movement in slowly transpiring plants and mass flow in the case of rapid transpiration. Hence it is commonly stated that osmosis contributes little to water uptake as compared to the transpiration pull. Driving forces are due to the tension or negative pressure in the xylem sap.

The forces involved in water absorption (A) can be described as follows:[684]

$$A = \frac{(\psi_m + \psi_o)\ \text{soil} - (\psi_p + \psi_o)\ \text{root}}{r_{soil} + r_{root}} \tag{5}$$

where ψ_m is the matric potential; ψ_o is the osmotic potential; ψ_p is the pressure potential (all expressed in MPa); and r is the resistance to water in s cm^{-1}. Absorption of water by roots can be osmotic (active) or passive.

2.3.1.2.1 Osmotic (Active) Absorption of Water

In optimum physical conditions of root environment, the root system often acts as an osmometer due to the difference in the water potential between the root xylem (rich in solutes) and soil (poorer in solutes). As a result, water diffuses into the root by the decreased pressure or tension produced in the xylem sap.

Root water potential depends on the osmotic potential of the xylem sap (which is usually higher than −0.2 MPa) and soil water potential. The relation between soil water potential and root water potential of loblolly pine seedlings is shown in Figure 11.

It was found[1080] that the water potential in tree roots growing in moist soil increased from the base toward the root apex and in dry soil the gradient was reversed.

Alternatively, the positive pressure is often called root pressure, which is also used to express mechanical growth forces of roots. This pressure can be explained in three manners. One explanation consists of theories that assume secretory activity of root cells; the second is based on various electroosmotic theories; the third considers the roots as osmometers. The above explanations are described in detail by Kramer[684] and Miller,[831] who indicate that a simple osmotic process is most reliable. Some authors[856,1347] indicate that while an osmotic component certainly exists, the xylem concentrations are inadequate to account for the total pressure and that therefore the remainder must result from the active water pumping.

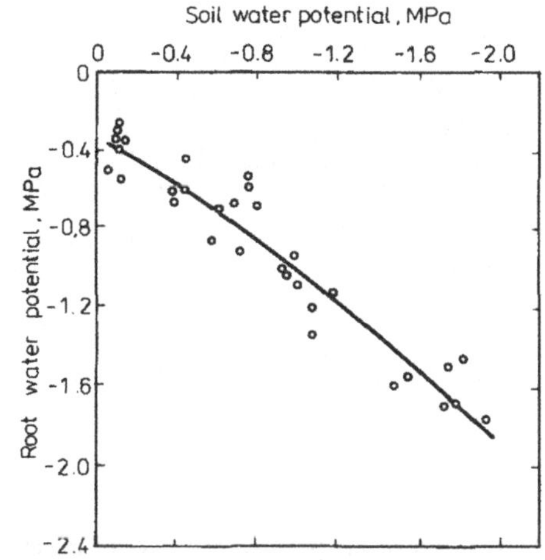

FIGURE 11. Relationship between soil water potential and root water potentials of loblolly pine seedlings sampled at noon. (Data from Kaufmann, M. R., *Plant Physiol.*, 43, 281, 1968a.)

Diurnal periodicity in water root pressure and exudation rates with the highest values at midday and the lowest at midnight is reported.[401,775,1222] The processes are influenced by deficient aeration, low temperature, and other soil physical conditions, the role of which will be discussed in Chapter 4.

2.3.1.2.2 Passive Absorption of Water

This kind of absorption occurs as the rate of transpiration increases. Increase of mass flow of water through the roots dilutes the root xylem sap, causing gradual transition from osmotic flow to nonosmotic. In this case, the root acts as an absorbing surface through which water moves by increasing tension in the xylem sap. Endotrophic mycorhizae decrease resistance to the entrance of water into roots.[684]

2.3.1.3 Water Movement in Roots

There are two paths of water movement within the root: radial and longitudinal.

2.3.1.3.1 Radial Water Movement

Water enters the roots through epidermal cells being in contact with moist soil. Then the water passes through cortical cells, the endodermis, pericycle cells, and into the xylem elements. There are three possible pathways for radial water movement from the epidermis to the xylem (Figure 12):

1. Across the cortical cells and their walls (transcellular)
2. Through the symplast (symplasmic)
3. Through the cell walls[684]

The path of radial movement in roots that have undergone secondary growth presumably occurs through parenchyma cells. In water movement, all three pathways may take part, but their contribution is not fully explained.[623,873] Recent studies of the hydraulic conductivities of epidermal, cortical, and endodermal root cells of wheat and maize showed that in freshly excised roots, the bulk of water movement is unlikely to be via the transcellular pathway.[1006]

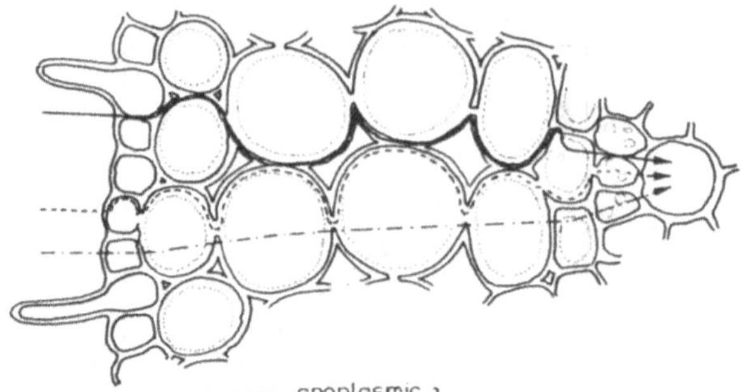

FIGURE 12. Diagram of the possible pathways of radial water transport in barley root. (Data from Steudle, E. and Jeschke, W. D., *Planta,* 158, 237, 1983.)

The radial water flow occurs between the root surface and the xylem elements. Their hydraulic conductance or resistance is calculated on the basis of the radii and lengths using Poisseuille's equation. This flow is limited by the hydraulic conductivity (conductance) of a root which is affected by the structure of the osmotic barrier made from the epidermis, cortex, Casparian strips, and endodermis.[1070] The Casparian strips are considered by Newman[872] to be the greatest barrier to radial water flow. Also, components of apoplasmic, symplasmic, and transcellular transport are involved. High resistance to the radial movement may prevent excessive dessication of roots in dry soils.

The driving force for the radial flow is a result of both the atmospheric demand and the root conductance and also a function of the osmotic phenomena opposing (or adding) flux at the root-water flow barrier.[400] It was found that the hysteresis phenomenon in the flow rate may occur when the roots are subjected to stepwise changes in hydrostatic pressure.[400,694]

Water transport through the root system or segments of roots is expressed by the relationship:[400]

$$J_v = L_p(\Delta P_h - \sigma\Delta\pi) \tag{6}$$

where J_v is the flow of volume per unit area of root surface ($cm^3\ cm^{-2}\ s^{-1}$), ΔP_h is the hydrostatic pressure difference between the outside and the inside xylem elements of the roots (MPa); σ is the dimensionless reflection coefficient or osmotic efficiency; $\Delta\pi$ is the osmotic pressure difference between the inside and the outside of the roots (MPa) and L_p is the so-called hydraulic conductivity coefficient ($cm^3\ cm^{-2}\ s^{-1}\ MPa^{-1}$).

The above equation used in plant physiology is based on irreversible volume flow through semipermeable membranes. Also, water transport in roots can be described by the equation including the hydraulic head concept, widely applied in soil science:[400]

$$J_v = L_p(\Delta H - \sigma\Delta\pi) \tag{7}$$

where ΔH is the difference in hydraulic head ($\Delta H = \Delta h_p + \Delta h_g$); Δh_p is the difference in pressure head, and Δh_g is the difference in gravitational head.

2.3.1.3.2 Longitudinal (Axial) Water Movement

Longitudinal water movement through the xylem to the leaf is closely related with the transpirational flux. It is usually assumed that the flow through the xylem has a large conductivity or negligible resistance. Xylem water flow decreases with age of the root or with the increase of root length.[186]

In some cases, water could be transported along water ducts from roots to leaves graviosmotically, along the xylem tracheal elements. An osmotic flow transport of water in membrane systems directed against the force of gravity was discovered by Przestalski and Kargol in 1972[960] and called graviosmosis. This phenomenon was further well described in other publications[344,555,641,642] and the results were summarized.[961]

2.3.1.4 Translocation of Water in the Soil by Roots

Some authors[846,1079,1081] reported translocation of water through the roots from moist to dry soil. Kirkham[666] showed by means of a physical model (wick simulating a plant) that water can move from a wetter side of a root system to a drier side or vice versa. The quantity of translocated water is small in relation to that being extracted. According to Klepper,[670] due to the backflow, more water is delivered into deep dry soil layers than in the case of upward direction from deep wet layers to drier surface.

Baker and Van Bavel[67] concluded from the experiment on bermuda grass that water can move from root to soil if the potential gradient is in that direction, although other factors such as soil hydraulic conductivity may become limiting. Slavikova[1080,1081] showed that the root system of *Fraxinus excelsior* grown in soil with differentiated humidity behaved as a hydrodynamic unit in respect to suction force. The individual parts do not respond to changes in soil humidity separately by changes in their suction force, but rather in a mutual relationship which is brought about by gradients in the suction force. These gradients are the cause of water translocation between individual branches of the root system.

2.3.2 SORPTION AND TRANSPORT OF NUTRIENTS

Various chemical and physical factors of the soil rhizosphere zone, as well as plant properties, are responsible for mineral nutrients sorption by the root system. According to Jungk and Claassen,[629] factors of great importance in this matter are

1. Chemical nature of a nutrient, its concentration in soil solution connected with the rhizosphere pH and aeration status
2. The location of the nutrient in the soil relative to plant roots
3. Mobility of the element under investigation in relation to the mass flow and diffusion which implies the ability of plants to create gradients of nutrient concentration and water potentials in the rhizosphere
4. Growth, distribution, and shape of the roots, as well as their physiological ability to mobilize and absorb nutrients
5. Root exudation of low-molecular-weight organic compounds (e.g., amino acids), high-molecular-weight gelatinous materials (mucillage), and sloughed-off cells and tissues may mobilize mineral nutrients either directly (e.g., by chelation) or indirectly by providing the energy for microbial activity in the rhizosphere.

Nutrients enter the roots in aqueous solution and the routes which they and water follow from the soil are similar. However, the rate of absorption of water and nutrients from external solution, as well as the individual ions, is different.

When the entire root system is placed in a solution of uniform concentration, the concentration in the sap of the xylem vessels is frequently much greater than in the external solution. During the physical movement of ions (by diffusion or mass flow) into the free space of the root (across the endodermis to the xylem), the influx of cations is greater than that of anions due to the predominance of negatively charged sites on the surfaces of cell walls.

The quantities of cations and of anions which enter roots are frequently not equal, but they are balanced by the outward movement.[1024] The amount of nutrients which enters a root from a solution of constant concentration depends moreover on the salt status of the plants and the rate of transpiration (metabolic demand).

The relationship between root fluorescence and their ion uptake (e.g., iron absorption by soybean) was shown by Dyer and Brown.[346] Ions inside the root may move through the "active" or "passive" transport.[1024] "Active transport", depending on metabolism, energy, provided normally by respiration, consists in moving ions from a zone of lower to a higher electrochemical potential across membranes, against concentration gradients. During "passive transport" (physical process), the movement of ions of one sign by active transport can cause ions of the opposite sign to move down a concentration gradient. These two kinds of transport are reflected in absorption of nutrients by intact roots when the external concentration of nutrients is low (Mechanism I, showing characteristics of an active process) or high (Mechanism II, showing characteristics of a passive process).

Most mineral nutrients enter roots in a free ionic form, but there are exceptions, e.g., urea or chelates of trace elements. To characterize the flux of nutrients across the root surface, the term "root absorbing power" is used.[888] It is expressed as the rate of nutrients arrival per unit surface area (x):

$$x = \frac{\text{Flux (g cm}^{-2}\text{ root surface s}^{-1})}{\text{External concentration (g cm}^{-3})} \tag{8}$$

Jungk and Claassen[629] described the phosphate and potassium flux into roots in relation to their concentration in the form of a saturation curve which can be expressed by the Michaelis-Menten equation as modified by Nielsen:[870]

$$I_n = \frac{I_{max}(C - C_{min})}{K_m + C - C_{min}} \tag{9}$$

where I_n is the net influx of a nutrient; I_{max} is the maximum influx (the parameter for the capacity of a root to absorb a nutrient); K_m is the Michaelis constant which describes the affinity of the uptake system for a nutrient; and C_{min} is the minimum concentration to which plants can deplete a solution. Torbert et al.[1195] calculated the magnesium uptake rates by tall fescue from the change in total Mg content and the change in fresh weight of fescue propagule roots using the following equation:

$$I_m = \frac{M_2 - M_1}{WR_2 - WR_1} \ln \frac{(WR_2/WR_1)}{t_2 - t_1} \tag{10}$$

where I_m is uptake rate per gram of fresh weight of root; M is total elemental content in tall fescue propagule (leaves and roots); WR is fresh root weight; and t is time (days).

An important factor for nutrient competition in plants is the cation exchange capacity of roots (RCEC). It is mainly due to the presence of carboxyl groups (–COOH) connected with chemical constituents of cell walls (uronic acid, pectin, pectic acid, cellulose, and other organic compounds)[336,852] and it is affected by plant species and variety, age, and growth conditions.[280,336,338] The RCEC values of major species of shrubs, forbs, and grasses have been listed in Table 2.

TABLE 2
Average Root-Cation Exchange Capacity of Major Plant Species

Species	Average root REC (meq/100 g)	Ref.
Shrubs		1334
Artemisia nova	36.8	
Artemisia tridentata	40.2	
Atriplex confertifolia	16.9	
Ceratoides lanata	12.6	
Chrysothamnus nauseosus	15.7	
Chrysothamnus viscidiflorus	29.4	
Leptodacytylon pungenus	22.2	
Purshia tridentata	36.6	
Qurecus gambelii	20.1	
Tetradymia canesceunus	14.3	
Xamthocephalum sarothare	21.8	
Forbs		1134
Agoseris	60.2	
Astragalus spp.	60.2	
Balsamorrhiza hookeri	29.4	
Cryptantha spp.	52.3	
Iva axillaris	36.5	
Lomatium spp.	27.9	
Pestemon spp.	36.5	
Phlox longfolia	19.2	
Spheaeralcea spp.	25.2	
Thelesperum subnudum	26.6	
Grasses		1334
Agropyron cristatum	15.2	
Agropyron spicatum	15.3	
Bouteloua gracillis	15.2	
Hilaria jamesii	8.4	
Poa longiligula	23.3	
Poa sandbergii	14.6	
Oryzopsis hymenoides	13.2	
Sitanium hystrix	11.5	
Sporobolus cryptanrus	13.2	
Pear	7.6	852
Peach	6.0	
Unshu (citrus)	15.7	
Grape	11.7	
Kaki	7.7	
Lotus	34.9	530
Clover	45.8	
Trifolium pratense	34.0	820
Lolium perenne	24.0	
Hordeum sativum	20—30	998

Drake et al.[324] pointed out that the RCEC of dicotyledons was higher than that of monocotyledons. For these reasons, dicotyledons often take up polyvalent cations more efficiently than monocotyledons do. In contrast, monocotyledons frequently take up monovalent cations in relatively greater quantities than dicotyledons.[539] Since the negative charge is conceived of as

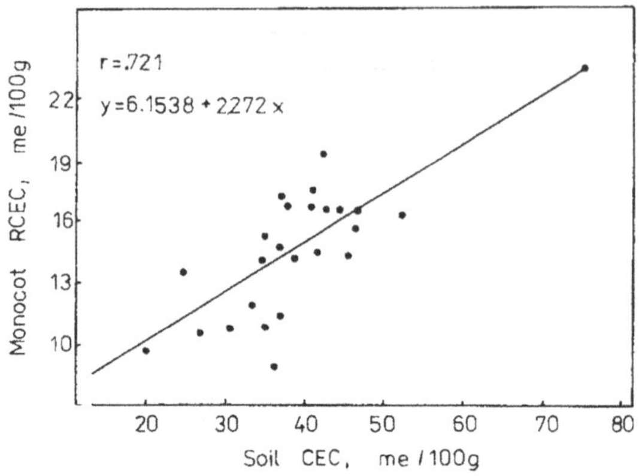

FIGURE 13. The relationship between the average monocotyledon RCEC and soil CEC. (Modified from Woodward, R. A., Harper, K. T., and Tiedemann, A. R., *Plant Soil*, 79, 169, 1984.)

being uniformly distributed over the root surface, polyvalent cations are more strongly drawn to the root surface than monovalent ions, assuming all other things to be equal.

Some authors[336,1334] have reported a positive correlation between RCEC and soil CEC (Figure 13). The increase in RCEC caused the increase in sorption of N,[51,810,852] Ca and Mg,[852,855] P,[855] and K[820] by roots or the decrease of K[324,852] and Na.[1263]

Kulkarni and Savant[696] reported the effect of soil compaction increase from 1.1 to 1.85 mg m^{-3} on RCEC increase of eight crop plants. They tried to explain it by the increase in percentage of the nitrogen content of the roots and carboxyl groups on the roots. Investigating the relationship between the RCEC and plant behavior, Chamucah and Dey[219] found that the RCEC of 15 genetically different tea clones were inversely proportional to the $K/(Ca)^{0.5}$ ratio of the plant tops. According to Crooke and Knight,[273] crop yields of 16 varieties of leeks were directly proportional to their RCEC.

Particular attention has been drawn to the susceptibility of some plant species to toxic elements, e.g., aluminum. It has been clearly demonstrated that adverse proportionality between plant resistance to toxic elements and RCEC existed.[857,1226,1262] Amory and Dufey[27] proposed a mathematical model which allows prediction the composition of root exchange complex in a wide range of soil solutions. Increased lime additions to soil increased the RCEC of lotus (from 35 to 40 meq/100 g) and clover (from 46 to 51 meq/100 g). Liming caused a decrease in P, Al, Mn, and N accumulation in the roots.[530]

In natural conditions, different parts of the root system have different abilities to absorb and translocate nutrients. Also root age,[693] volume, weight, length, surface area,[12,128,129,1157,1195] irregular root distribution in the soil and partial contact between roots and soil, interactions with mycorrhizal partners, soil moisture, aeration, temperature, and pH influence nutrient absorption by roots.[1236] The influence of the physicochemical soil properties on ion uptake by roots has been described in Chapter 3.2

When most of the ions absorbed by the roots affect its metabolism and growth positively, some of them, e.g., aluminum[113-117,532] and nitrite,[728] are toxic. The excess of Al resulted in stunted and coralloid root systems due to disruption of mitosis in root tips,[574] in decreased effectiveness of the roots,[1074] and in increased root rigidity by cross-linking pectins.[383] Dufey et al.[338] reported that the quantity of Al adsorbed by the roots is closely related to root cation exchange capacity of root species ($r^2 = 0.991$) and for this reason, this root characteristic might be used as a screening criterion for Al fixation capability.

Root uptake of nutrients from the soil is a function of the parameters divided into three categories: root, rhizosphere, and bulk soil. For example, in the rhizosphere soil, i.e., soil zone deeply modified by root activity, the roots may cause biological weathering of phosphates,[491] carbonates,[610] and silicates.[554] The complex interaction of these parameters needs mathematical models and quantitative data for these models.

Taking into consideration an equation of Nye and Marriott,[891] Claassen and Barber[233] developed a model to calculate nutrient uptake by plants from the soil. Ten parameters of soils and plant properties were used: soil solution concentration, differential diffusion coefficient in the soil, the maximum influx, the Michaelis constant, minimum concentration to which plants can deplete a solution, root radius, distance between roots, water influx, the root growth constant, and initial root length.

Models of nutrient uptake by plants include other root parameters such as rate of root extension, mean root radius, and mean root hair density and length.[598,599,890] Other authors also include environmental factors such as temperature, pH, and soil moisture in the model.[779,780,849] Cushman[278] introduced root age as a parameter to numerical study on nutrient uptake by plants.

Claassen et al.[236] verified the mathematical model by simulating K uptake from the soil. de Willigen[301] studied theoretical aspects of limited soil-root contact for the uptake of water and nutrients by roots.

Van Noordwijk[1233] suggests that root length per unit of soil volume can be used for theoretical estimates of the water and nutrients which can actually be taken up by the root system at the required rate while root dry matter per unit volume of soil is important in the description of soil life. The relative depletion capacity of roots of various diameters, when compared on a volume or weight basis, increased as root diameter decreased.[303]

2.4 RELATIONS BETWEEN ROOTS AND ABOVE-GROUND PARTS

Terrestrial plants have specialized organs (roots and shoots) to exploit the environmental conditions. Shoots use solar energy for photosynthesis and production of growth metabolites. Roots anchor the plants in the soil and absorb water and nutrients. In spite of these separated functions, both parts of the plant interact in the growth and function of the plant. In constant environment conditions, there is a logarithmic linear relationship between shoot and root weights during vegetative growth.[1024] This relationship is interferred with by variation in factors of the environment.

There are very close relationships between roots and shoots. Shoots supply roots with carbohydrates, growth regulators, and organic substances and receive from the roots water, nutrients, and growth regulators. The indicator of the balance between these two parts of plants is the root to shoot ratio (R:S). This ratio depends on the kind of plant, e.g., Beta (3.06), *Solanum tuberosum* (1.54), *Zea mays* (1.54), *Hordeum* (0.25),[684] genotype,[143] and the stage of plant development and growth conditions.[1024] Anderson-Taylor and Marshall[31] found that the spring barley root system reached maximum mass 6 weeks from sowing when the shoot weight was only 50% of its value at maturity.

Heen[533] reported a decrease of the R:S ratio during the first 2 months of vegetative growth for various cereals (barley, oat, and wheat) from 7.30 to 6.04. The R:S ratio of plants increases with water stress which reduces leaf growth more than root growth.[920,1212] The R:S ratio of grass in the vegetative phase is constant under constant conditions (water and fertilizers).[373] Stable R:S relationships are maintained by root systems able to compensate for root members that either do not develop or are lost. Removing a portion of a root system results in a faster relative growth rate.[586]

Jordan et al.[625] showed that a compensatory growth within the existing root member of grain sorghum was capable of maintaining root length and volume, but not root dry weight when crown root members were severely reduced. The rapid growth compensation was attributed to competition between shoot and root for available carbohydrate and N, with the organ nearest the source being the greatest in fulfilling its needs.

According to Yoshida et al.[1341] the deep root system of rice and sorghum is correlated with tall plant stature and a low tiller number. The weight of superficial roots of rice is linearly correlated with crop yields within a range of yield up to about 6000 kg,[1338] while beyond that range, any distinct relationship was not observed.[655]

Root-shoot relations are modified by nutrients. Barraclough[93] reported a considerable variability in R:S ratios for different crops during winter and early spring which resulted partly from differences in soil N availability and timing of application of fertilizer nitrogen. Synthetic plant growth regulators (PGR) influenced gibberellin synthesis and altered the proportion of roots to shoots and nutrient uptake when applied to plants.[494,627,692] Eavis and Taylor,[350] investigating the influence of soil water content and NPK fertilization on soybeans, found differentiated root length to leaf area ratio (RL:LA) in the range of 39 to 140 cm cm^{-2}. The R:S ratio can be reduced by a shortage of some elements in the root environment, e.g., Zn for wheat.[189]

Correlations between the root density of three upland rice varieties[792] and wheat[416] and leaf water potential were found. Some plants are characterized by a definite pattern of root and shoot development, e.g., the number of roots on any culm of wheat can be predicted from the number of leaves of the culm ($r^2 = 0.9$).[670a]

Fiscus[398] reported that the ratio of root absorbing surface to leaf transpiring remains constant in beans grown under two levels of irradiance. A relatively stable relationship exists among absorbing, transport, and transpiring tissues as a result of internal feedback processes.[1294] Taylor[1172] questioned if cotton and soybeans need as many roots for water absorption as they produce.

Varietal differences of the S:R ratio (3.9 to 7.8) for five wheat lines were reported by Troughton and Whittington.[1204] They also cited data showing that the R:S ratio of low heritability in maize and oats and relatively high in *Pinus taeda*. The close relationship between root length density due to irrigation and grain yield of maize was obtained by Aina and Fapohunda[18] (Figure 14). The decrease of root weight to overground part ratio with increase of soil compaction level has been reported for soybeans[1138] and oats.[1060] For maize this ratio was similar in loose and compacted soil.[1166]

The effect of the decreasing root growth in mechanically impeded conditions on shoot growth and finally the yield itself depend on which type of root and which of their parts is restricted. Barry and Miller[97] found, in controlled conditions, that mechanical restriction of the nodal root length (by 54%) of maize did not significantly reduce shoot growth and final yield (Table 3). Authors suggest that mechanical impedance of nodal roots does not cause imbalances of growth regulators in the shoot and that seminal roots are able to compensate for a restricted nodal root system. However, dry soil treatment resulted in significantly less shoot and root growth.

The drop of leaf water potential of the sunflower to around −1.5 MPa was accompanied by 25% contraction of root tissues compared to their turgid volume.[385] According to Van der Valk,[1229] variation in soil aeration status did not change the R:S ratio of a tulip. Some researchers[313,1254] imply that a reduction in root growth due to soil physical conditions may not cause a yield reduction if the plant is not stressed for water and nutrients. Thus, timely rainfall or irrigation reduces the possibility of yield decreases. In water-stressed conditions, the R:S ratio usually increases. This is due to both continued root growth to explore a larger soil volume and reduced shoot growth to limit transpirational area being important adaptations to drought.[577]

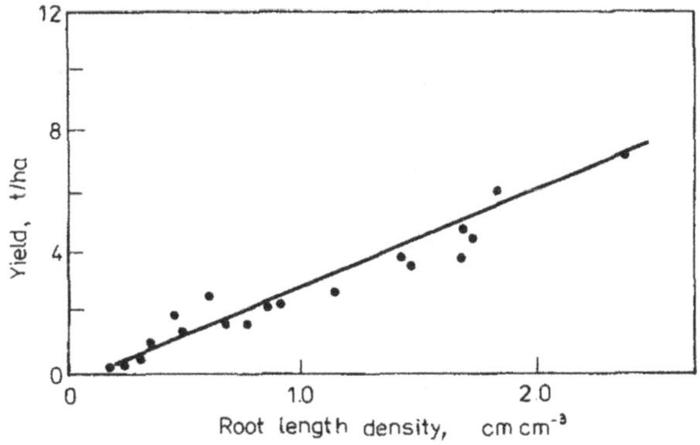

FIGURE 14. Relationship between maize grain yield and root length density. (Modified from Aina, P. O. and Fapohunda, H. O., *Plant Soil*, 94, 257, 1986.)

TABLE 3
Plant Height, Shoot Dry Weight, and Length and Fresh Weight of Nodal Root System of Corn Plants as Related to Various Soil Treatments

			Nodal roots	
Treatments	**Plant height (cm)**	**Shoot dry wt ($g\ plant^{-1}$)**	**Length ($m\ plant^{-1}$)**	**Fresh wt. ($g\ plant^{-1}$)**
Control	96.8	9.29	192.5	30.6
Compaction	93.2	8.20	88.1	18.1
Dry soil	82.8	7.56	18.2	15.3

Data from Barry, D. A. J. and Miller M. H., *Can. J. Soil Sci.*, 66, 689 1986.

Chapter 3

SOIL-ROOT INTERFACE

3.0 INTRODUCTION

Considering the phenomena occurring at the soil-root boundary, it is necessary to define some terms, such as "bulk soil", "soil-root interface", "rhizoplane", and "rhizosphere". "Bulk soil" denotes the zone of soil which lies outside the rhizosphere zone and which is not directly influenced by growing roots except by the withdrawal of water and nutrients. The term "soil-root interface" is defined as a boundary between the outermost root tissues and the particles of soil which are closest to them.

"Rhizosphere" is a zone of microbial activity within and around the roots. It is provided with substances rich in energy and minerals released by the roots. It may also be defined as the volume of interaction between a plant's rooting system and its immediate soil environment.[1035] The rhizosphere is sometimes divided into two regions:[772] endorizosphere, including the inner root region (stele with conducing elements: xylem and phloem, epidermis, cortex, endodermis, and root cap), and ectorhizosphere, including the outer root region (plant and bacterial mucigel, root hairs, plant mucilage, and sloughed root cap cells). Newman and Watson[876] proposed a mathematical model for the growth kinetics of the rhizosphere and rhizoplane regions including in it the changes in substrate concentration with time, considering the root as an idealized cylinder. This model was not proved by measurements and has many limitations. "Rhizoplane" refers to the zone of attachment of microorganisms to the root surface itself (it is the external root surface).

3.1 PHYSICAL ASPECTS

3.1.1 ROOT-SOIL CONTACT AREA

The main physical aspect of the root-soil interface is the root contact area between the root and the soil. The degree of this contact affects the movement of water and nutrients from the bulk soil to the root (see Chapter 4). It is known that this contact is not continuous and it can be impaired either by the enlargement of the cavities or by the decrease in diameter of the roots themselves.

Loss of water from the swelling soil due to both evaporation and its removal by the roots leads to shrinkage and thus cracking of the soil and in consequence rupture of the roots. A similar effect can be caused by soil movements due to frost action. On the other hand, roots may shrink at low water potentials and thus loose contact with part of the wall of their channel.[892]

Root contact area with the soil can be increased with time due to cambial increase in root diameter in the case of dicotyledons. The monocotyledons have no such progressive thickening, but the diameter of their roots can decrease as they age because of the degeneration of the cortex due to invasion by microorganisms or dessication. Also diurnal changes in the diameter of some plant roots were reported by Huck et al.[579] for roots of cotton plants and by Faize[384] for sunflower roots.

Lack of contact between root and soil may be reduced by the following effects:[892]

1. Root hairs tend to proliferate in the humid air of large soil pores and channels.
2. Even though pores of a diameter exceeding 10 μm are drained at –300 hPa water potential, there is always a film of water on the root and root hair surfaces, linking them to the soil moisture continuum.
3. Roots tend to excrete a layer of mucilage up to 5 μm thick, but on average it is more like 1 μm thick, which will bridge small gaps.[478,617]

The mechanical stresses at the root-soil interface prevent root growth. These stresses expressed by a penetrometer point resistance vary for different plants from 1.0 to 3.4 MPa.[73,85,245,248,249,1169]

Collis-George and Yoganathan[248,249] used the spherical cavity expansion theory of Vesić (see Chapter 1) to estimate quantitatively the stress at the root-soil interface that prevents germination, root and coleoptile growth, and emergence of spring wheat. They found 2.3 MPa as a limiting value for root growth. There are some suggestions concerning soil particles entering the root and thus forming the endorhizosphere after the cells of the epidermis and the cortex have been destroyed by the soil microbes. This assumption was proved by Weiming et al.[1285] on the basis of high concentration of Si observed in the tissues outside the root endodermis.

3.1.2 ROOT-SOIL WATER RELATIONS

As the roots extract water in wet soil, the rate of their water uptake decreases with the decrease of soil water content. The decrease is explained by the increase of hydraulic resistance in the soil-plant system. This resistance, named "the rhizosphere resistance" or "interfacial resistance" appearing in the soil immediately adjacent to each single root, is distinguished from the so-called "pararhizal resistance" which concerns the movement of water through the bulk soil to the rooting zone.[870,1024]

If the interfacial resistance is sufficiently high, the roots of a rapidly transpiring plant may reduce the water potential of the neighboring soil to the value near the wilting point even though the water potential at a short distance away is close to field capacity. Roots can effect water in their immediate vicinity and the quantity of water they can extract is closely related to the total density of roots. Some authors, e.g., Newman,[870] Cowan and Milthrope,[262] Tinker,[1191] and recently Hasegawa and Sato[525] and Hasegawa,[523] tried to determine the magnitude of interfacial resistance.

Newman[870] evaluated the relative magnitudes of plant and soil resistance and concluded that soil resistance is almost negligible when compared to plant resistance until a soil matric potential of about −1.5 MPa is reached. Cowan and Milthrope[262] and Tinker[1191] paid attention to the interfacial resistance between roots and the soil which is caused by an incomplete contact of roots with soil as a result of the decrease of the soil water-root contact area and the shrinkage of the roots and the soil.

Hasegawa and Sato[525] and Hasegawa[523] in their root plane experiment with soybean examined (1) the effect of soil water-root contact area on water uptake by roots, (2) the magnitude of soil hydraulic conductivity impeding water movement toward roots, and (3) the potential drop between roots and soil. They concluded that the area of contact between the roots and soil water did not significantly affect the water uptake because of its transfer into the vapor phase.[262] Water supply to plant roots was more affected by hydraulic conductivity of the soil. Enough water to plant roots was moved until the hydraulic conductivity of the soil fell to around the order of 10^{-5} cm d^{-1}. The water potential of the root surface increased with the increase of hydraulic conductivity of soil. An example of the soil water potential gradients in the vicinity of soybean roots is shown in Figure 1.

Examining water content and ion concentration gradients (using the root plane technique) in soil near a plane of onion seedling roots, Dunham and Nye[341] found that potential gradients near roots were steeper in drier than in wetter soil. The process of root growth, if rapid enough, can reduce the effect of the localized draw down of both matric and osmotic potential around each root. Schleiff,[1049] investigating the water uptake by barley roots in the rhizospheric saline soil, found that the water uptake rates were reduced to a greater extent by decreasing soil matric water potentials than by decreasing soil osmotic water potential.

3.2 PHYSICOCHEMICAL AND CHEMICAL ASPECTS

The physicochemical aspect of soil-root interface is connected mainly with the rhizosphere

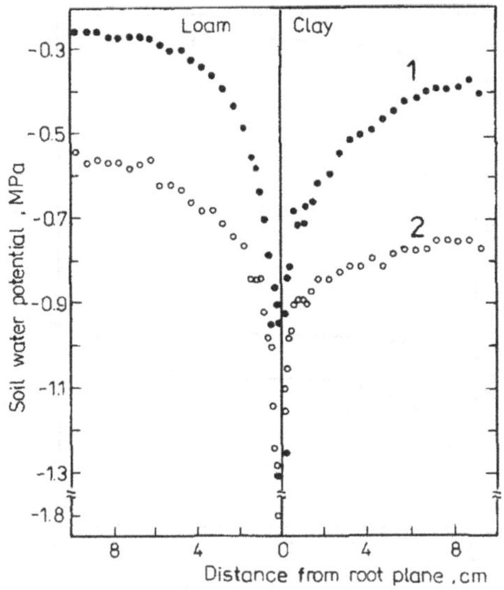

FIGURE 1. Soil water potential gradients in clay and loamy soils with a distance from soybean root plane. (1) Moist treatment; (2) dry treatment. (Modified from Hasegawa, S. and Sato, T., *Trans. Jpn. Soc. Irrig. Drain. Reclam. Eng. (Nogyo Doboku Gakkai Ronbunshu)*, 17, 22, 1985a.)

pH changes. Most plant species in their apical root zones increase the acidification, the reducing capacity, and the rate of iron and manganese reduction and their uptake.[800] Root-induced pH changes at the interface may also affect the uptake of some elements, e.g., phosphate.[491,557,988]

Depending on the ratio of cation to anion uptake, consumption or release of O_2 altering the redox potential in the rhizosphere, and pH buffering capacity of the soil, the rhizosphere pH and Eh may differ considerably from that in the bulk soil.[798,799,889,1046,1344] The differences in charge between total cation uptake and total anion uptake is balanced by the excretion of either H^+ or OH^- to the rhizosphere. This is mainly influenced by N uptake.[314,393,902] Usually, the NH_4^+ supply is correlated with preferential cation uptake and thus a higher net excretion rate of H^+ over HCO_3^- or OH^- and nitrate supply causes the reverse.

Smiley,[1085] examining the rhizosphere pH (pH_r) vs. bulk soil pH (pH_b), found that wheat growing in slightly alkaline or acidic soils treated with NH_4-N or NO_3-N fertilizers exhibited pH_r differences of up to 2.2 units and pH_r vs. pH_b differences of up to 1.2 units. These differences depended on the forms, concentrations, and placement of N in soil, the pH buffering properties of soils, and on the variety and age of wheat.

Marschner and Römheld[798] found that with the nitrate-fed maize plant, the increase in rhizosphere pH was usually most distinct along the main axes when the rhizosphere pH of lateral roots was either not increased or even decreased. Within the root system of individual plants and also between the rhizosphere pH and the pH of the bulk soil, distinct pH gradients exceeding two units are observed along the root axes.[798,889]

Schaller,[1046] using Sb microelectrodes, found that the root tips are particularly effective in changing rhizosphere pH and that these changes for peanut seedling roots in soils with differentiated pH between 3.9 and 7.7 decreased soil pH within 2 to 2.5 d to a minimum value. The greatest distance from the root surface where pH dropped significantly was found in soils with pH near 6 (Figure 2).

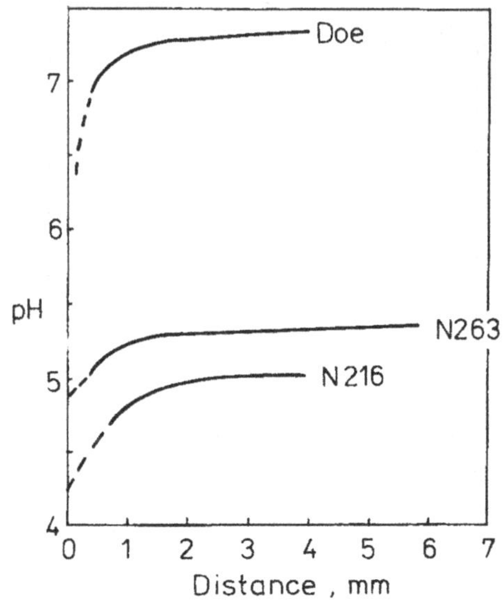

FIGURE 2. pH profiles of the rhizosphere of peanut roots in three soils. (Modified from Schaller, G., *Plant Soil*, 97, 439, 1987.)

A particular case of differences in rhizosphere pH within a root system can be observed in the so-called cluster-rooted plants.[424,715] The root system of these plants is characterized by clusters of finely divided, highly branched sections of roots, called the proteoid roots. Their capacity for acidification of the cluster-rhizosphere soil may drop pH to about 4.5 compared to about 7.0 in the other root zones.

Root biomass and distribution are strongly dependent on soil pH. This was shown by Erickson[375] for *Salix dasyclados* of which very few roots penetrated deeper layers of peat soil of pH below 4.0. The same was found for bean roots by Welbank.[1287] Similar to water and pH changes in soil, in the vicinity of roots, changes of ion concentrations are reported, e.g., by Claassen and Jungk[234] and Jungk and Claassen.[629] Scanning the film density of autoradiographs of maize root-soil interface, the above mentioned authors showed K and P depletion profiles which extended over 5 mm (for K).

Figure 3 shows potassium distribution in soil (Luvisol from loess) in the vicinity of ryegrass roots after different growing periods. The influence of this effect on the magnesium uptake by ryegrass, barley, maize, and rape was found. The rate of this uptake increased when the potassium concentration at the root surface decreased and was doubled when K concentration decreased below 20 μmol/l.

3.3 BIOLOGICAL ASPECTS

The inner boundary of the interface consists of the walls of the outermost root cells. In relatively young roots, such as those of annual plants, these tissues are the epidermis and root hairs (see Chapter 2). The biological aspects of the interface include water and nutrients absorption from the soil by root hairs and secretion by the roots of biological compounds such

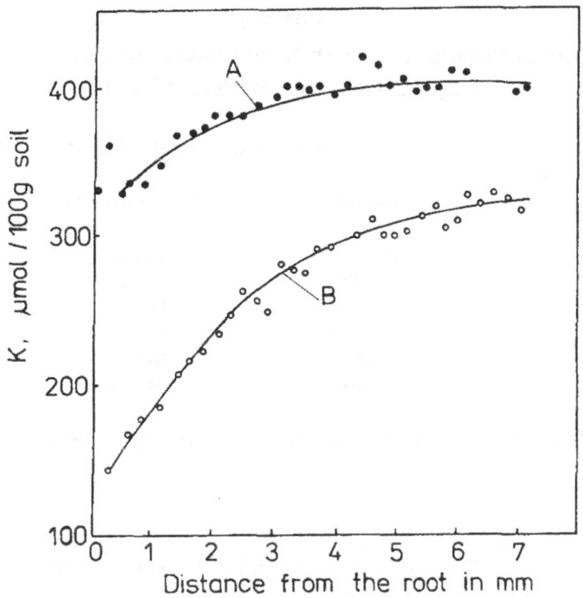

FIGURE 3. Potassium distribution in loess soil in the vicinity of ryegrass roots after (A) 2 and (B) 16 d. (Modified from Seggewiss, B. and Jungk, A., *Z. Pflanzenernaehr. Bodenkd.*, 151, 91, 1988.)

as mucigel, slimes, soluble exudates, cellular contents of living plant tissues which are the substrate for the microbiological flora, and material cementing soil particles into aggregates (see Chapter 5), and strengthening the attachment between root and adjacent soil particles.

All normal roots, in natural conditions, support a large population of microorganisms on their rhizoplane and in the rhizosphere as is shown in Table 1. This population includes bacteria, streptomycetes, fungi, yeasts, and protozoa. Some are free-living; others are symbiotic (e.g., root nodule bacteria and mycorrhizal fungi).

Each gram of root can support 36 mg of bacteria which is equal to 2×10^{10} cells. Typical populations are 10^9 bacteria, 10^7 actinomycetes, 10^6 fungi, 10^3 protozoa, and 10^3 algae per gram of rhizosphere soil being 50 to 100 times as many microorganisms as the bulk soil (the rhizosphere effect).[406] The distribution of microorganisms in the rhizoplane is not uniform.[1010] Microscopic pictures show that over many parts of a root, bacteria can be from 10 to 40 cells deep while other parts appear to be relatively free of microorganisms. Bacteria and fungi are often found arranged in lines along the axis of the root. Finally, some organisms, e.g., fungi, may traverse all of the rhizosphere and enter several cells deep into the root cortex.

Foster et al.,[406] by combining electron microscopy with the use of undisturbed soil and root samples either as ultrathin sections for transmission electron microscopy or as frozen, dehydrated specimens for scanning electron microscopy, developed a new visual concept of the ecology of microorganisms at the root-soil interface. The excellent photographs show the spatial structural and temporal relationships between the soil, root, and microorganisms. The results from ultrastructural studies on the relations of the root tips of maize mucigel-clay-microorganisms showed that the mucilaginous sheath of the apical zone of the roots possesses an outer

TABLE 1
Microorganism Populations at Different Distances From the Root For Lupine Seedlings by Plate Count

Distance from root (mm)	Microorganisms (1000s/g of oven-dried soil)		
	Bacteria	Streptomycetes	Fungi
Rhizoplane	159,000	46,700	355
0—3	49,000	15,500	176
3—6	38,000	11,400	170
9—12	37,400	11,800	130
15—18	34,170	10,100	117
Control soil	27,300	9,100	91

After Nye, P. H. and Tinker, P. B., in *Studies in Ecology,* Vol. 4, Blackwell Scientific, Oxford, 1977, 172.

fibrillar structure containing granules of polysaccharide, and the clay layers in contact with this part of the roots are either embedded in the mucigel or adsorbed at its surface.[173,495]

There is a large differentiation in the composition of the rhizosphere flora which is associated with different species of higher plants. From the viewpoint of plant growth, three groups of microbes are the most important: microbes of the nitrogen cycle, mycorrhizal fungi, and pathogens which cause specific diseases by invading the roots. In the nitrogen cycle, symbiotic nitrogen-fixing and free-living organisms can be distinguished.

Symbiotic nitrogen fixation associations between plants and microorganisms have been identified in more than ten families of higher plants, but fixation by *Rhizobium* spp. in the Leguminosae alone is of agricultural importance.[886] Atmospheric nitrogen may be also fixed by nearly 20 genera of free-living rhizosphere bacteria and a smaller number of blue-green algae.[886] Species of *Azotobacter, Beijerinckia,* and *Clostridium* are the main nitrogen-fixing bacteria.

Free-living organisms can bring about two other processes besides nitrogen fixation from atmospheric sources, such as the mineralization of organic nitrogen and denitrification. Rhizosphere nitrogen fixation by blue-green algae and photosynthetic, free-living (diazotroph), and associative heterotropic bacteria is of considerable significance in submerged rice soils.[222,970,1342]

The role of mycorrhizal association and root hairs in nutrient uptake by roots is widely discussed, especially P uptake. The symbiotic mycorrhizal association between fungi and plant root systems consists in entering the roots by the hyphae, either through the epidermis or root hairs, and extends out into the soil. This association exerts a considerable effect both on the growth of roots and on the uptake of nutrients.[996]

The mycorrhizal infection is referred to by several authors, e.g., Graham et al.[467] and Johnson et al.,[621] connecting it with sugar concentration in root exudates. However, Azcon and Ocampo[60] indicated that the amount of sugar exuded from six plant species is not a decisive factor for the initiation of VA mycorrhizal infection.

Sands et al.[1040] found that hydraulic conductance of loblolly pine seedlings was independent of the amount of mycorrhizal infection. There was no such clear statement on the role of root hairs in nutrient uptake. Bole,[147] Crush and Caradus,[276] and Caradus[203] showed different influences of root hairs on P uptake by mycorrhizal and nonmycorrhizal plants.

In some cases, an invasion of root tissues by rhizosphere organisms appears. Some bacteria can enter epidermal cells, either by perforating the cell wall or entering the middle lamellae between adjacent cells.[901] The invaded roots become decorticated; their diameter is appreciably reduced and their contact with the soil may be impaired. The bacteria on root surfaces are strongly nonrandomly distributed, but tend to be aggregated.[874,1012]

Some microorganisms, named nonparasitic pathogens or exopathogens, associated with the rhizosphere induce root damage by liberation of a toxin or other detrimental substances which are absorbed by roots, producing abnormal metabolic reactions.[1332] Aggressively colonizing root growth-inhibiting rhizobacteria (pseudomonads) were found among others by Bennett and Lynch,[119] Kloepper and Schroth,[671] Frederickson and Elliot,[408] and Elliot and Lynch.[366] The term "rhizobacteria" was defined by Schroth and Haricock[1053] to distinguish bacteria isolated from rhizosphere and rhizoplane regions with little regard to function or population from those that have been thoroughly studied and are known to heavily colonize these regions. In the case of the sugar beet, rhizobacteria reduce seed germination, cause root distortion lesions, reduce root elongation, increase infection by root-colonizing fungi, and significantly reduce plant growth.[1145]

Garret[427] and Zentmeyer,[1346] examining root infection by fungi at the interface, found that susceptibility to this infection is connected more with plant resistance to diseases than with soil-root interface. On the other hand, it is known[1010] that the proximity of plant roots stimulates the germination of the spores of many nonpathogenic fungal species lying dormant in soil.

Some authors pay attention to the role of plant mucilages in the retention of metallic cations in the rhizosphere.[170,573,734,850] Morel et al.[850] and Mench et al.,[815,816] examining the mucilage-binding capacity of maize roots for Pb, Cu, Zn and Cd ions, found that mucilages are able to bind these metals and that this binding depends on the nature of the cation, following the order Pb > Cu > Cd. These reactions can be due to exchange processes involving mucilage cations (Ca^{2+}, Mg^{2+}) and heavy metals. The authors state that mucilages through modification of the flux of heavy cations to the root act as selective filters for these cations, completing the role of cell walls from which they are derived.

Chapter 4

INFLUENCE OF SOIL PHYSICAL FACTORS ON ROOT GROWTH AND FUNCTIONS

4.0 TEXTURE

It is a general opinion that root growth is influenced more by a coarser soil texture than by finer one, which is clearly shown for *Phleum pratense* in Figure 1 and for wheat in Table 1. This is mainly related to low unsaturated hydraulic conductivity which may form a dry barrier layer at the root front and low moisture retention abilities of the coarse soils.[534,750,830]

Soils shown in Figure 1 are characterized by the same porosity and nutrients content which differs in both soils in four levels. However, it should be mentioned that the same amount of nutrients of various textures modifies root environment by differentiation of sorption capacity of soils. So, we can have, in this case, an indirect effect of soil texture on roots through modification of nutrient availability. For wheat (Table 1), it is seen that seminal, adventitious, and especially secondary roots prefer sandy soil over clayey soil. Soil of a silty texture is intermediate.

Prihar and Gajri,[958] examining the effect of irrigation on dryland crops, used two soils of loamy sand and sandy loam texture and found a higher progressive growth of roots and a shallower rooting depth of wheat in sandy than in loamy soil (Figure 2). They also received differentiation in wheat root density in respect to the texture of the above-mentioned soils in further experiments[416] (Figure 3).

Roots of *Pinus radiata*,[291] grown in soils of various texture, showed a greater concentration and a greater proportion of the root length in loose-textured when compared to heavy-textured soils (Table 2). On the other hand, Rasmussen[972] found that root development of barley cultivated on loamy soil was about three times more intensive (about 1.5 cm cm^{-3}) in deeper layers than in sandy soil (about 0.56 cm cm^{-3}). Also, Beukes[127] showed that the percentage of the distribution of apple tree roots was in the case of some diameter groups (e.g., 1 to 3 mm) even twice as high in more clayey and silty soils than in soil containing less of these size fractions. Maertens[788] obtained similar results for soybean (Figure 4). However, in the case of vineyards, Nagarajah[864] showed that maximum root densities were 0.4 mm cm^{-3} in coarser soil, 0.8 to 1.0 mm cm^{-3} in moderately coarse soil, and 0.7 to 1.7 mm cm^{-3} in fine soil. Ruark et al.,[1020] summarizing the literature on root growth of forest trees, indicate that in fine-textured soil they are shorter and more branched, but in poorer, well-aerated soils, they tend to be longer, thinner, and straighter. Madsen[786] reported that on pure sandy soils the root zone capacity (RZC) is very low and similar for grass, spring-sown cereals, and winter-sown cereals. In loamy areas, the RZC is higher than in pure sandy areas, and great differences occur in relation to the crop.

O'Brien[894] examined the effect of three root media consisting of different, artificially prepared, size sand fractions with particle sizes of 2 to 3, 1 to 2, 0.5 to 1, and less than 0.5 mm on seminal root growth of wheat. He found the maximum expression of most variables in a 1- to 2-mm-sized sand fraction and for the maximum depth of root penetration, number of lateral roots, and total length of lateral and seminal roots (Table 3). The examination of the influence of soil stratification (fine and coarse textural layers) in comparison to uniformly textured soil on the root development of 12 crops[1336] showed that none of the root systems of the plants tested could penetrate the stratified layers to any important extent, even with the application of adequate water and fertilizer, although all the layers were sufficiently porous to allow surplus water to drain through them freely.

The experiment with layered sand columns and tomato plants described by Flocker and Timm[403] confirmed that root penetration within these columns was affected by the textural

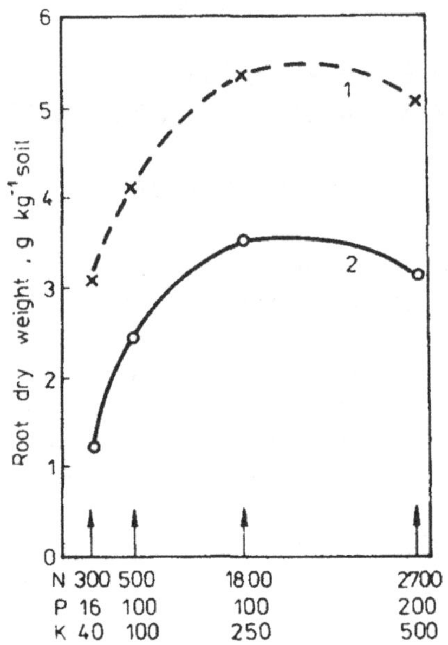

FIGURE 1. Root dry weight of *Phleum pratense* L. cv. B.51 in sandy (1) and loamy (2) soils as a function of NPK concentration in soils. (Modified from Callot, G., Chamayon, H., Maertens, C., and Salzac, L., *Soil-Root Interaction. Aspects of Mineral Nutrition,* INRA, Paris, 1982, 212 (in French).)

characteristics of the soil profile and was correlated with moisture retention of the respective layers. Effects similar to the above mentioned were stated by Babalola and Lal[61,62] in greenhouse experiments with the growth of maize seedling roots in drums filled with three soil materials (sand, clay, and sandy loam) layered with gravel material below the surface (Figure 5).

Soil texture differs significantly within rhizosphere microorganisms. Pera et al.,[935] examining the effect of organic matter on rhizosphere microorganisms and root development of sorghum plants in pliocenic clay and in alluvial sandy soils, found significant differentiation in the content of fungi, actinomycetes, aerobic and nitrite-oxidizing bacteria, and cellulolitic microorganisms (Figures 6 and 7).

The above review shows that soil texture influences root systems more indirectly, creating mechanical impedance of the soils, modifying their water and aeration status, nutrient content, and availability. Generally, root growth is influenced more in soil of coarser texture than in finer texture due to lower fertility, lower unsaturated hydraulic conductivity, and lower water storage capacity. These interrelationships are discussed in Chapter 4.7.

4.1 STRUCTURE

Misra et al.[837,839] investigated the effects of aggregate size in the range of 1.0 to 29.0 mm and various strengths on pea, cotton, and sunflower root penetration. They also presented the theory of axial root penetration into soil aggregates. This has been achieved through a comparison of theoretical results for ideal, cylindrical aggregates with experimental results from approximately spherical aggregates. They found the effect of aggregate size on root growth force was much larger than the effect of aggregate strength obtained by saturating aggregates with

TABLE 1
Root Characteristics of Wheat During Tillering in Three Soils of Various Texture (Units For Length in Millimeters)

Root variable	Soil texture: Clayey	Silty	Sandy
Seminal and adventitious roots			
Number	8	8	10
Mean length	42	127	129
Total length	340	1,021	1,290
Mean diameter	0.48	0.48	0.34
Total surface (in mm²)	510	1,500	1,380
Secondary roots			
Number	181	638	941
Mean length	10.2	12.8	17
Total length	960	7,200	16,000
Mean diameter	0.30	0.22	0.20
Total surface (in mm²)	904	5,000	10,080
Tertiary roots			
Number	697	1065	320
Mean length	1.74	1.9	1.49
Total length	1220	2,100	477
Mean diameter	0.24	0.17	0.12
Total surface (in mm²)	913	1,100	176
Total length	2520	10,300	17,466
Total surface (in mm²)	2329	7,600	11,636

Data from Callot, G., Chamayon, H., Maertens, C., and Salzac, L., *Soil-Root Interaction. Aspects of Mineral Nutrition*, INRA, Paris, 1982, 212 (in French).

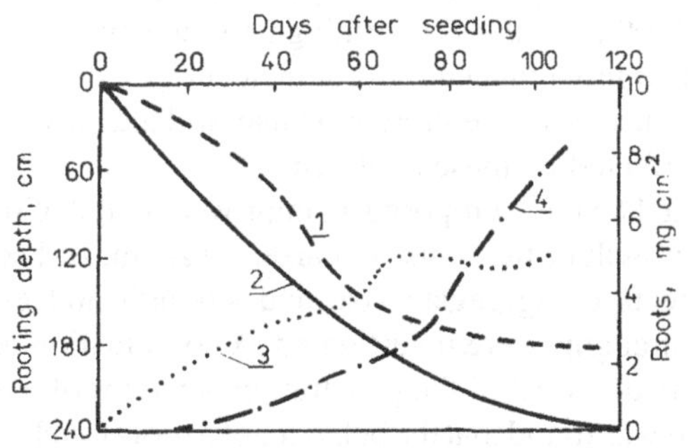

FIGURE 2. Maximum rooting depth of wheat in loamy sand (1) and sandy loam (2) irrigated with 60 mm of water at seedling and progressive root growth in loamy sand (3) and sandy loam (4). (Data from Prihar, S. S. and Gajri, P. R., in *Trans. Symp. Rainwater and Dryland Agriculture*, Indian National Science Academy, New Delhi, 1982, 115.)

polyvinyl alcohol. It has been found that large axial root growth pressure must be exerted by roots to penetrate large aggregates (Figure 8). With the increase in aggregate diameter, total root length decreased, while the length of root hairs increased.[838] The lower root growth in beds of

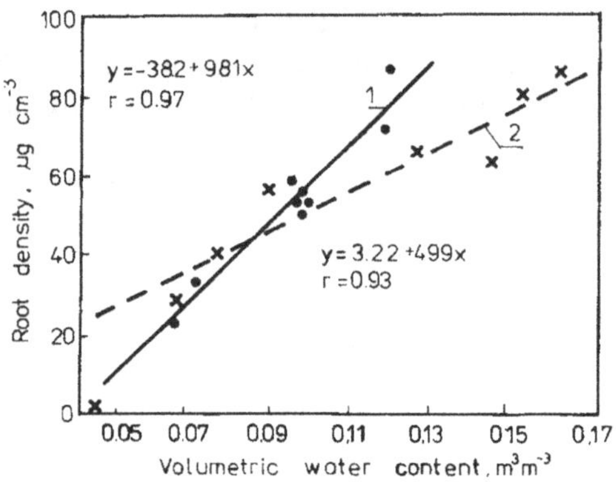

FIGURE 3. Wheat root weight density as a function of soil water content in surface 30-cm layer of loamy sand (1) and sandy loam (2) soils during a 30- to 60-d period, based on 2 years of data. (Modified from Gajri, P. R. and Prihar, S. S., *Field Crop Res.*, 12, 115, 1985.)

large compared with small aggregates is attributed to greater mechanical impedance to root penetration.[837] Agraval and Jhorar[14] found that root quantity was greater in ten 4- to 10-mm and <2-mm soil aggregate treatments than in 2- to 4-mm aggregates. The response of root system upon aggregate size is also related to the type of roots. Donald et al.[317] reported that the main axes of seminal and nodal roots were longer in the coarser aggregate system, the length of primary laterals was not affected, and the length of secondary laterals was lower in the coarser systems. The diameter of the main axes roots was greatest in the largest aggregate fractions.

Anderson and Kemper[32] have also observed that large aggregates can be an impedance to root growth, but Goss[455] showed no difference in barley root elongation when grown in glass beads of different sizes (100 to 1000 μm) unless confining pressures were applied. Two soils broken into units of 2 mm or less largely reduced root rot incidence of subterranean clover compared to intact soil. In clay loam, the roots were shorter in intact soil than in disturbed soil, but in sandy loam, root length was unaffected by these treatments.[1086]

Dexter[306] and Hewitt and Dexter[549] proposed an improved model of root growth in structured soil. They examined the growth of the roots of maize, sorghum, and soybean through beds of spherical aggregates. Effects of aggregate size and strength and effects of the spread or distribution of aggregate strengths were investigated. A combination of a statistical model for soil structure with a statistical model of the penetration behavior of a root at a void/aggregate interface was introduced. It was found that the behavior of a root at such an interface is dependent on the previous history of the root in its passage through the soil and that the smaller the aggregate size, the greater is the nutrient availability per unit length of root. Further, new information on the penetration behavior of roots has been incorporated into this model by Misra et al.[841]

Dexter[305] described axis root growth as a geotropic downward movement until the root encounters an aggregate, assuming continuation of its growth through the aggregate or its deflection around the aggregate (Figure 9). A displacement of the aggregate from its path is also possible.[1299] They include the possibility of deflection of both the root and the aggregate and predict that narrow roots are more easily deflected and that larger aggregates are less likely to be displaced by the root.

The modeling of the growth of roots in structured soil[306,312,549,1297] contains the following:

TABLE 2
Root Concentration of *Pinus radiata* in Two Soils of Various Texture

Forest site	Depth (cm)	Root conc. (mm/cm^3)	Clay (%)	Soil bulk density (Mg/m^3)	Air pores in soils (%)
Stoodley	0—10	2.60	17	0.95	47.3
	10—20	2.79	18	1.15	39.5
	20—30	1.11	20	1.15	37.4
	30—40	1.34	24	1.31	32.9
	40—50	1.22	25	1.36	31.4
	50—60	1.45	24	1.30	31.8
	60—70	1.61	25	1.38	29.4
	70—80	1.73	25	1.45	26.9
Strahan	0—10	11.56	1	0.58	55.6
	10—20	3.94	2	0.96	44.3
	20—30	2.91	1	1.27	41.9
	30—40	1.16	1	1.35	42.2
	40—50	1.26	2	1.42	38.9
	50—60	1.93	1	1.45	35.6
	60—70	1.97	2	1.48	36.6
	70—80	2.07	1	1.50	39.4

Data from Davis, G. R., Neilsen, W. A., and McDavitt, J. G., unpublished.

1. A decription of the proportion (C) of roots penetrating an aggregate is given by a Boltzmann-type expression:

$$C = \exp[-F(\theta)S/(S_{max} - S)] \qquad (1)$$

where $F(\theta)$ is an unspecified function of the angle of incidence (empirical values are given by Dexter[306]); S is the soil strength, and S_{max} is the maximum pressure beyond which a root will not penetrate. The maximum pressure is either the pressure at which a root will no longer elongate (S_e) or the strength which causes buckling of a root (S_b) which is calculated from the Equation:

$$S_b = 80.8EI/\pi d_r^2 Y^2 \qquad (2)$$

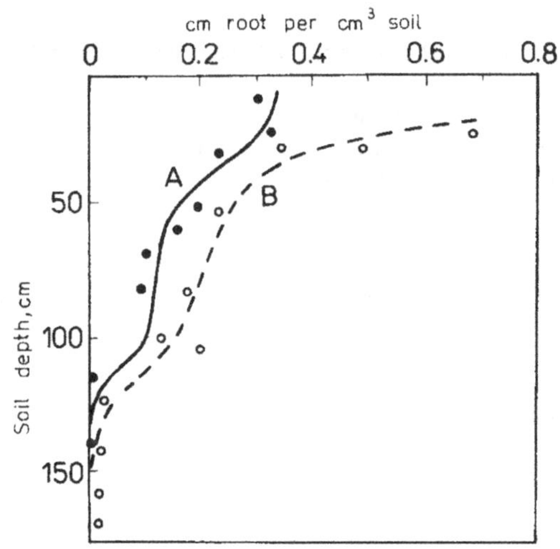

FIGURE 4. Root profile of soybean grown in loamy (A) and clay (B) soils. (Modified from Maertens, C., CETIOM INRA, Supplement à Informations Techniques, 1er trimestre, 1986, 34.)

TABLE 3
Mean Values of Root Variables For Wheat Plants Grown For 10 d in Four Different-Sized Sand Fractions

Root variable	Sand size fractions (mm) 2—3	1—2	0.5—1	<0.5
Root fresh weight (mg)	96a	108a	108a	107a
Rooot dry weight (mg)	17b	16b	16b	16b
Fresh weight shoot to root	0.98b	1.03b	0.95b	0.98b
Dry weight shoot to root	1.58a	1.61a	1.53a	1.60a
Number of seminal root axes	4.50b	4.80b	4.95b	5.20a
Total length of the seminal root axes (cm)	56.7a	55.0a	50.6a	45.5a
Maximum depth of penetration (cm)	16.0a	14.0b	14.2b	12.7b
Number of lateral roots	41.0b	49.8a	34.4b	34.8b
Total length of lateral root (cm)	22.5b	31.6a	18.8b	18.1b
Total length of seminal root (cm)	79.2a	86.6a	69.4b	63.6b
Length of root hair bearing root (cm)	16.8a	25.3a	21.0a	23.3a

Note: Means in the same row followed by the same letter are not significantly different at the 5% level according to Tuckey's procedure.

Data from O'Brien, L., *Crop Sci.*, 18, 685, 1978.

where d_r is root diameter, E is the Young's modulus of the root; I is inertia; and Y is the length of the void through which a root has grown.

$$E = fx^3 6.4/3\delta I \tag{3}$$

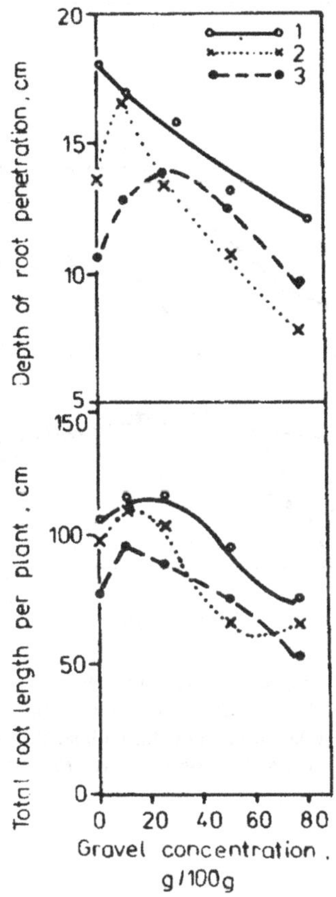

FIGURE 5. Effect of soil texture (1, clay; 2, sandy loam; and 3 beach sand) with various gravel concentrations on the depth of root penetration and the total root length of maize. (Data from Babalola, O. and Lal, R., *Plant Soil*, 46, 347, 1977b.)

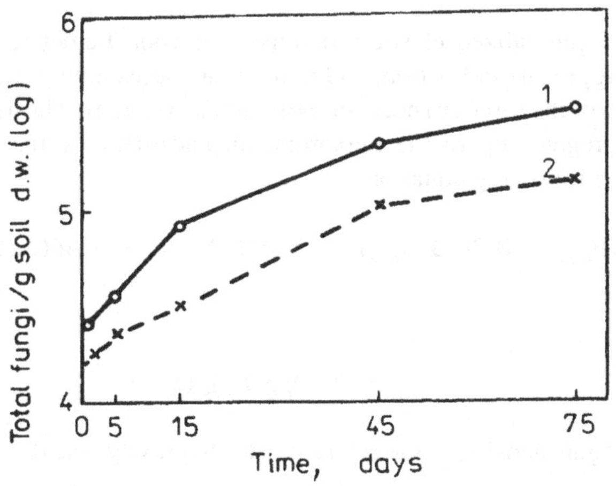

FIGURE 6. Total fungi of sorghum rhizosphere in sandy (1) and clay (2) soils. (Data from Pera, A., Vallina, G., Ines Sireno, Lorella Bianchin, M., and de Bertoldi, M., *Plant Soil*, 74, 3, 1983.)

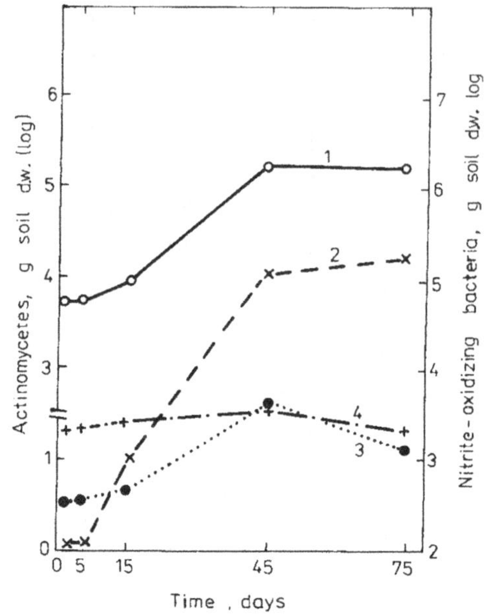

FIGURE 7. Actinomycetes of sorghum rhizosphere in sandy (1) and clay (2) soils and nitrite-oxidizing bacteria in sandy (3) and clay (4) soils. (Data from Pera, A., Vallina, G., Ines Sireno, Lorella Bianchin, M., and de Bertoldi, M., *Plant Soil,* 74, 3, 1983.)

where x is the distance between aggregates; δ is the deflection distance; and f is the force on the root.

2. The root force (f), if the distance between aggregates is assumed to be the same as aggregate diameter, is as follows:

$$f = Md_r^2(d_r/D)^{\alpha}[0.3D - \Delta)/d_r]^{\beta} \tag{4}$$

where M is the generalized elastic modulus; d_r is root diameter; D is the aggregate diameter; Δ is aggregate deflection; and α and β are adjustable parameters.

3. The relationship of the force (F) necessary to result in a specified horizontal displacement of dry soil aggregates up to the maximum displacement by the maximum force is expressed by the following equation:

$$F/F_{max} = 0.20(\Delta/\Delta_{max}) + 2.25(\Delta/\Delta_{max})^2 - 1.46(\Delta/\Delta_{max})^3 \tag{5}$$

where

$$F_{max} = 30.0\rho g D^3(d_s/D)^{1.98} \tag{6}$$

where ρ is aggregate density; g is acceleration due to gravity, and d_s is the depth below the soil surface.

Dexter's[306] system includes the influence of aggregate size, root diameter, and soil depth. It describes root contact with soil along a line, but does not include branching. A model is empirical for dry aggregates and may not apply to moist aggregates which have more adhesion for each other.

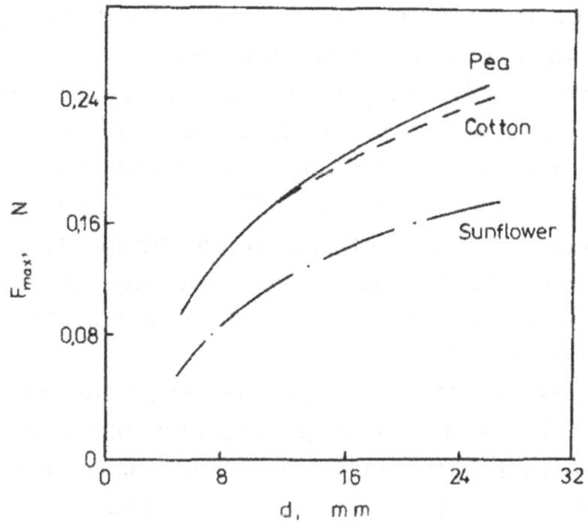

FIGURE 8. Maximum axial root growth force (F_{max}) of pea, cotton, and sunflower as a function of aggregate diameter (d). (Data from Misra, R. K., Dexter, A. R., and Alston, A. M., *Plant Soil*, 94, 59, 1986a.)

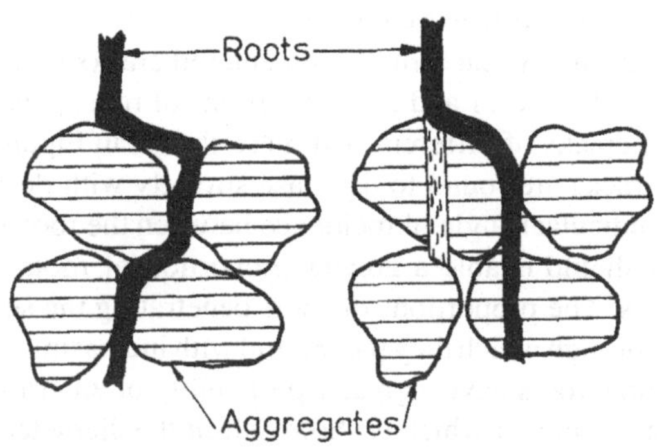

FIGURE 9. Representation of possible paths of a root axis through a bed of aggregate. (Data from Dexter, A. R. and Hewitt, J. S., *J. Agric. Eng. Res.*, 23, 17, 1978.)

Logsdon et al.,[755] studying the effect of aggregate size of silt loam soil on maize root impedance developed Dexter's equations to describe the root pressure necessary to avoid deflection around an aggregate. The critical root pressure was predicted to increase with increasing aggregate size, decreasing neither the diameter nor the deflection angle. They found that the larger aggregates did constitute a slight root impedance as roots were deflected around them. Diameters of roots grown in 3- to 6-mm aggregates increased significantly, whereas root lengths were not always significantly decreased. The smaller aggregates did not impede root growth and were readily displaced by roots. Large aggregates were more of an impedance to lateral than to main axes.

Soil profile is very often heterogeneous and shows structural interfaces or strength discontinuities especially between an aggregated seedbed of the tilled layer and the hard or compacted layer beneath.[307] If this compaction layer is firm, roots are deflected and grow horizontally along the base of the seedbed, not penetrating below a seedbed.

Compaction of the untilled subsoil layer often has cracks and biopores (wormholes and channels left by previous roots) which may be penetrated by roots.[308,550,1300] Cracks are natural or artificial. Natural cracks are usually formed by desiccation of swelling/shrinking soils during dry periods or during growth of the preceding crop. The usual desiccation crack pattern consists of somewhat irregular polygons. The crack patterns can be modified by plants. Cereal crops sown in rows tend to produce vertical linear cracks midway between the rows, whereas in grassland, cracks of a circular nature tend to surround the individual tussocks. Artificial cracks can be induced by tillage machinery.

Dexter,[308] Whiteley and Dexter,[1300] and Hewitt and Dexter[550] found in model experiments that when a soil contains cracks between macroscopic structural units (e.g., clods), it is common to find a preferential growth of roots on the clod surfaces rather than within the clods. The reduction in root elongation rate when growing through the clods is probably a result of reduced aeration. Roots are able to elongate more rapidly in cracks narrower than the root diameter than through undisturbed clods without cracks, provided that the crack was not oriented at an oblique angle to the preferred geotropic growth direction.[1298]

When the tip of an elongating root grows out of a soil clod and crosses a crack, it can either penetrate the next clod or it can buckle and be deflected along the surface of the clod (Figure 10). This is likely for soil aggregates (see Figure 9). Sugar beets grown in heavy silty clay-loam soil with the presence of 80-mm clods in the plow furrow were characterized by a high percentage of forked beets, irrespective of tillage treatments.[1345]

The proportions of seminal wheat roots which entered cracks of various widths and angles of incidence are shown in Figure 11 and the proportions of roots penetrating soil with various strengths after crossing cracks of different width are shown in Figure 12. The proportions of roots which enter the cracks are found to decrease strongly with decreasing crack width and increasing (more perpendicular) angle of incidence between the root and the crack. Moreover, parallel crack patterns should enable a greater proportion of roots to enter the cracks than hexagonal crack patterns. The proportions of roots penetrating the second clod decrease with increasing crack width through which they passed and with increasing soil strength. Lateral roots being thinner than seminal roots have a greater probability of entering narrower cracks. First-order lateral roots have a diameter which is about half of the diameter of the seminal root axes of the same plant.

McSweeney and Jansen[813] described soil structure patterns and associated behavior in reclaimed mine soils. They proposed a new term — fritted structure — which consists of rounded aggregates loosely compressed together. Subsoils with a fritted physical condition favor good rooting due to the presence of the extensive void spaces between aggregates. Operations using a mining wheel in combination with belt transportation favor the formation of the fritted structure, whereas operations using scrapers exclusively favor formation of the less desirable massive physical condition.

Aggregate size affects ion uptake by roots. It was shown[253,841,1307] that P-uptake is higher from small than from large aggregates, while N-uptake was slightly affected by aggregate size.[1256,1307] In soils in which the roots can only penetrate through interaggregate pores, cracks, and fissures, the accessibility of nutrients is related to mobility of the nutrient. This subject was discussed in more detail by De Willigen and Van Noordwijk.[303] Misra et al.[837] found that root hairs explored more soil volume in macrostructured soil and strongly influenced accessibility of phosphorus to roots.

Generally speaking, in beds of large compared with small aggregates, roots are characterized by lower growth, greater diameter, greater deflection, and increased percentage of forked beets. Large aggregates reduce greatly the proportion of roots penetrating into them. Nutrient uptake,

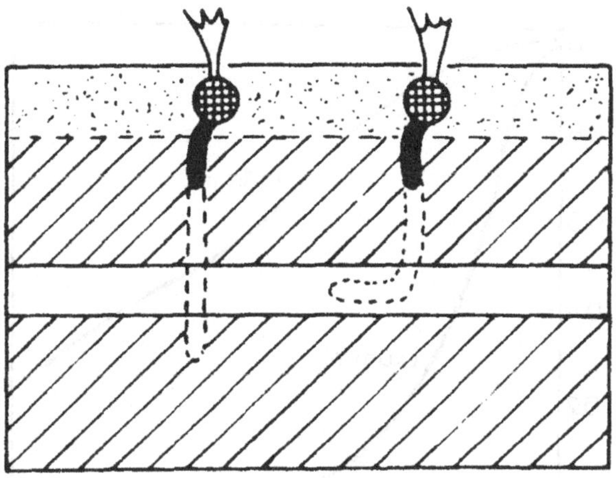

FIGURE 10. Possible behavior of seminal roots penetrating two clods separated with a crack. (Modified from Whiteley, G. M. and Dexter, A. R., *Plant Soil*, 77, 141, 1984b.)

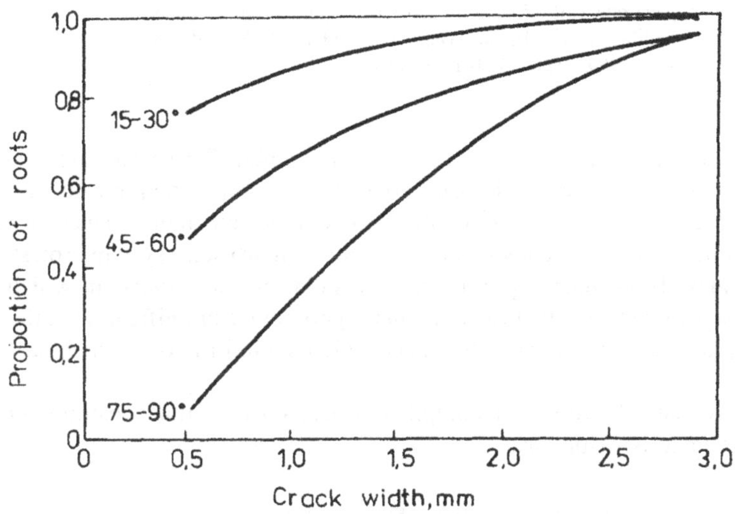

FIGURE 11. Proportion of seminal roots of wheat which enter cracks of various width and angle of incidence. (Data from Dexter, A. R., *Plant Soil*, 95, 235, 1986b.)

e.g., P, is lower from large aggregates. The response of roots to aggregate size is closely related to the type of roots. Main axes of seminal and nodal roots are longer in the coarser aggregate systems, while the length of secondary laterals is lower. Root hairs explore greater soil volume in macrostructured soil, thus increasing nutrient absorption. The effect of soil structure on root growth and functions in relation to other soil physical factors is discussed in Chapter 4.7.

4.2 MECHANICAL IMPEDANCE

4.2.1 ARTIFICIAL MEDIA

Root growth (elongation) can be reduced by a number of factors, in particular, increased

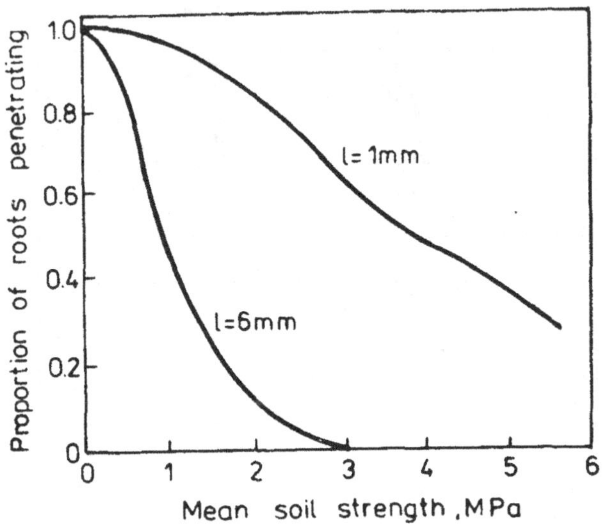

FIGURE 12. Proportions of seminal roots of wheat which penetrate soil of various resistance after crossing cracks of widths(l) of 1 and 6 mm. (Data from Whiteley, G. M. and Dexter, A. R., *Plant Soil*, 77, 141, 1984b.)

mechanical resistance, soil aeration, soil moisture, and others. These limiting conditions can occur in strong soil simultanously, making it difficult to distinguish unequivocally between their effects. For these reasons, many studies on the effect of mechanical impedance on root growth have been concerned with the behavior of roots grown in artificial systems (plaster of Paris, porous sinters, glass bead matrix, porous plate, capillaries, and others) in which effects of pressure can be separated from other factors. This experimental simplification of the real-world situation is justifiable as demonstrated by the considerable information which has come from it.

The pressure available for roots to accomplish work against external constraint is termed root growth pressure[442] and is written as:

$$\frac{F_t - F_{ew}}{A} \tag{7}$$

where F_t and F_{ew} are, respectively, the summation of the longitudinal forces in the root that arise from turgor pressure and the longitudinal forces in cell walls which resist cellular elongation and A is the cross-sectional area of the root at the plane where force is determined.

The results of maximum axial root growth pressure by virtue of hydrostatic pressure (turgor) are summarized by Taylor[1170] and Russell.[1024] The pressures ranged for a number of species from 0.9 to 1.3 MPa. Maximum radial growth pressures are around 0.5 MPa.[442] The values of maximum axial root growth pressure estimated from the maximum axial root growth force and root diameter were 497 (for pea), 289 (for cotton), and 238 kPa (for sunflower).[840] The radial root growth pressures were of similar magnitudes to the axial pressures.

In a study of the growth of various roots through porous sinters, Wiersum[1306] found that roots are incapable of decreasing in size to enter pores of smaller diameter. Results of the growth of maize in rigid systems of the glass bead matrix of different diameters[54] and of the growth of

TABLE 4
Effect of Mesh Size on the Growth of Perennial Ryegrass

Mesh size (μm)	Maximum thickness of root below mesh (mm)	Proportion of roots above mesh (wt%)
Control (no mesh)	970	52.7
712	950	44.8
572	940	48.3
542	950	48.4
401	960	50.9
315	880	45.5
210	530	61.3
160	450	72.3
120	300	91.1
100	230	87.6
76	220	97.3

Data from Scholefield D. and Hall, D. M., *Plant Soil*, 85, 153, 1985.

maize and barley through aerated beds of glass ballotini[455,1027] supported Wiersum's conclusion. Aubertin and Kardos[54] showed that in rigid systems, maize roots did not grow through pores of a diameter below 138 μm, but in nonrigid systems, they were able to grow through 46-μm pores.

The growth of roots in constricted media usually causes morphological changes.[605,657] It was reported that roots of barley,[747] *Pisum sativum* and *Zea mays*[84] grown in such media were characterized by larger diameter. This was mainly due to a great enlargement in the cortex in which cells became shorter in longitudinal direction and wider in the traverse, while the cell volume was unaffected.[84] Peterson and Barber[940] suggested that the shorter, wider cortical cells, with their greater absorptive surface area, would help in overcoming nutrient stress when roots are grown in the more resistant media. Roots which are unable to penetrate the strong model seedbed layer are deflected to grow horizontally along the interface between the loose and strong layer until they encounter vertical cracks.[307,308] The proportions of roots which entered cracks were found to decrease strongly with decreasing crack width and increasing (more particular) angle of incidence between the root and the crack.

It was observed that increased mechanical impedance caused distortion of the root apex and induced formation of nodal roots and increased lateral branching.[83,156,455,1056] In addition, mechanical restraint caused the root axes of barley to curve around the restraint, and the initiation of laterals occurred on the convex (tension) side, whereas root hairs were more numerous on the concave (compression) side.[455]

Recent studies in Japan and the U.K. showed that roots of Italian ryegrass[684] and *Lolium perenne*[1052] are able to penetrate into pores of a lower diameter than themselves (Table 4). The ratio of the thickness of the penetrating root to mesh size reached a maximum of 2.78 for primary roots of ryegrass and 2.89 for the first-order laterals.[1052] According to the present authors, the ability of grass roots to penetrate rigid pores should be considered to be a mechanism alternative to the exertion of radial pressure by which the plant may penetrate compressed soil layers.

In earlier papers, it was indicated[476,665] that reduced extension growth of roots was mainly caused by external pressures counteracting the normal cell turgor pressure necessary for cell extension growth. However, such osmoregulatory explanation was insufficient when it was shown[455,1027] that pressures of 20 to 50 kPa, being much smaller than those of cell turgor, considerably reduce root growth in a number of species. Clear-sighted discussion and interpretation of the incompatible results[459,473,1027] showed that complex metabolic processes are involved.

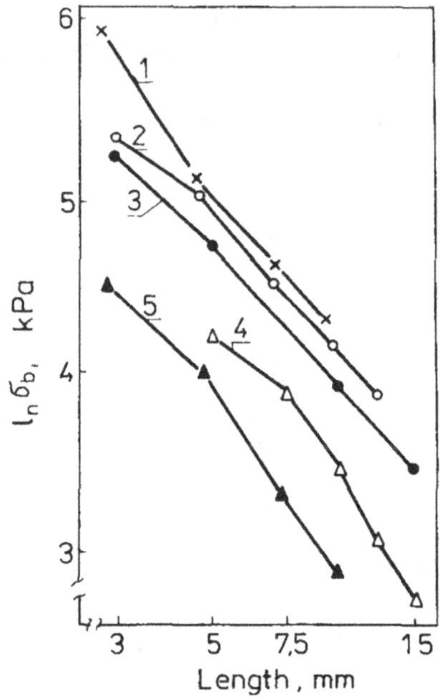

FIGURE 13. Buckling values (σ_b) against length (l) for seminal and primary lateral roots; (1) rape seminal; (2) soybean seminal; (3) soybean lateral; (4) linseed seminal; and (5) wheat seminal. (Modified from Whiteley, G. M., Hewitt, J. S., and Dexter, A. R., *Physiol. Plant*, 54, 333, 1982.)

Several workers[84,459,657] suggested that the response of roots to mechanical impedance can be hormone-mediated and indicated a role of ethylene which evolution significantly increased with impeded roots. This suggestion was supported by Dawkins[292] on the basis of evidence that no reduction in volume of root cells due to mechanical restriction has been observed in spite of decreased cell length and a decreased extent of the elongation zone.

Following the suggestions on a hormone-mediated mechanism in root extension, Lachno et al.[705] studied the effects of mechanical impedance on root morphology and the levels of abscisic acid (AbA), suggested as the root cap inhibitor, and indol-3 ylacetic acid (IAA), regarded as a root hormone in root tips of *Z. mays*. The AbA levels of impeded roots did not differ significantly from those of normal root tips. However, the levels of IAA were increased by about 3.5 times in impeded roots (176.3 as compared with 52.4 ng g^{-1} fresh weight) and it was concluded that this response is likely to be the main cause of the morphological and growth changes brought about by soil compaction. Authors also indicate that a possible role of ethylene has still to be explored, as well as the problem of whether the inhibition of root extension is due to the direct action of IAA or to its stimulation of ethylene production. Whiteley et al.[1301] indicated that the maximum pressure the root can exert to penetrate a solid surface is limited by the maximum buckling stress of the root tip, whenever this is less than the maximum growth pressure. The mechanical stress required to buckle the root tips of seminal and primary lateral roots of some plants is presented in Figure 13.

For measurements of maximum root growth pressure avoiding buckling,[1107-1109] the root was grown in capillary and the growth force was measured by means of steel plate deformations. It was found that the axial growth pressure of the seminal roots of maize is of the order of 250 kPa. The root elongation rate decreased rapidly with the increase of a mechanical obstacle.

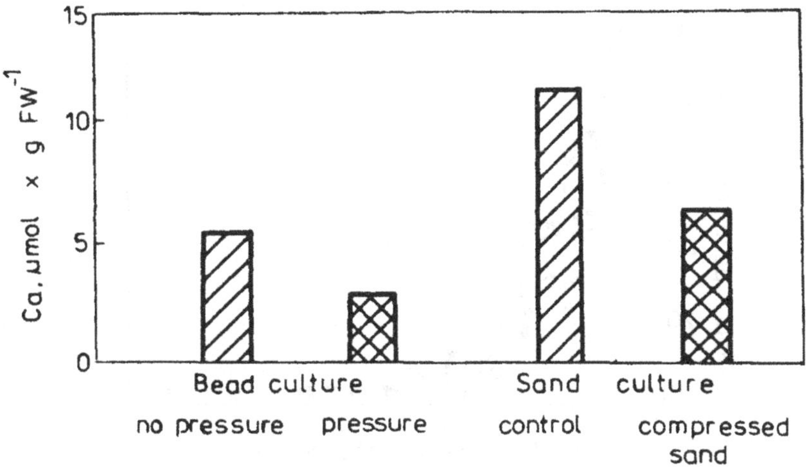

FIGURE 14. Effect of pressure on concentration of Ca in roots (fresh weight, FW) of barley grown in bead and sand cultures for 7 d. (Data from Lindberg, S. and Pettersson, S., *Plant Soil*, 83, 295, 1985.)

Dexter[309] studied root entry into artificial cylindrical biopores of different diameters and found that the proportion of roots which entered the holes decreased with decreasing hole diameters. There was no evidence for roots sensing and growing preferentially toward the holes (trematropism). Investigations of Whiteley et al.,[1301] Souty,[1107] and Souty and Faure[1109] have indicated a number of complications in attempting to define accurately the mechanical properties of plant root tips which are probably a reflection of the complex nature of the biological material involved.

As pressure in a bed of glass beads was applied,[747] the concentration of nitrogen and calcium in roots and shoots decreased. The largest effect of pressure was reflected in Ca uptake (Figure 14). At the same time, K and S uptake were little affected and it can be related to the fact that apoplastic ion uptake is more disturbed than symplastic uptake by mechanical stress. A similar effect for K and also for P was observed by Russell and Goss.[1027]

4.2.2 SOIL CONDITIONS

Parameters commonly used to relate the mechanical impedance of the soil to root growth is bulk density and penetrometer pressure (penetration resistance, cone index, and strength). Penetrometer pressure is defined as the ease with which a probe can be pushed into the soil and in a given soil it is a function of bulk density and soil water content.

4.2.2.1 Root Growth

It is generally accepted that when mechanical impedance increases, root penetration is reduced. The increase of mechanical impedance reduced root size of field bean,[635] soybean,[635,690,860] sorghum,[540] timothy,[415] spring barley,[752,754,939] spring oat,[752,939] winter rye,[752,1207] winter wheat,[95,805a,939,1207] cowpea,[1207] potatoes,[939] winter rape,[939] peas,[939] oil radish,[939] groundnuts,[121] tulip bulbs,[297] maize,[121,156,656,1044,1069] peanuts,[1181] and winter barley.[1318] The example for soybean root penetration is shown in Figure 15.

A field experiment showed that the depressive effect of soil compaction on rooting depth and root spatial arrangement for winter wheat appeared only at the beginning of plant growth,[805a]

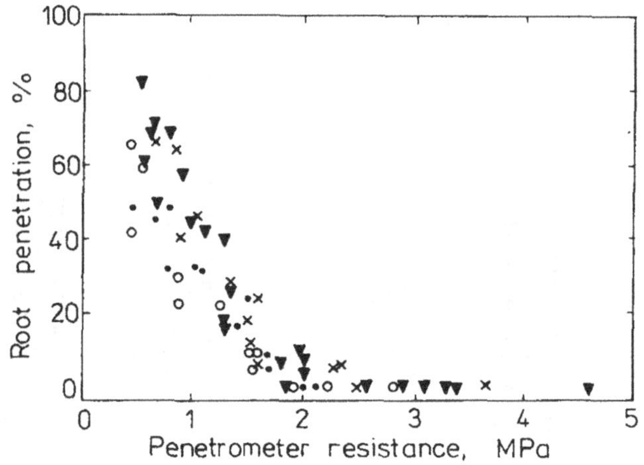

FIGURE 15. Relation between cotton root penetration and penetrometer resistance of various soils. (Modified from Taylor, H. M., Roberson, G. M., and Parker, J. J., Jr., *Soil Sci.*, 102, 18, 1966.)

while in spring cereals the effect may remain throughout the whole growing season.[751] This can be due to both various length of growing period and moisture conditions during the early growth of winter and spring-sown cereals.

It was shown that mechanically impeded root growth results from the reductions in cell proliferation and extension while cell numbers are unaffected.[1088] The effect of mechanical impedance on root growth depends on the soil conditions, plant species, and stage of development of the plant. For example, Boone and Veen[156] found that root elongation of maize in sandy loam at the same moisture content (18% w/w) was very sensitive to changes in mechanical resistance, especially at a moderate range between 0.9 and 1.6 MPa.

Taylor and Ratliff[1181] showed that roots of peanuts elongated at the rate of 2.5 to 2.7 mm h^{-1} in loose loamy sand and that increased penetration resistance to 1.9 MPa caused reduction of the elongation rate to 50% of the maximum. The same reduction of cotton root elongation was noted when penetrometer resistance increased to 0.72 MPa. In Collis-George's studies,[249] when the shear strength of a fine sand seed bed increased from 19 to 52 kPa, root elongation of seminal roots of spring wheat decreased from 43.5 to 0.2 cm d^{-1}.

Maize root length sharply decreased with increasing bulk density of loamy soil from 1.20 to 1.75 Mg m^{-3} (Figure 16).[1069] Trouse[1207] reported that bulk densities of 1.25, 1.35, and 1.45 Mg m^{-3} of clay loam reduced the daily root elongation rate by 57, 67, and 72%, respectively (Figure 17).

De Haan and van der Valk[297] studied the influence of four levels of soil compaction in sandy soil (beach bank soil) on the development of the root system of bulbs. They found that root growth was already strongly inhibited at the pore volume of 43% which corresponded to penetration resistance of about 4 kg cm^{-2}. The roots of bulbs were much more sensitive than those of cereals.

It was also found that too loose soil reduced root and shoot development of the plant.[23] For that reason, it was suggested that the soil, being more loose than optimum for plant growth, could be effectively compacted.[1278]

The ability of roots to overcome the mechanical resistance of the soil varies with species. The differences among plants may result from the diameter or anatomy of roots. For example, roots of Bahia grass (*Paspulum notatum* var. Flugge) possess the ability to penetrate hard layers due

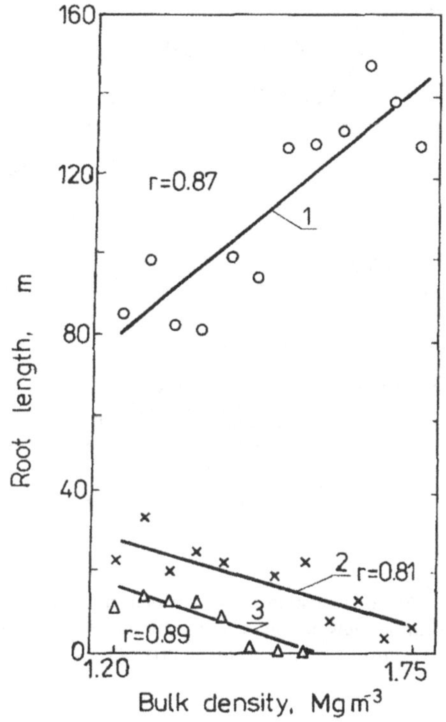

FIGURE 16. Total root length of maize in the upper (1) central compacted (2), and lower (3) layers of a soil. (Modified from Shierlaw, J. and Alston, A. M., *Plant Soil*, 77, 15, 1984.)

to the presence of a fibrous sheath beneath the epidermis.[365] These roots increased the number of pores of a diameter greater than 1 mm which enhanced cotton root penetration. Heinonen[535,536] indicated in his reviews that also the taproots of lucerne, red clover, lupin, pigeon pea, *Leucaena*, and to a lesser degree roots of oil seed rape[1298] have this ability. Whiteley and Dexter[1298] indicated that the ability of roots to penetrate a compact soil increased with increasing diameter. On the other hand, thin roots can use narrow channels in rigid soils and can grow more rapidly with a limited supply of assimilates which occurs at an early stage of development.[535]

Bennie and Burger[121] reported that the relative decrease in rooting density due to mechanical resistance was the same for maize, wheat, cotton, and groundnuts, indicating that mechanical resistance had the same limiting effect on root elongation for all plants. Petelkau et al.[939] studied the influence of different bulk density of loamy sand soil on the rooting depth of seven plants of the same age (Figure 18). The differences between rooting depths of particular plants decreased as soil bulk density increased. In loose, loamy, sand soil of bulk density of 1.35 to 1.38 Mg m^{-3}, about 90% of the roots of winter wheat and spring barley were found in the layer from 0 to 50 cm while in soil compacted to bulk density of 1.65 to 1.66 Mg m^{-3}, they were found in the layer from 0 to 15 cm (Figure 19).[939]

The influence of the layer with high mechanical impedance for root growth depends on its place and dimensions. A less harmful effect of such a layer appears when it is situated in the soil profile and the horizontal dimension is smaller than the lateral spread of the root system[151] and when its thickness decreases.[71]

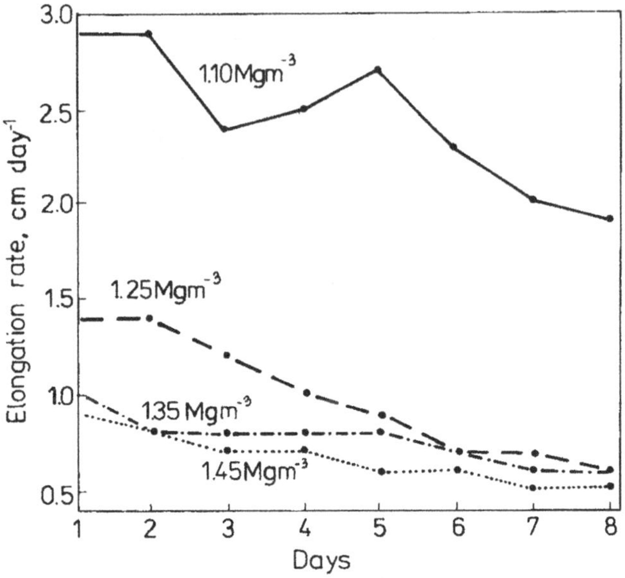

FIGURE 17. Mean elongation rates of winter-grown cowpea secondary roots at various bulk density of soil. (Modified from Trouse, A. C., Jr., *Zesz. Probl. Postepow Nauk Roln.*, 220(I),349, 1979.)

In field conditions, pan layers may restrict root penetration. Taylor and Burnett[1174] and Taylor[1170] reported an excessive soil strength of the pans (southern Great Plains, U.S.) when they dried below –0.1 or –0.2 MPa; water potential was the main factor distorting rooting patterns, but they could become nonrestricting due to rain or irrigation. The critical values of penetration resistance of soil for root growth of various plants are shown in Table 5. The above discussion indicates that the critical impedance of the soil for root growth differs as affected by soil moisture status and stage of plant development. Lack of such information in the literature makes the critical values incomparable and difficult to interpret.

4.2.2.2 Morphology of Roots

In most cases, roots encountering mechanical impedance increase their diameter. Higher mechanical impedance resulted in thicker roots of potatoes,[153] maize,[155,156,705,756,1069,1248] pea,[216,348] bean and soybean,[635] and wheat.[248]

Figure 20 shows substantial changes in root diameters of maize seedlings with increasing loamy soil bulk density in a pot experiment. It is worth noting that the diameters also changed below the compacted layer. A similar response of maize roots to soil compaction was found in a field experiment on sandy soils[155] (Table 6). Thicker roots indicated that in compacted soil, pores, with a size equal to or larger than the root, were lost. Usually the increased diameter of the roots is accompanied by the decreased length of roots.[756,1069] Such a response for maize roots is shown in Figure 21. The studies of Collis-George and Yoganathan[248] showed that the diameter of seminal roots of wheat increased significantly with a relatively small increase of shear strength of soil at the seed zone (Figure 22).

It was shown that severe compaction of the soil increased the degree of flattening (width to thickness ratio) of the roots to 1.5 to 2.0/1 or more.[612,751,1148,1205] The presence of mechanical constraint in the soil also resulted in the zigzaggy growth of roots.[751,1170] Figure 23 shows such

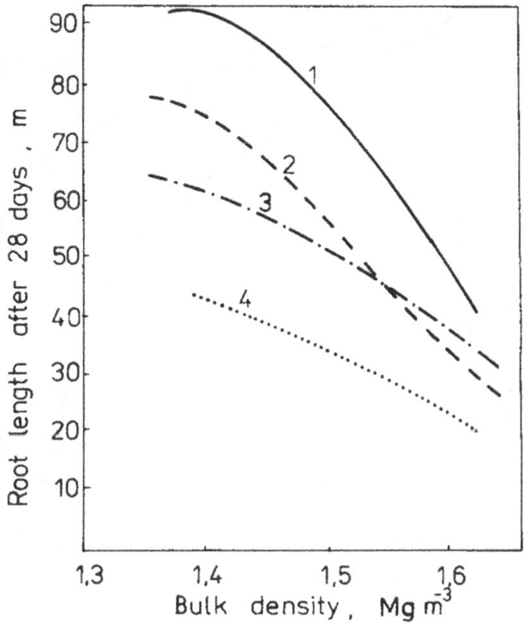

FIGURE 18. Root length of oil radish (1), winter wheat (2), winter rye (3), and peas (4) as a function of bulk density of soil. (Modified from Petelkau, H., Gatke, R., Dannowski, M., and Seidel, M., Report of Soil Science Research Centre, Müncheberg, East Germany, 1985, 24.)

a response in spring barley roots grown in field loess soil. It was suggested that the flattened regions of roots grown in dense soil occur from elongating cells being allowed to expand ahead and parallel to the fracture, but not perpendicular to the fracture,[951,1205] and zigzaggy roots are due to growth-conforming structural ped surfaces.[1146]

A characteristic morphological response of sugar beets[718,824,887,1345] and carrots[663,1291] to mechanical impedance is their forking and fanging. Roots of sugar beets grown in mechanically impeded soil contained more harmful nonsugars.[1150] Hallmark and Barber[508] reported that under low bulk density, in contrast to higher, low soil K fertility increased root surface area per unit of root weight and per unit of shoot weight.

4.2.2.2.1 Root Branching

It was shown that branching of roots grown in mechanically impeded soil is restricted.[156,1002,1059] Boone and Veen[156] reported that the total length of laterals per centimeter of main axis of maize decreased as the mechanical resistance increased and was mainly caused by a decreased lateral root number, whereas the length of the individual lateral roots was not influenced. In rhizotron studies, increasing bulk density of the soil from 1.1 to 1.3 Mg m^{-3} decreased the angle of deflection of maize roots.[74]

4.2.2.2.2 Root Hairs

Mechanically restricted elongation of pea roots increased the number of root hairs per unit length of root.[348] This response is in contrast to the one found in studies with nutrient solution (see Chapter 4.2.1).

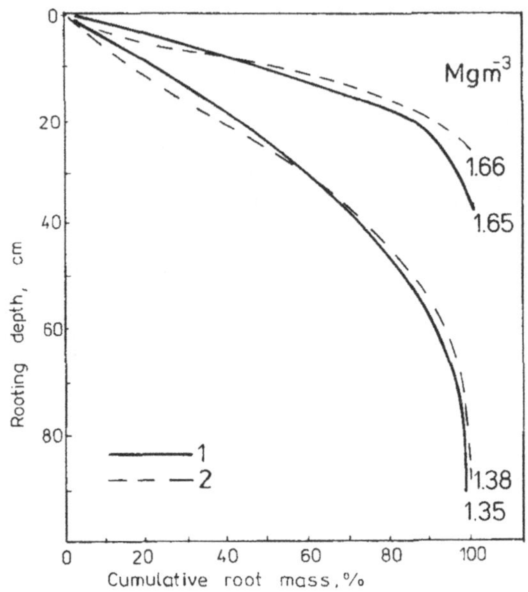

FIGURE 19. Cumulative root mass of winter wheat (1) and spring barley (2) with depth of soil at various bulk densities. (Modified from Petelkau, H., Gatke, R., Dannowski, M., and Seidel, M., Report of Soil Science Research Centre, Müncheberg, East Germany, 1985, 24.)

4.2.2.2.3. Root Distribution

Greater mechanical impedance of the soil in a surface zone resulted in surface accumulation of the roots of Kentucky bluegrass,[13] maize,[156,752] and spring barley[754] and also affected vertical and horizontal distribution of maize roots over the whole profile of soil.[1167] This accumulation can develop an uneven distribution of nutrients and give rise to exhaustion of the surface layers.[156] Mechanical obstacles located at the base of the plowed layer caused a reduction of maize root density, not only at the depths of compacted soil, but also in the nontilled layer and caused a "shadow effect" in the nontilled layers.[1164]

4.2.2.3 Root Anatomy

In mechanically impeded roots, cellular distortion has been observed by Barley and Greacen,[86] Goss and Walter,[460] and Schumacher and Smucker.[1056] This distortion may cause compression of the intercellular air spaces, resulting in the loss of internal air spaces and thus reduction in root porosity. As a consequence, an increase of the tortuosity of the path of oxygen diffusion can lead to the reduction of the radial aeration through the root and greater accumulation of ethylene and other volatile growth regulators within the root tissue.[657]

The influence of soil compaction on the anatomy of plant roots was studied in a limited range. The results obtained hitherto indicate that the cells of cotton[191] and barley roots[751] were radially enlarged (Figure 24). In roots of the soybean,[70] more oval vascular cylinder and thickening of the Casparian strips and cell walls in the xylem vessels were observed. Thickening of the Casparian strips and cortex did not affect the movement of P and K which follow a symplastic pathway,[72] but that of Ca which follows the apoplastic pathway was barred by the Casparian band.[992] Soil compaction also resulted in an increase of the root porosity of Kentucky bluegrass,

TABLE 5
The Values of Soil Penetration Resistance Causing Reduction of Root Growth of Various Plants

Plant	Soil texture	Critical mechanical impedance (MPa)	Degree of reduction	Kind of experiment	Measure of resistance	Ref.
Cotton	Fine sand, sandy loam, fine sandy loam, loam	1.0—2.5 2.5	Considerable Stopped			1182
Cotton	Sandy loam	6.0—7.0	Stopped	Pot	60° cone, 2.7- to 3.5-mm shaft	434
Maize	Loam	5.0	50%	Pot		1069
Maize	Sandy loam	0.9—1.6	50%	Pot		156
Winter barley	Sandy clay loam	1.3—3.7	50%	Field		1319
Sorghum	Clay-sand mixture	2.0	Stopped	Pot	6.0-mm-diameter probe	540
Oats	Silt (tilled)	3.6	Stopped	Field	60° cone, 11-mm diameter	360
Oats	Silt (untilled)	4.6—5.1	Stopped	Field	60° cone, 11-mm diameter	360
Rice		0.3—0.5	75%	Greenhouse	30° cone, 3.23-cm^2 area	527
Rice		2.5	47%	Field	30° cone, 3.23-cm^2 area	527
Pea	Silt loam	2.0			60° probe, 3.8-mm diameter	138
Pea	Clay	3.0	Stopped			433
Peanut	Loamy sand	1.9	50%			1181
Pine		2.5	Restricted			1350
Pinus sylvestris		3.0	Restricted			1041
Tobacco	Loam	2.5	50%			1249
Tomato	Clay, silt, loam sandy loam	2.0—3.5	Restricted			453
Italian ryegrass	Clay, silt, loam sandy loam	2.0—3.5	Restricted			453
Ryegrass	Loam	8.0	80%			1069

to a higher degree in water-stressed conditions (Figure 25). This effect is related to reduced soil aeration in compacted soil.

4.2.2.4 Root Functions

Alterations in root growth and proliferation within the soil profile due to mechanical impedance affect water and nutrient uptake.

4.2.2.4.1 Water Uptake

The influence of mechanical impedance on water uptake by plant roots is not uniform.[13,102,152,753,1339] Agnew and Carrow[13] reported that compaction of silt loam decreased total water use, but increased water use per gram of roots. However, in studies of Bar-Yosef and Lambert,[102] increased bulk density of sandy loam resulted in lower water use efficiency. Lipiec et al.[753] reported that an increase of bulk density of clay soil resulted in both significantly higher total water use and water use per gram of maize roots grown under soil water pressure of –97 or –243 hPa (Figure 26). Compaction also increased the water use efficiency of rice.[924]

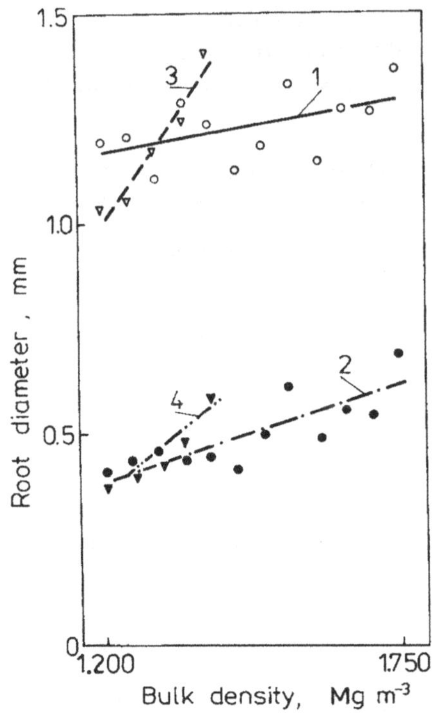

FIGURE 20. Diameters of axes and first lateral roots of maize in compacted layer (1, 2) and underlying loose soil (3, 4). (Modified from Shierlaw, J. and Alston, A. M., *Plant Soil*, 77, 15, 1984.)

TABLE 6
Diameter of Maize Roots as Affected by Compaction of Soils (Fine Sand, Loamy Fine Sand, Fine Sandy Loam)

Treatments	Main root axes x (μm)	First-order laterals x (μm)
Loose soils	1037	314
Lightly compacted (with packer)	1607	442
Severely compacted (with three tractor passes)	2208	640

Modified from Boone, F. R., van der Werf, H. M. G., Kroesbergen, B., ten Hag, B. A., and Boers, A., *Neth. J. Agric. Sci.*, 34, 155, 1986.

The higher water use efficiency by roots in compacted soil is attributed to higher unsaturated hydraulic conductivity[153,753] and decreasing percolation losses,[924] but lower water use in compacted soil was related to low O_2 levels.[13] In the field experiment of van Loon and Bouma,[1232] compacted topsoil retarded vertical root growth of potatoes and this resulted in the lack of significant capillary transport through dense soil between the groundwater and the lower boundary of the root zone.

Supplying a crop with adequate water and minerals in the resisted root system in compacted soil

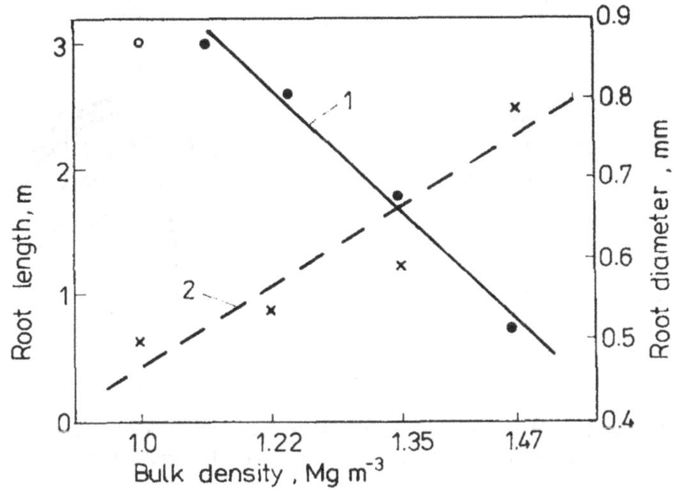

FIGURE 21. Root length (1) and root diameter (2) of 6-d-old maize seedlings as influenced by bulk density. (Data from Logsdon, S. D., Reneau, R. B., Jr., and Parker, J. C., *Agron. J.*, 79, 221, 1987b.)

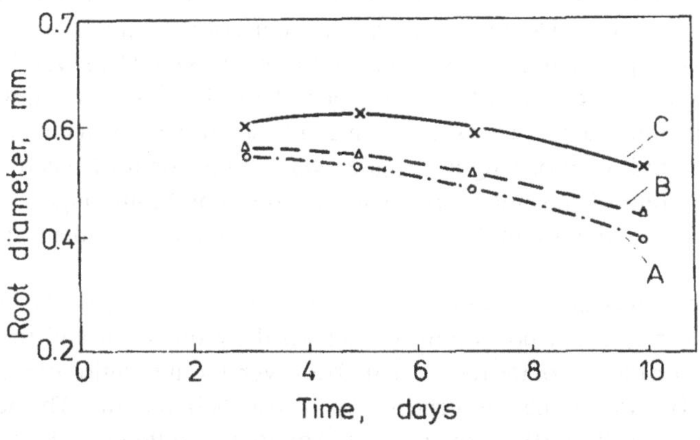

FIGURE 22. Root diameter of wheat as affected by initial strength at the seed zone; A, 1.9; B, 2.2; and C, 9.3 kPa. (Modified from Collis-George, N. and Yoganathan, P., *Aust. J. Soil Res.*, 23, 577, 1985a.)

depends on rainfall intensity and distribution during growing season.[1314] In a dry year, soil water via soil strength may restrict root penetration into deeper layers and thus limit water absorption. In a wet year, mechanical impedance does not restrict root penetration, but if rainfall is excessive, insufficient aeration may cause reduction of water absorption. Mild stress conditions affect yields of plants with a resisted root system more than unresisted ones.[1169]

4.2.2.4.2 Nutrient Uptake and Chemical Composition of Roots

Usually soil compaction reduces nutrient uptake, especially immobile phosphorus.[156,664,917,959,1069] Shierlaw and Alston[1069] grew ryegrass and maize under the same conditions

FIGURE 23. The growth of spring barley roots in loose (1) in comparison to root in compacted (2) loess soil. (Data from Lipiec, J., unpublished.)

and showed that bulk density affected phosphorus uptake by the former plant and did not by the latter (Figure 27)

A lower uptake rate of phosphorus than other nutrients resulted in an increased nitrogen/phosphorus ratio in the plant.[917] However, the uptake of P and K expressed per unit of root length of maize, wheat, cotton, and groundnut roots increased with a decrease in rooting density due to mechanical resistance.[121] Decreases in the concentration of nutrients in a crop due to reductions of the rooting zone because of compaction were reported by several authors.[148,593,1311]

Boone and Veen[156] indicated that the influence of mechanical constraints on ion uptake depends on the nutritional factor of the soil. At a high phosphate supply in sandy loam, the uptake of N, P, and K per unit of root length of maize was independent of mechanical resistance, suggesting that the rate of ion uptake is related to root length, although as shown in other papers,[660,948] soil compaction might be expected to increase the movement of ions toward the roots by diffusion.

This reaction depends on the kind of elements and the way in which the elements are absorbed.[76,1311] Phosphate and potassium are absorbed by diffusion and their absorption is strongly dependent on the size of the root system. However, the absorption of nitrate and calcium is by mass flow and the rate of their absorption depends on their quantity. The decreased volume of compacted loamy soil (to 1.30 Mg m^{-3}) explored by roots resulted in reduced Ca, K, Mg, and Mn uptake, while no significant effect on B, Fe, and P uptake was observed.[216] It was also reported that a considerable proportion of the root system situated in the upper layer of mechanically impeded soil did not participate in the ion uptake.[1248]

Lower nutrient uptake in compacted soil may be compensated by higher absorption in other parts with favorable conditions due to greater production of roots and thus total nutrients need not necessarily be decreased. This has been shown by Kubota and Williams[691] in field experiments for globe beet and barley, although crop yields were reduced.

The increase of bulk density of clay loam from 1.1 to 1.8 Mg m^{-3} resulted in significantly higher cation exchange capacity (RCEC) of eight plants (Table 7). This was apparently due to the increase in the percent of N of roots and –COOH groups of the roots. Compaction increased the amino acid content in roots of timothy when carried out in the wetter soil condition, and the increase of the root length tended to decrease the amino acid content and vice versa.[415] Root xylem accumulations of toxic anaerobic metabolites were directly correlated with soil bulk density and inversely correlated to oxygen diffusion rate in the soil.[50]

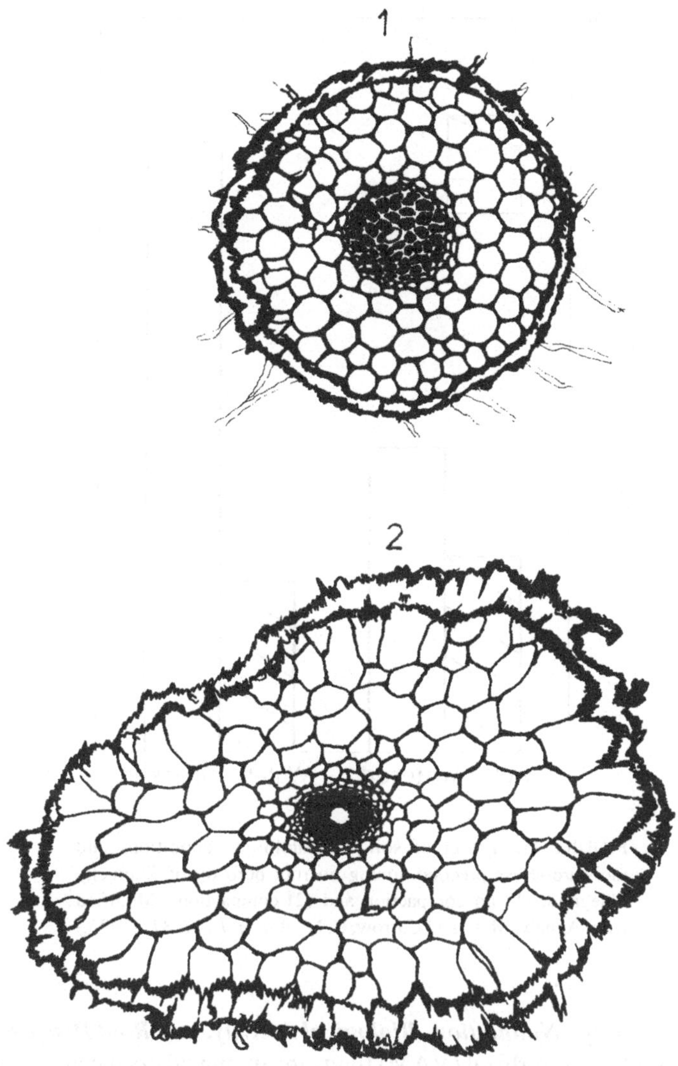

FIGURE 24. Cross section of spring barley root grown in loose (1) and compacted (2) loess soil. (Data from Lipiec, J., unpublished.)

4.2.2.4.3 Photosynthate Requirements and Oxygen Supply and Uptake

Studies of Sauerbeck and Helal[1044] showed that as mechanical impedance of soil increases, the amount of photosynthetic energy required to form and sustain a unit of root length strongly increases. Mechanical impedance through changes in other physical soil properties and roots predisposes them to oxygen defficiency. In compacted soil, the lower rate of oxygen diffusion to the surface of roots and through compressed cells of the roots was reported by Rickman et al.[986] and Schumacher and Smucker.[1056] However, the roots subjected to mechanical impedance consumed more oxygen per unit of fresh weight.[1056] It means that a greater oxygen supply is required at the root surface to prevent anoxia in mechanically impeded roots. It may result from the greater supply of assimilates available per fresh weight of stressed root. Soil compaction reduced the oxygen supply for roots of winter cereals[136,468] and tomato.[986]

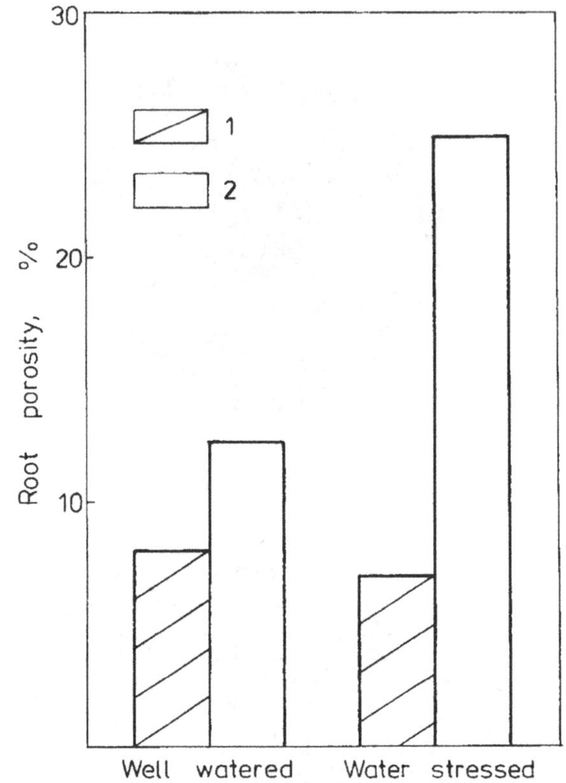

FIGURE 25. Effect of soil compaction preconditions and moisture stress preconditioning on root porosity of Kentucky bluegrass: (1) no compaction and (2) compaction. (Modified from Agnew, M. L. and Carrow, R. N., *Agron. J.*, 77, 872, 1985.)

4.2.2.4.4 Symbiotic Fungi, Nodulation, Biological Activity, and Root Diseases

Vesicular-arbuscular mycorrhizal (VAM) fungi are among others the most common symbiotic fungi in the soil environment. It was shown[1147] that the highest fungal colonization accompanied by the greatest demand for nutrients occurred when the root system of dry edible beans (*Phaseolus vulgaris*) was growing in mechanically impeded media.[1056] Mulligan et al.[858] investigated the influence of soil compaction by secondary tillage and traffic on root and VAM fungal colonization of dry edible beans grown on clay and sandy loam soils. The colonization decreased with increasing bulk density and depth within the soil profile (Figure 28). The cultivars having vigorous root systems which penetrated below the A_p horizon of compacted soil maintained high levels of colonization by the VAM fungi. The authors suggest that the decrease in colonization of roots in deeper layers may have resulted from the lower soil aeration as the oxygen diffusion readings during the first 2 months of the growing season were lower than 0.25 μg of O_2 cm^{-2} s^{-1}. Also the penetration of the soil by mycelial strands of mycorrhizal fungi associated with *radiata* pine roots is restricted in compacted soil.[1078]

The extent to which soil compaction can affect nodulation on the roots of soybeans (*Glycine max*) was studied in a field experiment on silty clay loam.[1258] Controlled wheel traffic produced soybean rows with no wheel traffic on either side and wheel traffic on only one or both sides. Measurements of nodule number and mass from undisturbed soil monoliths showed that rows

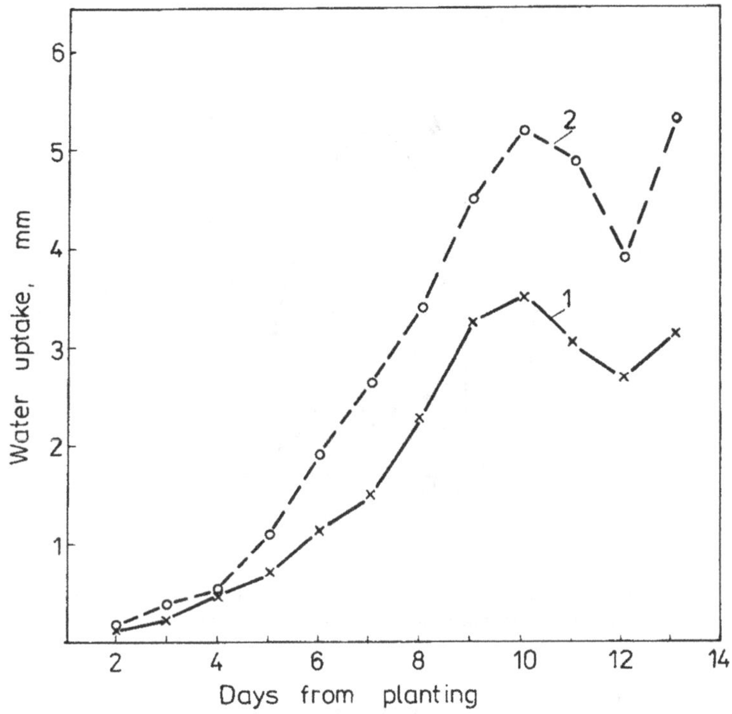

FIGURE 26. Daily water uptake by roots of maize from uncompacted (1) and compacted (2) soil at soil moisture tension 243 kPa. (Modified from Lipiec, J., Kubota, T., Iwama, H., and Hirose, J., *Soil Sci. Plant Nutr.*, 34, 417, 1988.)

with wheel traffic on both sides had from 20 to 30% fewer nodules and about 36% less total nodule mass than rows with wheel traffic on only one side. The influence of the controlled wheel traffic on soybean nodulation was also reported by Lindemann et al.[749] The experiments show that by controlling the wheel traffic of normal farming operations, the quantity and distribution of nodules can be altered in ways which have implications for increasing nitrogen fixation.

Fulton et al.[412] and Gray and Pope[471] reported that the incidence of *Phytophtora* root rot of soybeans increased in fine-textured soils subjected to compaction. The mechanism by which soil compaction increases the disease incidence is probably through changes in soil physical properties that either influence the host root system and/or provide an environment for rapid development of the fungus.[471,1088] The effect of mechanical impedance on root growth and functions in relation to other soil physical factors is discussed in Chapter 4.7.

4.2.2.5 Root Growth and Penetrometer Pressure

It is evident that the values of penetrometer pressure at which elongation of the root stopped or was reduced are much higher compared to the maximum pressure exerted by roots. Whiteley et al.[1302] and Eavis and Payne[348] reported that penetrometers had to exert a pressure several times (two to eight) greater than a root tip in order to penetrate the soil. The relation between the penetrometer pressure strength and root pressure for peas is shown in Table 8.

The differences result from different conditions in penetrometer resistance and the resistance encountered by plant roots. It was indicated that, compared with metal probes, root tips may penetrate the soil more efficiently because of a more advantageous distribution of the stress applied to the soil.[1,84,172,245,474,983,984,1302] The results strongly suggest that the displacement of soil

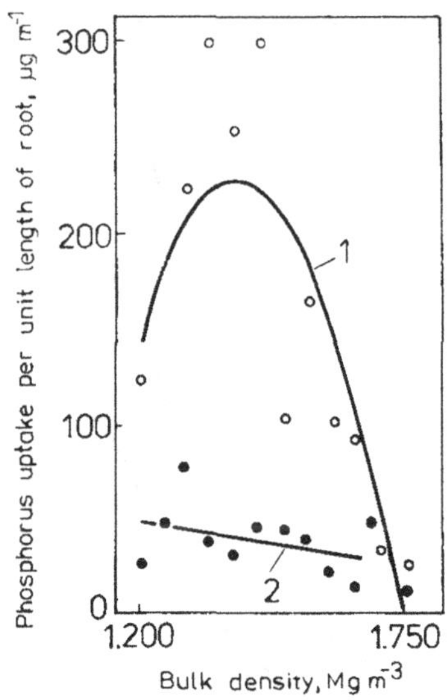

FIGURE 27. Phosphorus uptake per unit length of root of ryegrass (1) and maize (2) as a function of bulk density of soil. (Modified from Shierlaw, J. and Alston, A. M., *Plant Soil*, 77, 15, 1984.)

TABLE 7
Effect of Soil Bulk Density on Root-Cation Exchange Capacity and Nitrogen Content of Roots

Crop	Root-cation exchange capacity (meq 100 g⁻¹ dry roots)		Nitrogen content of roots (%)	
	Bulk density (Mg m⁻³)			
	1.10	1.80	1.10	1.80
Gram	34.57	33.27	1.73	1.81
Sunflower	34.66	44.46	1.32	2.22
Safflower	43.95	47.67	2.14	2.97
Maize	15.01	18.18	1.98	2.22
Pigeon pea	37.46	45.69	2.39	3.05
Cowpea	26.07	34.31	1.06	1.59
Soybean	29.85	34.22	1.65	1.89
Lentil	33.95	38.45	0.98	1.23

Data from Kulkarni, B. K. and Savant, N. K., *Plant Soil*, 48, 269, 1984.

particles by root tips was more nearly consistent with a cylindrical than with a spherical expansion. The cylindrical expansion of the tip weakens the soil in front of the tip and thus makes root elongation easier. This "inverse peristaltic" form of root elongation might enable a root to grow through dense soils five times stronger than it is by a uniform growth mechanism.[548] In contrast to penetrometer tips, the roots have also mucigel layers[729] which decrease friction between the root surface and soil particles during growth.

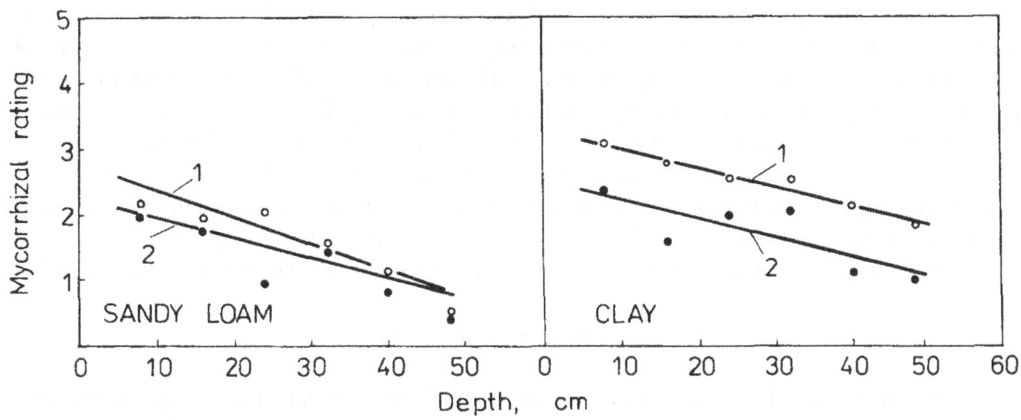

FIGURE 28. The influence of minimum (1) and excessive secondary (2) tillage and depth on the mycorrhizal rating of black turtle soup dry edible bean roots grown on clay and sandy loam soils. (Modified from Mulligan, M. F., Smucker, A. J. M., and Safir, G. E., *Agron. J.*, 77, 140, 1985.)

TABLE 8
Comparison of the Values of Pea Root Penetration Pressure (P_r) at Various Penetrometer Pressures (P_p)

P_r (kPa)	P_p (kPa)	R
155	563	3.83
198	732	3.83
329	1777	5.32

Note: $R = P_p/P_r$..

Modified from Whiteley, G. M., Utomo, W. H., and Dexter, A. R., *Plant Soil*, 61, 351, 1981.

The penetrating ability of plant roots is often assessed from the resistance to blunt and sharp penetrometer probes. However, soil resistance to probe penetration is much greater than the resistance to plant roots. The difference is due to the flexible nature of the roots in contrast to metal probes which are rigid. Blunt penetrometer probes (total tip angle of 60°) deform the soil almost spherically, whereas plant roots and sharp probes (total tip angle of 10°) deform the soil almost cylindrically.[111,840] Misra et al.[840] found a relationship between the axial pressure exerted by the root (P_x) to that exerted by a 1-mm-diameter blunt probe (P′),

$$P_x/P' = 0.7424 - 0.0228c - 0.0063d \quad (8)$$

where d (in millimeters) is the aggregate diameter and c (in kilopascals) is the shear cohesion of the aggregate.

The response of roots to mechanical impedance involves reduced elongation and increased radial growth and is hormonally modulated. In addition, the rigidity of the metal probes of penetrometers prevents them from seeking out planes of weakness; biopores and paths offering less than average resistance give an overestimate of the minimum soil pressure which appreciably reduces root elongation.[357,360,1084] It is confirmed by the lack of correlation between the rate of tunneling in soil by earthworms and penetrometer pressure.[305] The importance of the spatial varibility increases in forest soils, especially in unevenly aged stands, where mechanized operations are less evenly spread over the area.[457] The degree of correlation between root growth and penetrometer pressure also depends on the factors affecting the soil penetration resistance: size and geometry of a probe, skin friction, rate of advance, soil type, and soil depth.

Voorhees et al.[1259] found that root elongation was more closely correlated with soil resistance to a 10° probe than to a 60° probe. It was also more closely correlated with normal soil resistance (σ_n) to the probe which does not include soil-metal friction than with total point resistance (q) which includes the friction component.[86,386,1259] The total point resistance is related to σ_n by:

$$q = \sigma_n(1 + \tan\phi' \cot\alpha) \tag{9}$$

where α is the included semiangle of the probe and ϕ' is the coefficient of soil-probe friction.

Groenevelt[492] and Groenevelt et al.[493] analyzed the tip resistance (without frictional resistance) data to produce "penetrability characteristics" of the soil, relating percentage linear penetrability PLP (the percentage of a linear trajectory through the soil, for which the actual tip resistance is lower than a critical value) to a "maximum attainable" tip pressure (critical pressure). Assuming that a root can, at least temporarily, exert a certain maximum tip pressure, one can find what percentage of a certain soil at a certain bulk density is penetrable by that root.

It has been shown that a 1-mm-diameter probe had to exert a pressure about 50% greater than a 2-mm probe in soils of various textures,[453,1295] No significant differences in penetrometer pressure were found between 1.8- and 2.2-mm-diameter probes in remolded clay or in sandy loam soils[1259] and for probes of 3.8- and 5.1-mm diameter, although a close relationship with root length was found for a 3.8-mm probe.[161]

In numerous papers, the rate of probe penetration in the range corresponding to the rate of root elongation had no significant effect on penetrometer resistance.[86,138,161,245] Bradford[161] also found little effect of varying the penetration rate from 0.1 to over 100 cm d^{-1} on the resistance to penetration. It was found that in uniform cohesive soil, penetrometer pressure reaches the maximum value at a depth of three to four probe diameters and it remains constant with an increasing depth of penetration,[453] and the friction between soil and metal probe skin contribute more to total probe resistance at more acute angles.[1259]

Taking into consideration the importance of the factors influencing the relationship between penetration resistance and root elongation, Gooderhaam[454] and Jung La et al.[630] indicated that standarization of the techniques may significantly increase the value of penetration resistance in studies of root growth. It has been demonstrated[630] that the larger the base area of the probe, the lower the cone index and that a cone angle of 30° shows the greatest effect (lowest penetration resistance) on the cone index, compared to 60 and 90° cones. However, in some soils, the formation of soil bodies which are pushed ahead of the probe may limit standardization.[859] Bradford[161] indicates that standardization should include probe diameter and angle, speed of penetration, and method of probe advance into the soil. Taylor[1170] suggested using probes about equal in diameter to the roots. It was also suggested that the introduction of lubrication and axial and radial expansion systems in future penetrometers would simulate root growth much better.[493,859,1027,1296]

Although the penetrometers do not measure the forces encountered by growing roots, they do provide useful measures of soil to which root growth may be referred.[138,152,161,310,1069] Dexter[310] indicated that penetrometry with all its limitations will continue to be used for a long time to come, and Collis-George and Yoganathan[248,249] pointed out that if cone resistance measurements could be satisfactorily related to shear strength, they could also be used as rapid methods for measuring seedbed strength.

4.2.3 MODELING OF ROOT GROWTH IN RELATION TO MECHANICAL IMPEDANCE

Mathematical simulation of axial root growth to increasing mechanical impedance expressed in terms of the displacement velocity and tissue strain[808] showed that turgor pressure initially decreased behind the root apex, reaching a minimum within the elongation zone, and then increased to the base of the elongation zone. The simulation confirmed several root responses to mechanical impedance, such as decreased root elongation rate, decreased length of mature cell, and a decreased extent of the elongation zone itself. Recent papers[984,1250] described how new techniques for nonlinear modeling can be used to predict the strain-stress-displacement relationships in the soil-root interactive system.

Studies of Hewitt and Dexter[551] have shown that root maximum growth pressures, root buckling stresses, and soil penetration strengths are all nonnormally distributed and beta-distribution is recommended to be used in the development of mathematical models for root growth in soil.

Dexter[310] proposed the equation describing relative rate of root elongation as:

$$\frac{R}{R_{max}} = -\frac{\psi_o}{\psi_w} + e - 0.6931(Q_p/Q_{1/2}) \qquad (10)$$

where ψ_o is the external water potential; ψ_w is the wilting point water potential of the plant species; Q_p is the penetrometer strength; $Q_{1/2}$ is the value of penetrometer resistance that reduces relative elongation rate to one half; and the exponent −0.6931 is log e 0.5 and results from the use of $Q_{1/2}$. Figure 29 shows values of water stress and penetrometer strength and soil penetrometer strength at which root growth will cease.

According to Hillel and Tolpaz,[553] root elongation rate can be described by the equation:

$$dL/dt = CL \qquad (11)$$

where L is the root length capable of growth (in centimeters); t is the time (in days); and C is the specific root elongation rate per day. Bar-Yosef and Lambert102 tested the equation under controlled laboratory conditions and related C to soil probe resistance (Figure 30) and concluded that root growth of maize and cotton is satisfactorily described by the equation. Bennie and van Antwerpen[122] found the empirical equations used by Gerwitz and Page[436] and Rowse and Barnes[1017] (see Chapter 2.0.3.) to characterize the change in rooting density with depth in uncompacted soils, can be suitable to describe the rooting density distribution of wheat, peanuts, and cotton with depth of subsurface compacted soils and that after modification they can also be used for maize and peas.

Summarizing, it can be concluded that the characteristic responses of the root system of most plants grown in mechanically impeded soil are reduced root size; diminishing root elongation rate; uneven root distribution; thickening of roots and greater lateral branching; increased amount of assimilates and oxygen per unit of roots; decreased VAM fungi colonization; increased root rot incidence; and reduction of water and less mobile nutrient uptake. Roots of a few plants (such as ryegrass) are capable of penetrating into pores of lower diameter than themselves by decreasing their own diameter. Roots of certain plants, e.g., Bahia grass, are able to penetrate the strong layers thanks to the presence of a fibrous root sheath. Moderate mechanical impedance can increase water use efficiency due to higher unsaturated hydraulic conductivity and better root-soil contact area. Equations have been developed to describe root growth in relation to mechanical impedance of soil. The response of plant roots to mechanical impedance is closely interrelated with other soil factors (see Chapter 4.7).

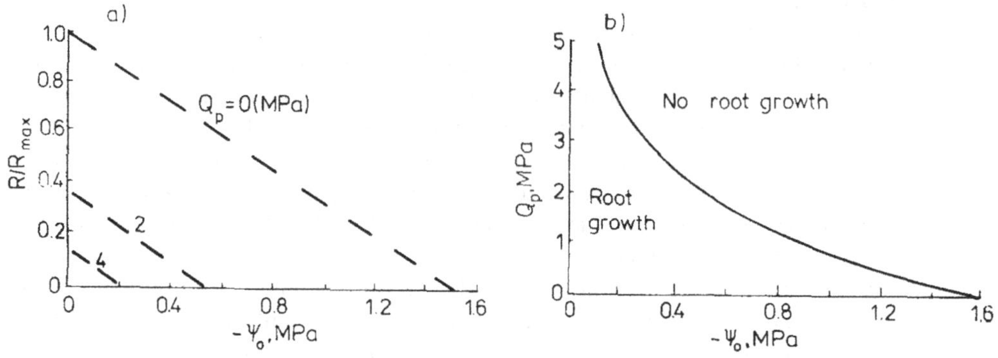

FIGURE 29. Relative rate of root elongation, R/R'_{max}, under combinations of soil water potential, $-\psi_o$, and penetrometer strength, Q_p (a) and curve indicating values of $-\psi_o$ and Q_p at which root growth will cease (b), both predicted from Equation 10. (Modified from Dexter A. R., *Plant Soil*, 98, 303, 1987a.)

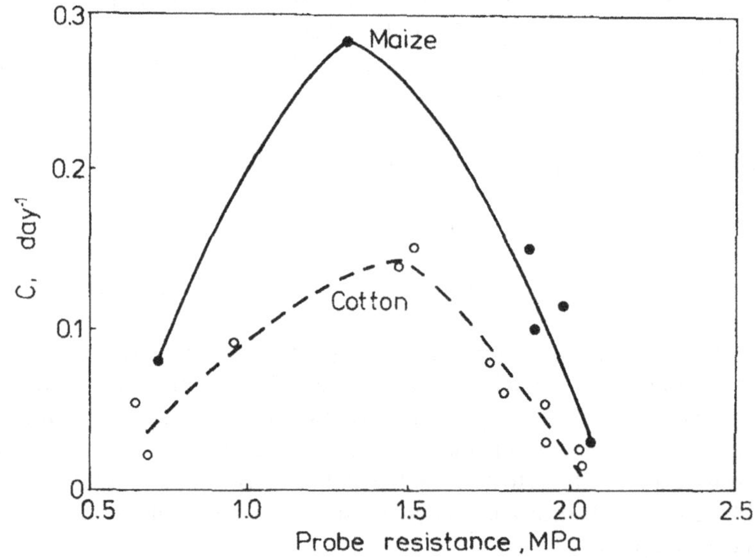

FIGURE 30. Maize and cotton root growth rate coefficient (C) as a function of soil probe resistance. Bulk density (1.45 Mg m^{-3}) was held constant at different matric potentials. (Modified from Bar-Yosef, B. and Lambert, J. R., *Soil Sci. Soc. Am. J.*, 45, 930, 1981.)

4.3 WATER

Soil water status is one of the most important parameters affecting root growth and functions. Rooting of a plant is directly affected by the influence of water and indirectly affected by other physical factors, such as aeration, mechanical impedance, temperature, and transport of nutrients toward the root in the soil.

4.3.1 ROOT GROWTH

It is known that the drier the soil, the smaller the root quantity.[372,929,957,1005,1127] It is also known that the root elongation rate of cereals becomes slower during prolonged periods when the soil is wet and faster, i.e., to greater depths, during dry conditions.[368] There is a relationship between root depth, available soil water, and soil texture. Dwyer et al.[345] found that coarse-textured soils with a smaller water-holding capacity contain deeper roots (maize, soybeans, and barley) than those of fine texture with a larger capacity.

In the case of upland rice varieties,[793] there was no significant relationship between soil moisture and total root dry weight, but a soil moisture regime resulting from different depths of the water table significantly affected the angle of root penetration. Soil moisture data have been used by some authors to elaborate a quick means of estimating root system configuration.[794]

It is generally considered that the water potential in the root zone of about –50 kPa causes the root growth to start to decrease and until about 1500 kPa, it may slowly continue.[509,724,867] Such relationships were noted for oats,[359] forage species,[18,318] maize,[279,780,955] chickpea,[57] grapefruit,[132] tomato,[955] and soybean.[1135] Roots of maize and tomato can grow in soil with the water potential less than –40 MPa if a part of them is well supplied with water.[955] In wet soil, root growth can be limited unless aeration is adequate.[1205] A relationship between soil water potential and root water potential was also found (Figure 31). Logsdon et al.[756] found that the root length of maize was not reduced by low soil moisture equal to a matric potential of –1.09 MPa.

Most crops proliferate more in high-moisture soil zones,[49,1129] and the contribution of roots developing deep in the profile increases as the surface soil becomes dry.[805] Supplying a root system with water in the soil depends not only on the water potential at the different depths, but also on the rate at which water moves through the soil to the roots. The amount of water available to the roots varies widely in soils as affected by their texture, organic matter content, soil density, soil temperature, and soil aeration (see Chapter 1). In general, soils with a smaller water-holding capacity have deeper roots, while those with a larger capacity have a shallower root zone. The rate of movement of water from soil to roots depends on the water potential gradient and on the hydraulic conductivity of the soil. They decrease with decreasing soil water content (see Chapter 1.6). Increased salt concentration in the soil water in saline soils or due to intensive fertilization leads to increased osmotic forces, and the limitation of root growth can be greater than that caused by matric forces alone.

4.3.2 ROOT MORPHOLOGY

Soil moisture significantly influences root morphology. The results obtained for maize grown on silt loam by Mackay and Barber[780] can serve as an example. The increase of soil moisture from 22 ($\psi = -173$ kPa) to 32% ($\psi = -7.8$ kPa) significantly increased total root length with root hairs, density of root hairs, and root surface area due to root hairs (Figure 32). Too dry or too wet soil conditions inhibit development of root hairs.[684]

Plant dehydration causes root shrinkage. A drop of water potential resulted in shrinkage of wheat roots to about 40% of their original diameter,[246] shrinkage of cotton roots to about 60%,[582] and shrinkage of broad bean roots about 50%.[1018] Taylor and Willatt[1185] indicated that the roots of soybean seedlings shrank to a much greater extent if the water potential was lowered at the root surface than if it was lowered in the leaves by transpiration when the roots were in close contact with moist soil. Shrinkage and subsequent rehydration of roots growing in the soil cause significant variations in root-soil contact resistances to water and ion movements.

4.3.3 ROOT POROSITY

For better growth under flooded conditions, some plants increase root porosity, especially in the adventitious roots[75,288,289,615,762,1240,1343] (Figure 33). According to Das and Jat,[289] high root

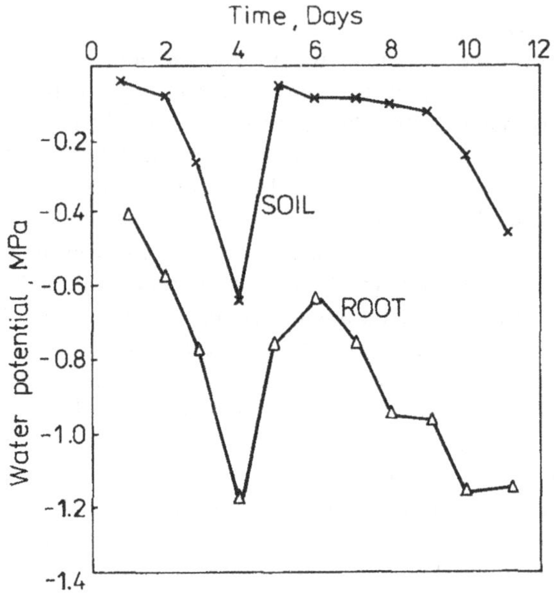

FIGURE 31. Relation between soil water potential and root water potential of maize seedlings. (Modified from Adeoye, K. B. and Rawlins, S. L., *Plant Physiol.*, 68, 44, 1981.)

porosity of rice due to continuous flooding was associated with greater root length and increased dry weight of roots. Other adaptation features to waterlogging are described in Chapter 4.4.

4.3.4 ROOT FUNCTIONS

Soil moisture influences water and nutrient uptake by roots. The rate of root water uptake from the soil depends on the root length density (cm cm^{-3} of soil), soil hydraulic conductivity, and the difference between soil water suction and root suction. A nonuniform root system can produce different suction gradients in the soil profile and induce water movement from one layer to another. The range of this movement is relatively small compared to the water uptake by roots, but in some cases, it can be considerable.

An analysis of the time of water withdrawal from different soil depth zones in relation to soil wetness and hydraulic properties and of the rate of plant-root uptake was presented by Rose and Stern.[1000] The water conservation equation for a given depth of soil (assuming the flow to be vertical only) for a given period of time (from t_1 to t_2) can be written as follows:

$$\int_{t_1}^{t_2}(i - v_z - q)dt - \int_0^z\int_{t_1}^{t_2}\frac{\delta\theta}{\delta t}\,dzdt = \int_0^z\int_{t_1}^{t_2}r_2dzdt \qquad (12)$$

where i is the rate of water supply; q is the evaporation rate from the soil surface; v_z is the vertical flux of water at a depth z; θ is the volumetric soil wetness; and r_z is the rate of decrease of soil wetness due to water uptake by roots.

The average rate of uptake by roots at the depth z is

$$r_z = \int_{t_1}^{t_2}r_2dt/(t_2 - t_1) \qquad (13)$$

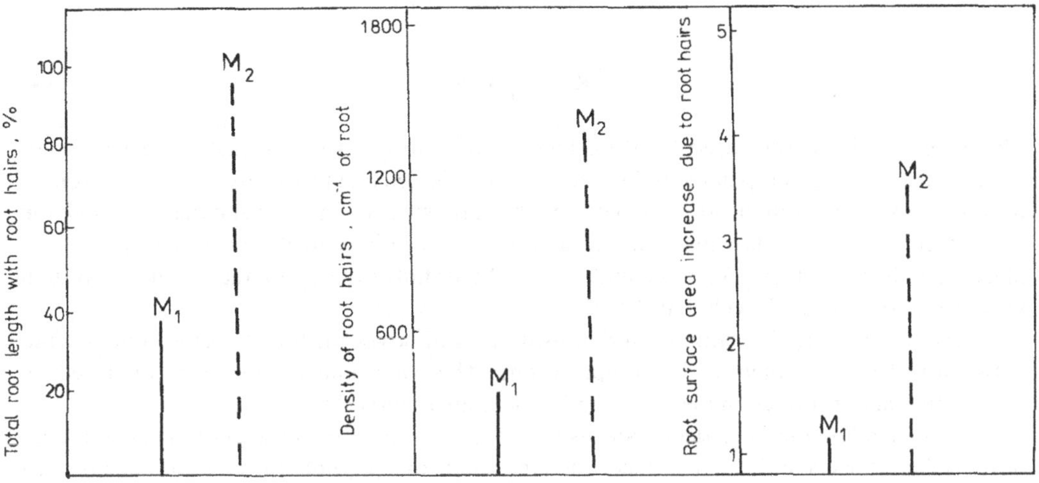

FIGURE 32. Root with root hair characteristics under the influence of soil moisture held at water potentials M_1 = –173 kPa and M_2 = –7.5 kPa and P level of 200 mg kg^{-1} of soil. (Data from Mackay, A. D. and Barber, S. A., *Plant Soil*, 86, 321, 1985.)

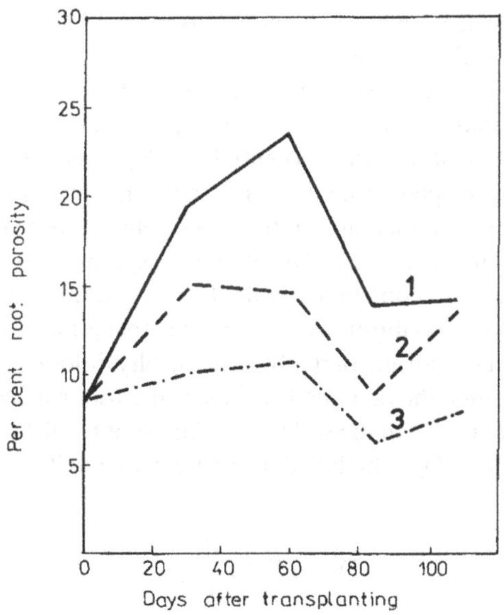

FIGURE 33. Changes in root porosity of rice at different stages of growth under different soil moisture regime: (1) flooding, (2) alternate flooding, and (3) no flooding. (Data from Das, D. K. and Jat, R. L., *Agron. J.*, 69, 197, 1977.)

The pattern of soil water extraction by a root system can be determined by repeated calculations based on the preceding equations for successive small intervals of time and depth. The total (cumulative) water uptake by the roots, R_z, is given by

$$R_z = \int_0^z r_z dz \qquad (14)$$

These relationships were used by Rose and Stern[1000] to describe the pattern of soil-water extraction by a cotton crop in the field. The results indicated that nearly all water extraction by the crop took place from the top 30 cm during the early stages of growth and that, although the zone of uptake extended downward during later stages, essentially all of the seasonal uptake took place from the top 100 cm. A similar and detailed field study of water extraction of root system was carried out by van Bavel et al.[1224,1225]

Eavis and Taylor,[350] examining the transpiration of soybean, found a close relationship between soil water content and root uptake rate. The water uptake rates were modified by fertilization and its effect on the root length to leaf area ratios (Figure 34).

It is a general opinion that deep roots usually extract water more efficiently (per centimeter of root) than roots near the soil surface when the soil is dry (Figure 35). This has been proven for soybeans,[1135,1321] and citrus trees[217] and was explained by the fact the deeper roots were younger, less crowded, and often located in wetter soil with higher hydraulic conductivity. This was indicated by Taylor and Klepper for cotton[1177] and maize.[1178]

Opinions on the relationships between the water uptake rate and the water potential in the soil are not uniform, especially in the range of high matric potentials. This is due to many interactions with other factors, such as oxygen supply, soil temperature, and nitrate concentration. Bar-Yosef and Lambert[102] found under laboratory conditions that the flux of water uptake by cotton and maize root systems had a peak at a matric potential between –13 and –20 kPa and was about 0.2 and 0.03 ml cm^{-1} of root per day, respectively.

When the whole profile is wet, roots can extract almost of all the water from the shallow layers. Hasegawa and Yoshida[528] showed that the water uptake rate by the dryland rice root system is not a simple function of soil water potential and depends on the depth in the soil profile. In the upper soil layer, water uptake rate decreases with the decrease of soil water potential, whereas in the deeper layers, the increase in the water uptake rate may occur with decreasing soil water potential as shown in Figure 36. The above is explained by the presence of axial root resistance in rice roots. The water content of a soil affects ion uptake by roots. This is done partly by influencing the transport of ions through the soil to the root surface by diffusion and mass flow induced by the transpiration stream and partly by causing physiological effects on the plant (e.g., water stress which may change the demand for ions of the root surface).[139,341,757]

The mobility of ions moving from the soil to the plant root by diffusion is determined by the effective diffusion coefficient, D_c, which is defined by Nye and Tinker:[892]

$$D_c = D_l \theta f \frac{dC_l}{dC} \qquad (15)$$

where D_l is the diffusion coefficient in water; θ is the volumetric water content; f is the impedance factor; C_l is the concentration of the nutrient in the solution; C is the quantity of the nutrient participating in diffusion; and dC_l/dC is the reprocical of the buffer capacity.

4.3.5 DROUGHT TOLERANCE

From the physiological point of view, water stress occurs when the water potential of cells falls below zero. In water-stressed roots, solute accumulation was enhanced to allow maintenance of the turgor and continuous growth.[476,1051,1066] The osmotic adjustment of the root tips was greater than for the root as a whole.[1051] In these conditions, the solute concentration increased. Also the decrease of root hydraulic conductivity occurs,[906,969] increasing root resistance. In the

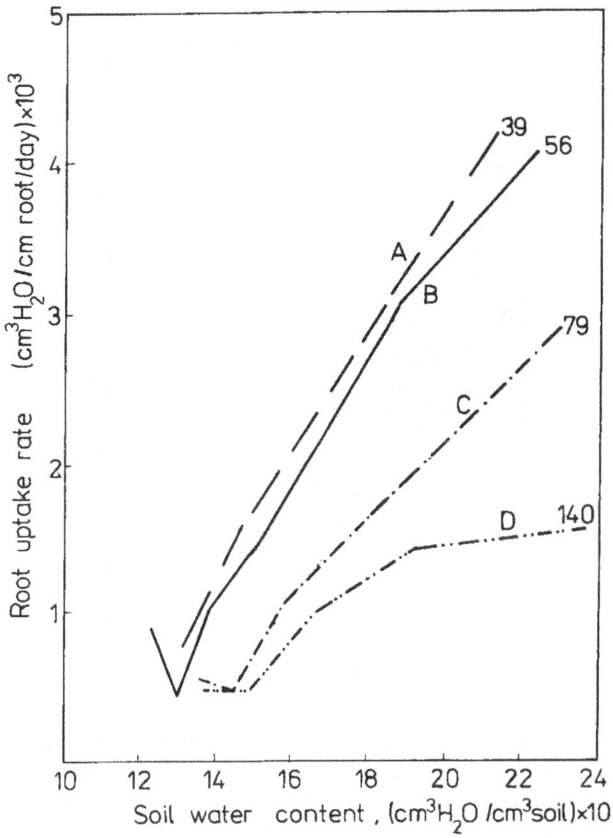

FIGURE 34. Effect of soil water content on water uptake by soybean roots at various fertilization: A and C, watered plants at high (A) and low (C) NPK fertilization, and B and D, water-stressed plants at high (B) and low (D) NPK fertilization. Numerals near the lines are the root length to leaf area ratios (cm roots/cm leaf area). (Modified from Eavis, B. W. and Taylor, H. M., *Agron. J.*, 71, 441, 1979.)

ecological or agronomic sense, this is concerned with a significant restriction in plant survival growth.[1024] The ability of the roots to survive dry conditions is attributed to the rapid extension of their axes to maintain continuing contact with the wet soil, extensive root branching, and development of short tuberized roots.[390,1242] The suberization of plant roots in dry soil is considered to be one plant mechanism for surviving drought by reducing water loss from plant roots.[991,1242]

Some succulent perennials, e.g., *Agave deserti* and *Ferocactus acanthodes,* lose little water through their roots during drought, yet they respond rapidly to light rainfall.[624] During 12 to 15 d after rainfall, new root production increased total root length by 47% for *A. deserti* and by 27% for *F. acanthodes*. Gradually increasing drought stress caused death of a root cortex of *Lolium perenne* at –2 to –10 MPa and death of the root tips at potentials below –10 MPa.[633]

Physiological and metabolic changes caused by water stress were discussed by Hsiao and Bradford.[578] In drought conditions, some plants, e.g., *A. deserti,* extend rain-induced roots in which conductance to water flow per unit of dry mass is 11-fold higher than that of established roots.[588]

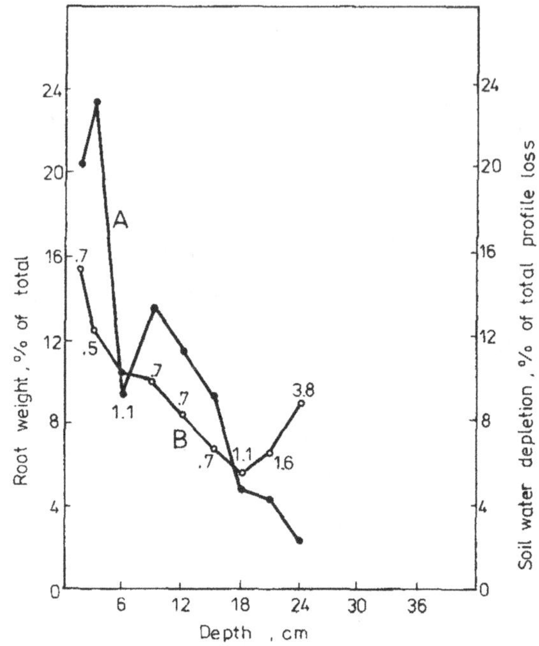

FIGURE 35. Root weight (A), soil water depletion (B), and apparent root efficiencies of sweet orange. Numerals at each depth are the ratios of water depletion to root weight. (Modified from Castle, W. S. and Kresdorn, A. H., *J. Am. Soc. Hortic. Sci.*, 102, 403, 1977.)

Marshall,[802] studying the influence of drought on the mortality of the finer roots of Douglas fir seedlings, found that an unwatered condition influenced root mortality (Figure 37), but drought was not directly responsible for death of fine roots, but was indirectly responsible by increasing root temperature and maintenance of respiration and by inhibiting photosynthate transport to the root system. Massive die-off of roots in response to drought was apparent only when carbohydrates were nearly depleted. Hoogenboom et al.[565] and Krizek et al.[688] indicated that soybean plants subjected to drought stress apportioned more fixed carbon into new root growth than did a plant with ample soil moisture. The same tendency was noted by Sheehy and Peacock[1067] for ryegrass.

Shallow-rooted crops, such as lettuce, wetland rice, and potatoes, are less drought tolerant than deep-rooted species such as alfalfa, maize, tomato, and dryland rice. Elkins et al.[365] indicated that in the sandy soils of Alabama confining the roots of cotton, maize, or soybeans to the top 15 cm of the soil will result in water stress of the crops after only 3 d without rain. However, when roots penetrate to a depth of 100 cm, then the period without rain must last longer than 20 d to cause water stress. The only disadvantage of an extensive root system is that it might remove a large quantity of soil water in the early season of growth causing its deficit during later growth. In some cases a limited root system and high root resistance by the slowing absorption of water in early seasons leave more water in the soil when drought appears.[913,919]

Ogata et al.[896-899] carried out a field study to evaluate the drought-resistance abilities of eight grass forage species: Dallis grass, Rhodes grass, Bahia grass, sweet sorghum, maize, African millet, Job's tears, and rancer grown in irrigated and water-stressed conditions. They found

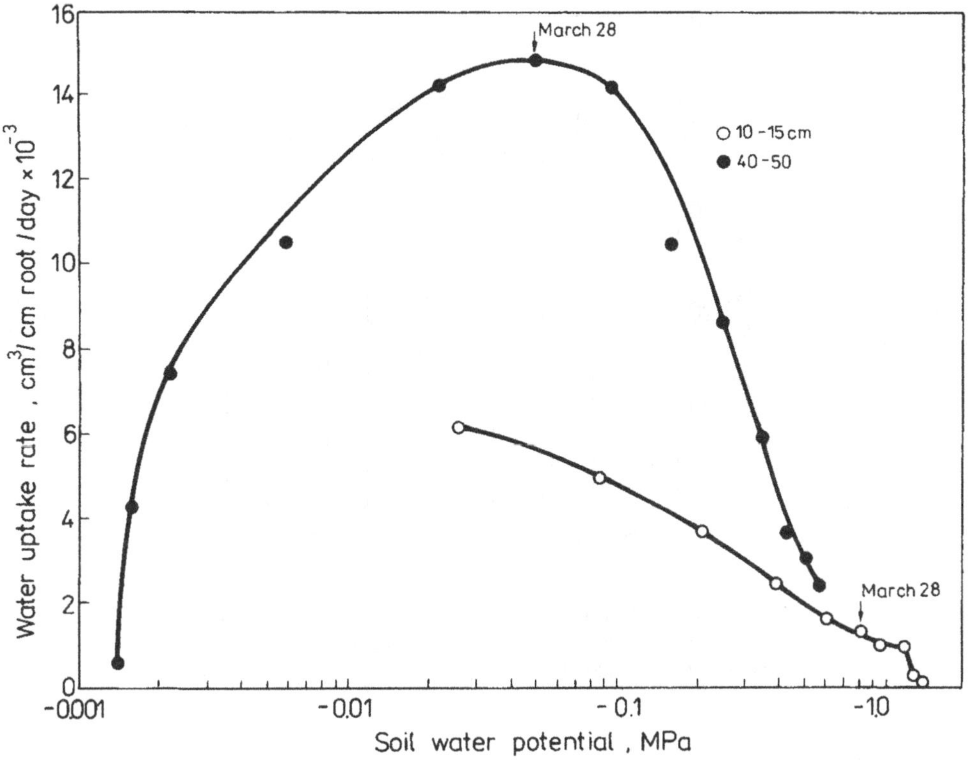

FIGURE 36. Water uptake rate by rice roots as a function of soil water potential for two soil layers. (Modified from Hasegawa, S. and Yoshida, S., *Soil Sci. Plant Nutr.*, 28, 191, 1982.)

that Dallis grass, Bahia grass, and Rhodes grass were most tolerable to water stress, but African millet and Job's tears were very sensitive to water stress. The roots of the former plants were characterized by higher dry weight, total length, root surface area, and nutrient uptake as compared to the latter ones. The authors indicated that good root development of the drought-resistant species was particularly due to efficient translocation of assimilated carbohydrates from leaves into roots or higher molecular root constitutes such as ligniocellulose. Root characteristics of two species (most and least drought resistant) are shown in Figures 38 and 39 and Tables 9 and 10.

Deeper root penetration on direct-drilled soil than on the plowed soil resulted in higher water extraction and heavier grain yield in dry conditions.[458] Water stress slows maize root elongation and increases root diameter.[231] Hoogenboom et al.[562] found in rhizotron a decrease in soybean root growth rate during the later stages of its vegetative growth or early reproductive development.

Diurnal patterns of water status are also reflected in diurnal variations in root diameter.[246,582] The roots of cotton[582] shrank in the middle of the day, with a minimum diameter on dry, sunny days being about 60% of the maximum diameter. Shrinkage occurred when roots lost water faster than they absorbed it. This loss was related both to soil water content and to transpiration rate. Consequently, the shrinkage may cause a poorer soil-root contact. In clay soil, the effect can be greater when roots grow along a ped surface than through the pore inside a ped where the pore itself may shrink as well.[1290]

The relation between soil moisture stress and root moisture stress of tobacco is shown in

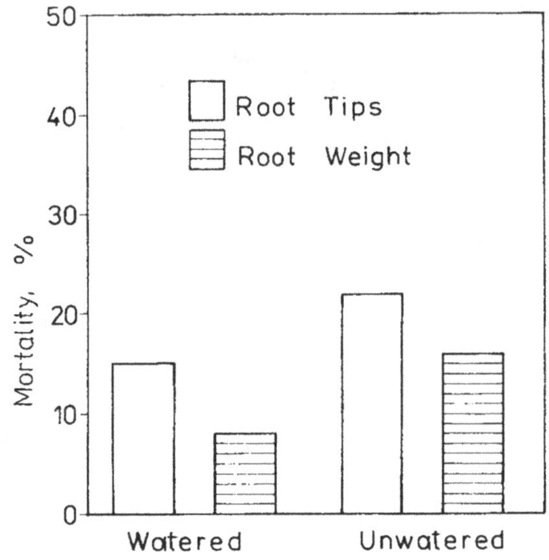

FIGURE 37. Fine root mortality of Douglas fir seedlings in watered and unwatered conditions. (Modified from Marshall, J. D., *Plant Soil,* 91, 51, 1986.)

Figure 40. Weir and Barraclough[1286] examining root growth of winter wheat in irrigated and drought-stressed soil, found a reduction of the total root length and weight by 16 and 8%, respectively, in drought conditions. The reduction occurred only in the top 40 cm of soil. The difference in water inflow to the root system between the same treatments is shown in Figure 41.

Water stress preconditioning, lasting for 2 to 3 d, of cotton roots decreased root pressure flux by 51% as compared to the well-watered control and resulted in significantly less water flowing from the roots to the soil. This flow was small compared to water flow in the normal direction (1:30).[906] When the upper roots are water stressed, the plant maintains its rate of transpiration by compensatory increase in water uptake from lower unstressed roots.

Water-stressed upper roots absorb phosphate more slowly. Rewetting induces new roots to grow from nodes at the stem base into the soil. These nodal roots absorb nutrient minerals effectively.

4.3.6 IRRIGATION

If drought periods are encountered, the increased water supply may increase root exploration in the soil profile. Irrigation generally causes roots to be distributed closer to the surface than under dry conditions. At shallow soil layers, irrigation promoted root growth and root concentration of sorghum,[143,144,823] apple trees,[743] alfalfa and sunflower,[789] maize,[18,789] grapefruit,[131] soybean,[583,789] and lettuce.[1016]

Irrigation of sugarbeets in a 21-year experiment in Czechoslovakia led to an increased root yield by 51% and an increased yield of raw sugar by 41% and a decrease of harmful α-amino nitrogen by 13%.[1150] Irrigation at different intervals of St. Augustine grass increased root mass, but not root length.[930] In the case of irrigated lettuce,[1016] the roots were thinner and less concentrated at depth. Romo and Diaz,[997] comparing the effect of drip and flood irrigation on peach trees, found a lower total number of roots, a higher concentration in the upper layer and

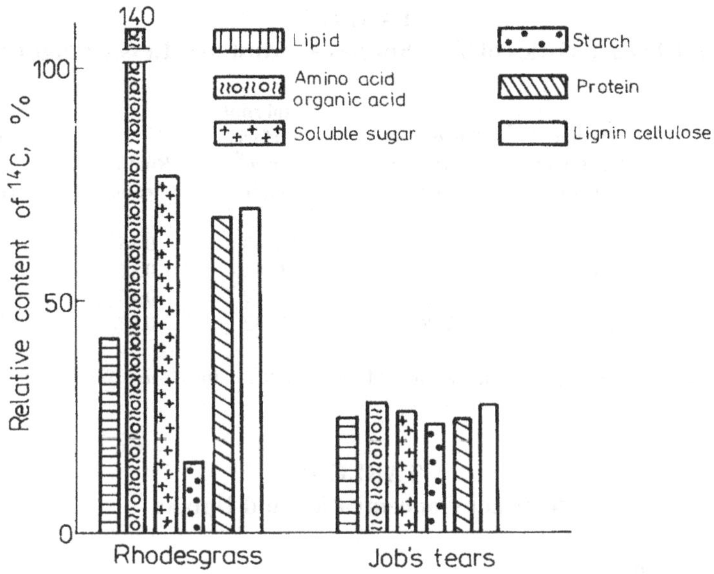

FIGURE 38. The relative content of ^{14}C in various root constituents of Rhodes grass and Job's tears under low soil moisture (pF of 2.8 to 3.0). (Modified from Ogata, S., Saneoka, H., Fujita, K., and Matsumoto, K., *J. Jpn. Grass. Sci.*, 31, 159, 1985a.)

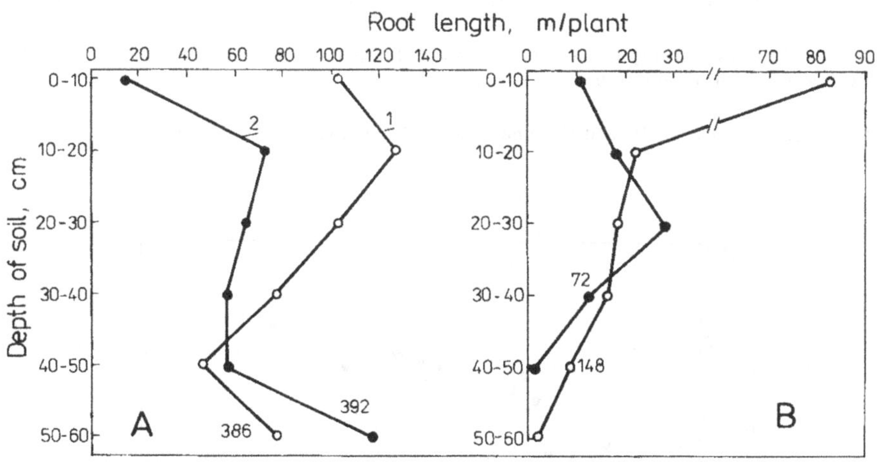

FIGURE 39. Effect of water stress on root length of Rhodes grass and Job's tears: (1) irrigated and (2) water stressed. (A) Rhodes grass; (B) Job's tears. Numerals at each line indicate the total root length in meters per plant. (Modified from Ogata, S., Saneoka, H., Fujita, K., and Matsumoto, K., *J. Jpn. Grass. Sci.*, 31, 263, 1985b.)

a thinner concentration in deeper horizons, on deeply irrigated trees than on flooded ones. Root and tuber crops and vegetables with a shallow root system are most likely to benefit from drip irrigation.[957]

Hoogenboom et al.[563] studied short-term changes in the rate of root growth of soybeans in

TABLE 9
Some Root Characteristics of Two Species — Most and Less Drought Resistant

Species	Water treatment (pF)	Mean root diameter (mm)	Total root Surface area (cm^2)	Root/top ratio	Water uptake rate (H_2O mg g^{-1} d^{-1})
Rhodes grass	1.7—2.0	1.74	24,643	0.35	2,043
(Chloris gayana)	2.8—3.2	1.11	13,515	0.41	483
Job's tears	1.7—2.0	2.09	9,719	0.46	2,647
(Coix lacryma-jobi)	2.8—3.2	1.76	3,732	0.62	36

Modified from Ogata, S., Saneoka, H., Fujita, K., and Matsumoto, K., *J. Jpn. Grass. Sci.*, 31, 263 1985b.

TABLE 10
Nutrient Concentration and Uptake

Species	Soil moisture treatment (pF)	N		P		K		Ca		Mg	
		Conc.	Up.	Conc.	Up.	Conc.	Up.	Conc.	Up.	Conc.	Up.
Rhodes grass	1.7—2.0	1.07	100	0.216	100	0.93	100	0.386	100	0.121	100
	2.8—3.0	1.44	67	0.148	35	0.67	49	0.521	54	0.124	22
Job's tears	1.7—2.0	1.86	100	0.226	100	2.22	100	0.850	100	0.610	100
	2.8—3.0	2.12	17	0.111	7	1.18	8	1.390	25	0.360	9

Note: Conc. % in dry matter; Up. expressed as the percentage of plant root with pF of 2.8 to 3.0 to 1.7 to 2.0 in roots of two species — most and less drought resistant.

Modified from Ogata, S., Saneoka, H., and Matsumoto, K., *J. Jpn. Grass. Sci.*, 31, 43, 1985d.

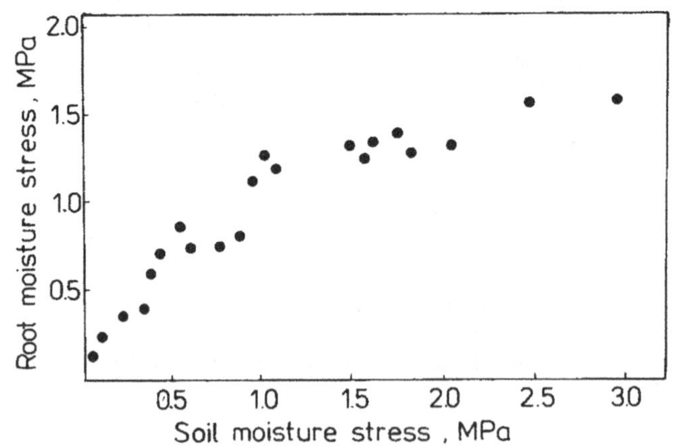

FIGURE 40. Root moisture stresses of tobacco plants as related to soil moisture stresses in sandy loam. (Modified from De Roo, H. C., *Agron. J.*, 61, 511, 1969a.)

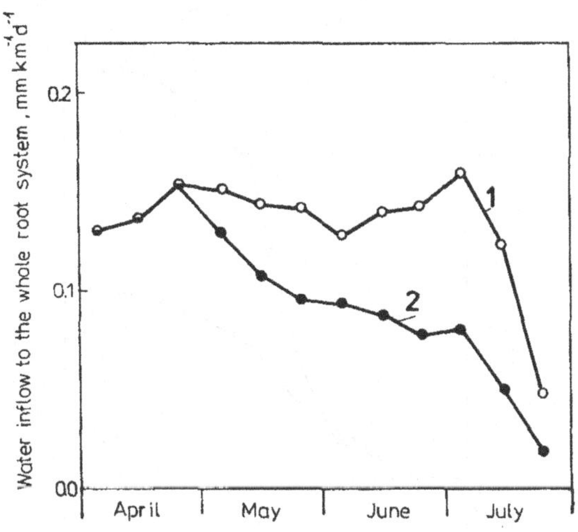

FIGURE 41. Changes with time of inflow to whole root systems of winter wheat grown in clay soil irrigated (1) and drought stressed (2). (Modified from Weir, A. H. and Barraclough, P. B., *Soil Use Manage.*, 2, 91, 1986.)

irrigated and nonirrigated rhizotron loamy sand soil and determined the differences in root growth rate and root length throughout the growing season. Examples of the results obtained are shown in Figure 42.

Heinonen[535] reported that when mechanical impedance of the topsoil becomes critical for root growth, a single supplementary irrigation, also called "a life-saving irrigation", may have a marked effect by softening up the soil, by washing down the nitrates whereby their evaporation flow is to the root-free top layer and by increasing turgor pressure and thus penetrability of the roots. Uppal et al.[1220] reported that in the growing season of low rainfalls, greater root penetration was obtained when the limited amount of water was split into light presowing and postsowing irrigations than when the entire amount was applied before sowing.

4.3.7 WATERLOGGING

It was shown that waterlogging via soil aeration decreased both extraction of water by roots, especially from deeper soil layers,[107,417,519] and root growth.[137,198,200,201,369] Root systems of conifer species show a different sensitivity to poor drainage. Levan and Riha,[742] examining the response of root systems of *Pinus strobus, P. resinosa, Picea glauca* and *P. mariana* to saturated soil conditions in a growth chamber experiment, found that none of the species appeared to be adapted to growth of roots into flooded soil. None of them had a significant internal air-filled pore space. These two genera responded to flooding differently. No root tips of spruce species survived flooding longer than 1 d, whereas 35% or more root tips of pine species survived even 7 d, although the spruce in the field is considered more tolerant to flooding.[1133] The phenomenon associated with the influence of flooding of roots is a rapid blocking of their acidification mechanism[325] and a severe reduction of the gibberellin from the main root to the shoot in the xylem sap.[976] This may be symptomatic of biophysical changes caused by flooding. The influence of waterlogging on root growth and functions is closely related to oxygen shortage. This aspect is discussed at greater length in Chapter 4.4.

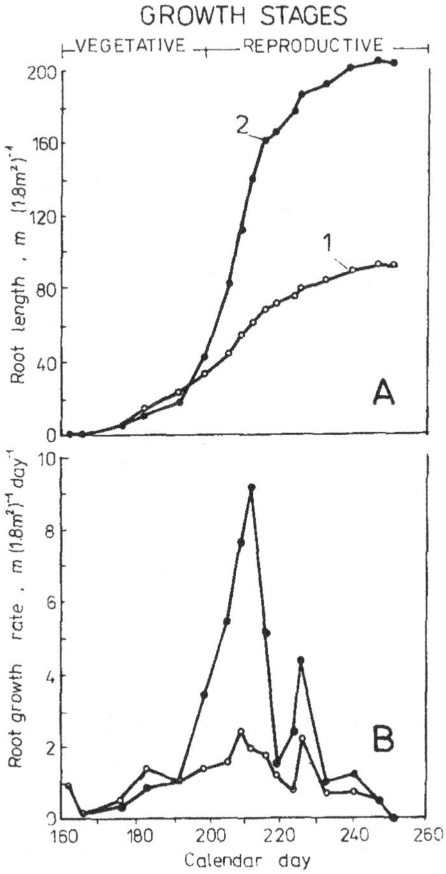

FIGURE 42. The influence of irrigation on the soybean root length (A) and root growth rate (B) per rhizotron window area: (1) irrigated and (2) nonirrigated. (Modified from Hoogenboom, G., Huck, M. G., and Peterson, C. M., *Agron. J.*, 79, 607, 1987b.)

4.3.8 DRAINAGE

Drainage changes soil moisture distribution in the soil profile, influencing root penetration and the pattern of water extraction. Drying of the soil surface limits the crown root number per plant and causes compensatory growth in the existing roots which subsequently reach deeper soil layers.[144,903] It allows for deep roots to be especially effective in absorbing water in a dry period.[139]

When a soil dries, the resistance to water uptake might increase either by a change in root permeability or by a contact resistance that originates from weaker wetting at the root-soil contact area.[545,1179] Under field conditions, the frequent variations in water supply of the soil profile are major reasons for the root distribution in soil.

4.3.9 ROOT DISEASES

The rhizobial nodulation of alfalfa is affected by the *Fusarium oxysporum* pathogen especially in a warm and moist climate.[1045] This fungus is the most widely distributed soil-borne pathogen in croplands and grasslands of Japan. Some soils are suppressive and others are conducive to soil-borne disease, but the reasons for that are not fully explained. Generally, most root diseases are favored by wet soil conditions.[250] The effect of soil water is by pathogen growth,[8,1106] sporulation,[342,942] and spore dispersal and motility.[343,774]

Root rot caused by *Phytophtora megasperma* f. sp. *medicagnis* was found by Kuan and Erwin[689] to be more severe on alfalfa seedlings grown in water-saturated soil than in unsaturated soil. Water saturation of soil predisposed alfalfa roots to *Phytophtora* root rot by increasing root damage and the exudation of nutrients which in turn increased chemostatic attraction of zoospores to the roots.

4.3.10 MODELING OF WATER UPTAKE BY ROOTS

The modeling of water uptake by roots is difficult because a number of factors influence this process. These factors are soil hydraulic conductivity, root growth, root depth, root density, root distribution, soil moisture potential, soil aeration, soil mechanical impedance, soil temperature, distribution of solutes, depth of the water table, soil and plant resistances, the presence of soil-borne diseases, and root-soil contact area.

Molz,[845] Hoogland et al.,[566] and Feddes[387] give reviews of different models proposed to describe water uptake in the root zone. The models differ in concept, in complexity, and in volume of input data and parameters needed.

Most models which describe water movement toward roots and root uptake have been developed by using the soil water potential concept[261,425,552,847,1179] which relates changes in root uptake to changes in soil hydraulic conductivity or soil moisture potential. This relation is not valid when the root-soil contact area decreases.[544,545]

The relative importance of the resistance of water movement in the soil and in roots in moisture extraction is different in various models.[183,512,533,846,881,882,999,1223] There are different opinions on where the predominant resistance is localized: in the roots[142,844,869,870,1179,1323] or in the soil.[261,426,585] Some authors suggest that both resistances are important. It is also an open question as to where the main resistance is localized within a plant: in the root or in the stem.[975] Reviews on resistance to water flow among plants were done by Tinker,[1191] Taylor and Klepper,[1180] and Molz.[845]

The models under controlled conditions are difficult to use under field conditions. Hasegawa[522] proposed a model of water uptake by the dryland rice root system in field conditions. He introduced an idea on soil water-root contact to this model and assumed the existence of large axial resistance along roots. The model confirmed experimental results that plants with shallow roots suffer from water deficits earlier than plants with deep roots even though the root length is identical.

There are two alternative approaches to modeling the uptake of soil water by roots.[552] The first approach proposed by Gardner,[425] called "a single root model", can be used to describe water uptake for individual soil layers. In this model, the root is assumed to be a line sink of an infinite length, uniform radius, and uniform absorption properties. The use of the model can be limited when the root spatial arrangement is significantly uneven.[69,1162,1163,1165] The second approach, based on the "sink term model", termed the macroscopic or root system model, describes the water uptake by the whole root system from the soil profile. One of the shortcomings of this model is the omission of root growth as a process affecting soil moisture changes in the soil profile and root responses to water stress. This factor has been introduced by Hillel and Tolpaz[553] to the dynamic simulation model of moisture extraction by a nonuniform root system.

Many models describing water uptake by the root system are based on an analog of Ohm's law.[1227] One-dimensional steady-state water flow in terms of water potential (pressure head) and hydraulic resistance of plant and soil is described as:

$$T = \frac{\psi_s - \psi_r}{R_s} = \frac{\psi_r - \psi_l}{R_p} \tag{16}$$

where T is the transpiration rate, ψ_s, ψ_r, and ψ_l are the water potentials in the soil, at the root surface and in plant leaves, respectively; and R_s and R_p are the flow resistances in soil and plant, respectively. Because it is difficult to measure ψ_r, the equation has been rewritten as:

$$T = \frac{\psi_s - \psi_l}{R_s + R_p} \tag{17}$$

On the basis of the above equation, it is possible to calculate the combined value of R_s and R_p. To calculate these characteristics separately, Gardner[425] proposed the single root model or microscopic model in which the root is viewed as a uniform cylinder of a radius r (in centimeters) extracting water at a steady-state q_r (cubic centimeters per centimeter of root per day). Water is assumed to be taken up only from a distance c (in centimeters), which is half the distance to another root, assuming uniform rooting. Then, $c = (\pi L_z)^{-1/2}$, where L_z is the root length density (centimeter of root per cubic centimeter of soil). The Gardner's formula can be written as:

$$\psi_s - \psi_r = \frac{q_r}{4\pi k} = \ln(c^2/r^2) \tag{18}$$

where k is the soil hydraulic conductivity (in centimeters per day).

Assuming uniform distribution of roots growing to depth d (in centimeters) and their equal effectiveness in water absorption, we can calculate:

$$q_r = \frac{T}{L_z d} \tag{19}$$

Reicosky and Ritchie,[975] substituting Equation 19 into Equation 18 and taking R_s from Equation 16 obtained:

$$R_s = \frac{\ln \dfrac{1}{r^2 \pi L_z}}{4\pi k L_z d} \tag{20}$$

The soil resistance can be also written[425a] as:

$$R_s = \frac{1}{BLK} \tag{21}$$

where B represents a dimensionless geometric and root distribution factor; L is the length of root per unit volume; and K is the hydraulic conductivity of the soil (in centimeters per day). Having the above data, it is postulated to calculate plant resistance to water flow:

$$R_p = \frac{\psi_s - \psi_l}{T} - R_s \tag{22}$$

The root extraction term, the so-called "sink term" (S), was added to the continuity equation for water flow through soil. This term represents the volume of water taken up by the roots per unit of bulk volume of the soil in the unit of time ($cm^3\ cm^{-3}\ d^{-1}$). The integral of the sink term over the rooting depth z_r (in centimeters) yields the actual transpiration (T) (in centimeters per day).

$$T = \int_0^{z_r} S d_z \tag{23}$$

Feddes et al.[388] used the sink term in the model, assuming the maximum sink term (S_{max}) as the ratio of the potential transpiration over the whole rooting depth. Hoogland et al.[566] modified the model, expressing S_{max} as a function of soil depth rather than rooting depth. The potential extraction rate decreases linearly with depth, from a constant, maximum potential extraction rate at the surface. They also tabulated S_{max} values for different crops and depths, which varied from 0.0035 to 0.084 cm^3 H_2O soil per day .

Alaerts et al.[19] compared four sink terms proposed by Nimah and Hanks,[881,882] Feddes et al.,[388] and Hoogland et al.[566] for simulating field water budget. It was difficult to conclude which term could be used as "a general extraction term simple and economical in use satisfying in concept and results." This is due to little knowledge on plant transpiration and water extraction patterns. A new attempt to define the volumetric sink term added to the continuity equation for soil water flow and to describe plant water uptake by roots has been proposed by Perrochet.[937]

Hasegawa and Sato[526] proposed a simple model to evaluate daily soil water uptake by roots growing in subsoil cracks. The input data of the model are crack spacing, depth of the cracks, water uptake rate per unit length of root, and root length density in the cracks. The model showed that daily water uptake by soybean roots growing through the crack of the clayey subsoil was about one order higher than the upward soil water fluxes toward the surface soil. Recently Novak[885] elaborated a simple method for estimation of soil water extraction distribution by maize roots, S(z), which is detrmined mainly by the distribution of the surface area of the roots. A good agreement between S(z) determined from field measurements and that calculated by means of the suggested method was found. For predicting water uptake by A. *deserti* and *F. acanthodes,* two-dimensional simulation models were proposed.[587]

Samui and Kar[1037] examined one of the sink root models for the simulation of the extraction rate, soil moisture distribution, and soil evaporation and evapotranspiration of groundnuts under dry and wet soil water condtions of lateritic sandy loam soil with deep water in humid subtropics. The input data are hydraulic conductivity function and water retention characteristics of the soil, time changes of rooting depth, soil cover, water pressure at the maximum rooting depth, and meteorological parameters.

Computer simulation models which describe the complex interactions between root systems of a crop growing vegetatively in a soil-plant-atmosphere continuum were proposed by some authors.[562-564,581] The last-modified version of the model ROOTSIMU(4.0), proposed by Hoogenboom et al.,[563] also contains a water-balance algorithm to account for water movement through both the plant and bulk soil. A section of this model which calculates root growth was revised.

A hypothesis of the model for water and nitrate nitrogen transfer in a soil and their absorption was elaborated and tested experimentally with young peach trees by de Cockborne et al.[296] A theoretical approach to predict soil/root water flows is made in the two-dimensional case by Lafolie and Bruckler.[706] The root water potential is estimated at each time step using an interactive procedure which calculates the best root water potential estimate in order to minimize the plant water stress. An original procedure is given to automatically compute the finite elements grid layout according to the given position of the roots. Results for various soils, climatic conditions, and root position are presented. Maps of the soil water potential after several days of simulation showed that a strong relationship between the root position and water uptakes exists.

Summing up of these results indicates that the influence of soil water on root growth is closely related to the plant species and the depth in the soil profile. It is generally accepted that with decreasing soil water potential from –50 kPa, root growth decreases. Prolonged wet soil conditions also result in slower root growth and favor the development of root diseases. The efficiency of water use by plants depends on soil conditions that affect water infiltration, distribution, and storage in the soil. Usually deep roots extract water more efficiently than roots grown near the soil surface. Numerous models which describe water movement toward the roots and water uptake by the roots by means of a soil water potential concept have been elaborated.

4.4 OXYGEN

Soil aeration status, described in detail by Gliński and Stepniewski,[447,448] is one of the most important factors influencing plant roots, which is clearly shown in Figure 43. As can be seen, soil oxygen plays a fundamental role for plant root functioning, among many components of the scheme, through its availability by the root system, its direct effects connected with the physiological events occurring within the plants, and through its indirect effects changing soil properties.

Most of the negative phenomena connected with oxygen deficiency in soil are caused solely and directly by the lack of oxygen for metabolic processes. Indirect effects appear only when there is a long-lasting shortage of oxygen in the soil. Plants vary widely in response to oxygen stress in the soil. In general, the critical oxygen level in soil for plants begins in the range of 5 to 10% O_2 by volume. Root growth of most plants is limited when less than 0.1 of the pore spaces in soil is filled with air and the oxygen diffusion rate (ODR) is less than 35 $\mu g\ m^{-2}\ s^{-1}$.[447]

Although few plants are subjected to complex anoxia, many are exposed to periods of oxygen deficiency caused by water-saturated (waterlogged or submerged) or compacted soils. The extreme case of oxygen deficiency for plant roots is that in which the soil air is almost completely replaced by water. Such conditions may be permanent (e.g., in bog or marsh soils), intermittent (e.g., in rice soils), or sporadic and of short duration.[447]

As to the plant's tolerance to soil waterlogging, there is a whole spectrum of plants — from hygrophites adapted to permanent flooding (e.g., rice) to mesophytes — among which significant differentiation exists. Hence, among these plants there are those which are moderately tolerant to flooding (most grasses and cultivated plants), as well as those which are very sensitive to it so that after only a few days of flooding, they die away (e.g., tobacco and peas). Xerophytes have been investigated only to a limited extent. Even if their system is sensitive to oxygen deficiency, the fact is that some of them can live with a reduced root system, being capable of increasing their resistance to all stresses, among them to an oxygen deficiency stress.[447]

4.4.1 DIFFUSION OF OXYGEN TO AND WITHIN PLANT ROOTS

The final stage in the transport of oxygen from the soil to the respiring tissue of the plant root is the radial diffusion of oxygen from the root surface to its center. In turn, part of the carbon dioxide produced in the roots diffuses outward, in a direction to the radial opposite to the radial movement of oxygen, and part is transported upward with the transpiration stream and also by longitudinal diffusion to the shoot where it can either be assimilated or released to the atmosphere.

4.4.1.1 Radial Diffusion

Radial diffusion of oxygen from the soil into a root of radius R having uniform respiratory activity (q_i) and diffusion coefficient inside the root tissue (D_i) and surrounded by a uniform layer of water-saturated soil of effective thickness (d) with an effective diffusion coefficient (D_e) and respiratory activity (q_e) (Figure 44) has lately been considered under steady-state conditions by de Willigen and van Noordwijk[302] and by van Noordwijk and de Willigen.[1235] Some more simplified situations without respiration within the saturated layer surrounding the root ($q_e = 0$) have been considered previously by Wanner,[1277] Lemon,[730] Lemon and Wiegand,[733] and Wiegand and Lemon.[1304]

The equations presented below are based on those reports (especially on those of de Willigen and van Noordwijk[302] and Van Noordwijk and de Willigen[1235]) giving the most general solutions. These authors have also presented a nonsteady-state solution for radial oxygen distribution within the root without a water-saturated layer:

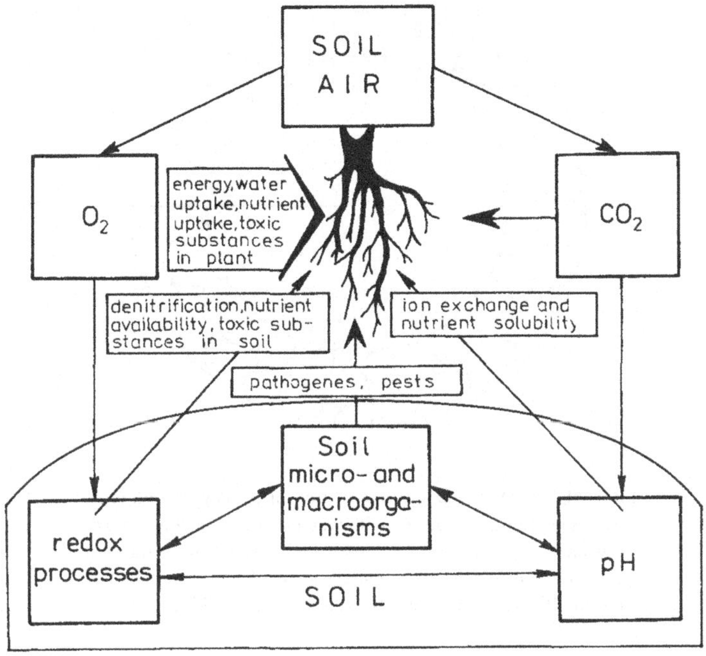

FIGURE 43. A scheme of direct and indirect effects of soil air on plants. (From Gliński, J. and Stepniewski, W., *Soil Aeration and Its Role for Plants,* CRC Press, Boca Raton, FL, 1985, 138.)

$$C_{i(t)} = C_o - \frac{q_i}{4D_i}(R^2 - r^2) + \frac{2q_iR^2}{D_i}\sum_{n=1}\frac{\exp(-\alpha_n t)J_0\left(\alpha_n \frac{r}{R}\right)}{\alpha_n^3 J_4(\alpha_n)} \qquad (24)$$

where $C_{i(t)}$ is the concentration of oxygen within the root tissue at a distance r from the root center after time t (assuming initial oxygen concentration within entire root as 0); C_o is external oxygen concentration; α_n is the nth positive root of $J_0(\alpha) = 0$; and J_0 and J_1 are Bessel functions of the first kind zero and first order, respectively.

The discussion of this solution by de Willigen and van Noordwijk[302] leads to the conclusion that a steady-state situation in roots is approximated within about 2 min. Thus, further considerations will concern only steady-state conditions:

$$\frac{\partial C}{\partial t} = D\nabla^2 C \pm q \qquad (25)$$

where C is the gas concentration; q is the respiration rate per unit volume; and D is the diffusion coefficient.

The general diffusion equation for oxygen (see Chapter 1), assuming the absence of longitudinal gradients in the cylindrical system for steady-state conditions, takes the following forms for the water film surrounding the root and for the root tissue, respectively:

$$D_e\left[\frac{d^2C_e}{dr^2} + \frac{1}{r}\right]\frac{dC_e}{dr} - q_e = 0 \qquad (26)$$

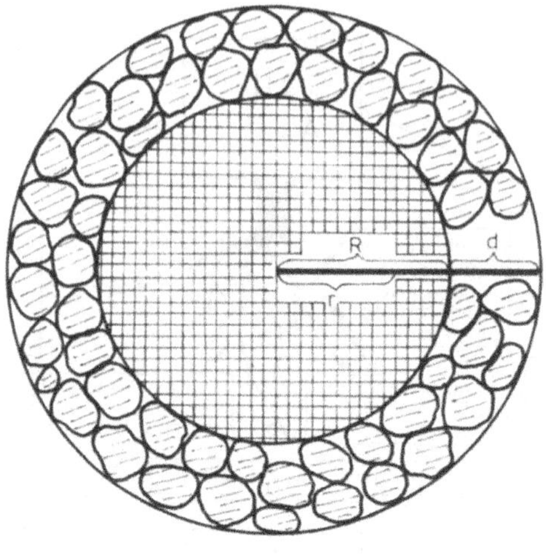

FIGURE 44. Schematic cross section of a plant of radius R, surrounded by soil saturated with water of effective thickness d: r, distance of a root element under consideration from the root axis. (From Gliński, J. and Stepniewski, W., *Soil Aeration and its Role for Plants*, CRC Press, Boca Raton, FL, 1985, 81.)

and

$$D_i \left[\frac{d^2C_e}{dr^2} + \frac{1}{r}\frac{dC_e}{dr} \right] - q_i = 0 \tag{27}$$

The boundary conditions in this situation are as follows:

$$r = 0, \qquad \frac{dC_i}{dr} = 0$$

$$r = R, \qquad C_i = C_e = C_r \qquad \text{and} \qquad D_e \frac{dC_e}{dr} = D_i \frac{dC_i}{dr}$$

$$r = R + d, \qquad C_e = C_o \tag{28}$$

The solution of Equations 26 and 27, when q_i and q_e are not dependent on the oxygen concentration, is given by two equations:[302]

$$C_e = C_o - \frac{q_e}{4D_e}[(R + d)^2 - r^2] - \frac{R^2(q_i - q_e)}{2D_e} \ln \frac{R + d}{r} \tag{29}$$

$$C_i = C_o - \frac{q_i}{4D_i}(R^2 - r^2) - \frac{R^2(q_i - q_e)}{2D_e} \ln \frac{R + d}{R} - \frac{q_e}{4D_e}[(R + d)^2 - R^2] \tag{30}$$

The concentration of oxygen on the root-water film interface (C_R) can be obtained by putting r = R to either of the equations. Then we get:

$$C_R = C_o - \frac{R^2(q_i - q_e)}{2D_e} \ln \frac{R + d}{R} - \frac{q_e}{4D_e} [(R + d)^2 - R^2] \tag{31}$$

Combining Equations 30 and 31, we obtain:

$$C_i = C_R - \frac{q_i(R^2 - r^2)}{4D_i} \tag{32}$$

The values of q_i for roots may vary from 10 to 300 mg m^{-3} s^{-1} and those of q_e may vary from 0.1 to 10 mg m^{-3} s^{-1}, while D_c is between 0.3 and 3.3×10^{-10} $m^2 s^{-1}$.[451] The values of D_i for the roots are from 2.2 to 9×10^{-10} m^2 s^{-1}.[733]

Oxygen concentration distribution, both within the root and outside, is calculated on the basis of Equations 29 and 30 for minimum respiration rate values and for maximum diffusion coefficient values likely to occur in roots and in soil and is presented in Figure 45. It should be emphasized that the above equations are valid only when the respiratory activity of the root does not depend on oxygen concentration, i.e., when $C_R > C'_R$ where C'_R is a critical concentration of oxygen on the root surface.

Equation 30 permits the calculation of the oxygen concentration at the center of the root ($C_{r=0}$), by substituting r = 0; then,

$$C_{r=0} = C_o - \frac{q_iR^2}{4D_i} - \frac{R^2(q_i - q_e)}{2D_e} \ln \frac{R + d}{R} - \frac{q_e}{4D_e} [(R + d)^2 - R^2] \tag{33}$$

and having included Equation 31 we get:

$$C_{r=0} = C_R - \frac{q_iR^2}{4D_i} \tag{34}$$

If the oxygen concentration at the root axis reaches the critical value of $C'_{r=0}$, below which the respiration rate decreases, the critical concentration on the root surface C'_R will be

$$C'_R = C'_{r=0} + \frac{qR^2}{4D_i} \tag{35}$$

i.e., it is increased in relation to $C'_{r=0}$ by the concentration drop inside the root.

The critical oxygen concentration at the exterior of the water or soil-water film surrounding the root, C'_{R+d} is, obviously, C'_R increased by the difference in concentration across the film (compare to Equation 31), i.e.,

$$C'_{R+d} = C'_{r=0} + \frac{q_iR^2}{4D_i} + \frac{R^2(q_i - q_e)}{2D_e} \ln \frac{R + d}{R} + \frac{q_e}{4D_e} [(R + d)^2 - R^2] \tag{36}$$

When the respiratory activity within the saturated layer around the root can be neglected, i.e., $q_e = 0$, then Equations 29 to 31, 33, and 36 are simplified as follows:

$$C_e = C_o - \frac{R^2q_i}{2D_e} \ln \frac{R + d}{r} \tag{37}$$

$$C_i = C_o - \frac{q_i}{4D_i} (R^2 - r^2) - \frac{R^2q_i}{2D_e} \ln \frac{R + d}{R} \tag{38}$$

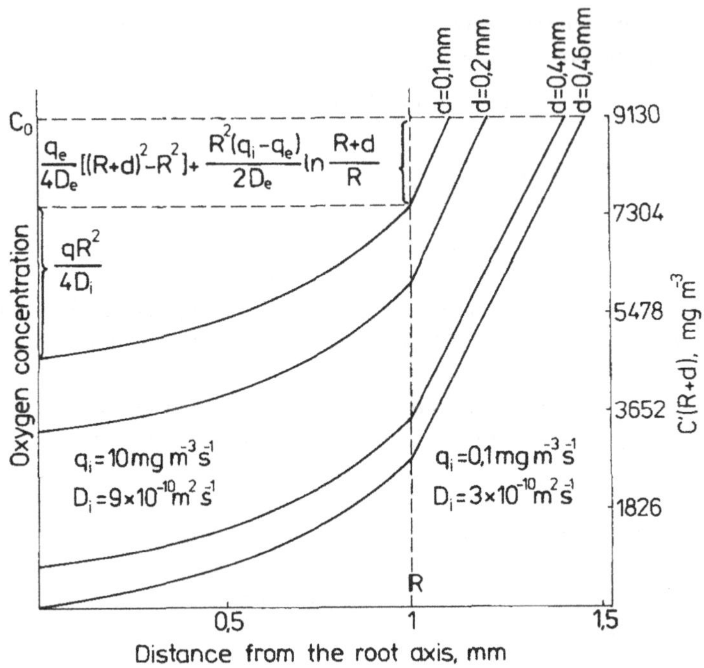

FIGURE 45. The effect of water film thickness on oxygen distribution inside and outside the roots as a function of distance from the root center for the minimum respiration rates and for the maximum diffusion coefficients likely to occur in the root and in the soil, calculated from Equations 27 and 28.

$$C_R = C_o - \frac{R^2 q_i}{2D_e} \ln \frac{R + d}{R} \tag{39}$$

$$C_{r=0} = C_o - \frac{q_i R^2}{4D_i} - \frac{R^2 q_i}{2D_e} \tag{40}$$

$$C'_{R+d} = C'_{r=0} + \frac{q_i R^2}{4D_i} + \frac{R^2 q_i}{2D_e} \ln \frac{R + d}{R} \tag{41}$$

As the affinity of cytochrome oxidase to oxygen is extremely high ($K_M = 0.1\ \mu M$, i.e., 0.008 kPa), we can assume that the critical oxygen concentration inside the root, i.e., on the cell level, is close to zero. Then we can put $C'_{r=0} = 0$ to Equations 35, 36, and 41.

Having considered this, we get the formula for critical oxygen concentration on the root surface,

$$C'_R = \frac{q_i R^2}{4D_i} \tag{42}$$

and on the exterior of the soil-water film surrounding the root,

$$C'_{R+d} = \frac{q_i R^2}{4D_i} + \frac{R^2(q_i - q_e)}{2D_e} \ln \frac{R + d}{R} + \frac{q_e}{4D_e} [(R + d)^2 - R^2] \tag{43}$$

which in the case of $q_e = 0$ (negligible respiration outside the root) is simplified to:

$$C'_{R+d} = \frac{q_i R^2}{4D_i} + \frac{R^2 q_i}{2D_e} \ln \frac{R + d}{R} \tag{44}$$

The values of critical external concentration of oxygen (C'_{R+d}) calculated from Equations 43 and 44 for different values of D_e, D_i, q_e and q_i are plotted vs. the d-value in Figure 46. Knowledge of oxygen consumption in a unit of its volume (q_i) permits the calculation of a unit oxygen flow (f_r) related to a unit of root surface. If the root has a length (l) and a radius (R),

$$f_r = \frac{R^2 l q_i}{2Rl} = \frac{R q_i}{2} \tag{45}$$

which, after combining with Equation 39, gives:

$$f_r = D_e \frac{C_{R+d} - C_R}{R \ln\left(1 + \frac{d}{R}\right)} \tag{46}$$

This has been presented by numerous authors, e.g., Lemon,[730] Lemon and Erickson,[731,732] and Stolzy et al.[1132]

The method of measuring the f_r value, using the polarographic method, which is also named ODR, has been elaborated by Lemon and Erickson.[731,732] It should be emphasized that during the measurement of the value of flow f_r by means of the polarographic method, using a platinum electrode (simulating the root), it is assumed that C_R equals zero at a surface of the electrode. In the case of a root, this is a simplification, as the critical concentrations of oxygen on the root surface are not lower than 0.02 to 0.03 $m^3\ m^{-3}$.

In view of the analogy of the mathematical description of diffusion flow to electrical current flow, where the intensity of diffusive flow corresponds to the intensity of electric current, and the difference between concentrations corresponds to the difference in the electric potential, in the design of electrical analogues of diffusion the concepts of diffusive permeability of a medium (P_D)[766] and its reciprocal known as diffusive resistance (R_D)[46] are introduced. These parameters are frequently used in the description of gas exchange in roots and in photosynthesis research. In the case of uniaxial diffusion, diffusive permeability (P_D) is equal to a ratio of the diffusion (D) to the length of the diffusion path (Δx),

$$P_D = \frac{D}{\Delta x} \tag{47}$$

and is expressed in meters per second.

Fick's equation,

$$f_x = -D \frac{dC}{dx} \tag{48}$$

for the diffusion flow (f_x) may therefore be written as follows:

$$f_x = P_D \Delta C \tag{49}$$

where ΔC is a difference in the concentration over the distance Δx and P_D is a diffusive permeability of the medium. Considering that

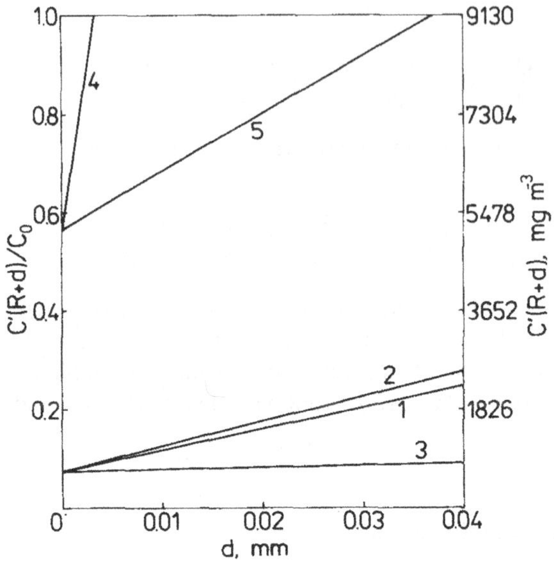

FIGURE 46. The effect of water film thickness on actual oxygen concentration outside the water film surrounding the root $C'_{(R+d)}$ at extreme values of q_i, D_i, and D_e likely to occur in roots and soils for the root of radius R = 0.25 mm, calculated from Equations 41 and 42.

1. $q_{i\,min} = 10$ mg m^{-3} s^{-1}, $D_{i\,min} = 2.2\ 10^{-10}$ m^2 s^{-1}
   $q_{e\,min} = 0.1$ mg m^{-3} s^{-1}, $D_{e\,min} = 0.3\ 10^{-10}$ m^2 s^{-1}
2. $q_i = 10$ mg m^{-3} s^{-1}, $D_{i\,min} = 2.2\ 10^{-10}$ m^2 s^{-1}
   $q_{e\,max} = 10$ mg m^{-3} s^{-1}, $D_{e\,min} = 0.3\ 10^{-10}$ m^2 s^{-1}
3. $q_i = 10$ mg m^{-3} s^{-1}, $D_{i\,min} = 2.2\ 10^{-10}$ m^2 s^{-1}
   $q_{e\,max} = 10$ mg m^{-3} s^{-1}, $D_{e\,max} = 3.3\ 10^{-10}$ m^2 s^{-1}
4. $q_{i\,max} = 300$ mg m^{-3} s^{-1}, $D_{i\,max} = 9.0\ 10^{-10}$ m^2 s^{-1}
   $q_{e\,max} = 10$ mg m^{-3} s^{-1}, $D_{e\,min} = 0.3\ 10^{-10}$ m^2 s^{-1}
5. $q_i = 300$ mg m^{-3} s^{-1}, $D_{i\,max} = 9.0\ 10^{-10}$ m^2 s^{-1}
   $q_e = 10$ mg m^{-3} s^{-1}, $D_{e\,max} = 3.3\ 10^{-10}$ m^2 s^{-1}

$$P_D = \frac{1}{R_D} \tag{50}$$

Fick's equation may be expressed in a form analogous to Ohm's law, i.e.,

$$f_x = \frac{\Delta C}{R_D} \tag{51}$$

where R_D is expressed in meters per second.

The use of the concept of diffusion resistance is convenient in that it permits the summation of diffusion resistance values along the diffusion path, in a manner analogous to the summation of electrical resistances. This is important, e.g., when determining the diffusion resistance of the whole plant, the complete root system, or individual roots, where we do not have an exact distribution of the diffusion coefficient which surely differs between the root epidermis and the internal tissues. In the literature, there is a lack of empirical data concerning the values of the radial diffusion coefficient in the roots.

In roots with no air-filled spaces, the oxygen diffusion coefficient is undoubtedly less than

2.1×10^{-9} $m^2 s^{-1}$ (at 20°C), i.e., below that for pure water, due to obstacles in the form of cell walls. Estimated values of this coefficient adopted by various authors are from 2.2×10^{-10} to 9×10^{-10} $m^2 s^{-1}$.[733] As it is known, plant root porosity may be considered and occasionally even reaches 0.6 $m^3 m^{-3}$.[46] It is difficult, however, to say whether and to what extent it affects an increase in the radial diffusion coefficient in the roots because of the lack of data from direct measurements. Radial diffusion permeability of adventitious roots of maize, as measured by Luxmoore et al.,[766] was 4.8×10^{-6} m s^{-1} , and for rice it was 1 to 11×10^{-6} m s^{-1}, and in both cases, radial diffusion permeability decreased with an increase in the distance from the root tips (Figure 47). In this Figure 47, it can also be seen that, contrary to expectations, an increase in porosity with an increase in the distance from the root tips is accompanied not by an increase in the diffusive permeability, but by a decrease. This probably follows from aging of the root epidermis and from a decrease in their permeability.

4.4.1.2 Longitudinal Diffusion

It is well known that plant roots contain free spaces filled with air,[17,37,38,46,644,1243] the volume of which, in the case of aquatic plants, is comparable to the air-filled porosity of soils. These are intercellular spaces having continuity along the root down to a distance of 2 to 3 mm from the root tip. They are found in the suberized root tissue known as the aerenchyma.[328]

The porosity of mesophyte roots is slight (of the order of several per cent of the root volume); however, under conditions of soil innudation with water, even with these plants it is quite common to find the development of adventitious roots containing aerenchyma[334] and having a porosity much greater than in the primary roots.[762] This results in an increase in root porosity under conditions of insufficient oxygen supply from the soil (Table 11).

The existance of a continuous system of air spaces in root tissue makes an internal gas exchange possible both through diffusion and through mass flow. As to the relative contribution of the two mechanisms, in relation to roots all the considerations presented for soils[447] are applicable, including the general conclusion that diffusion is the fundamental mechanism of gas transport. This is true in relation to the transport of oxygen downward in the roots, but the upward transport of carbon dioxide is also facilitated by the transpiration stream. The existence of the diffusive transport of oxygen to the roots has been shown experimentally by means of labeled oxygen for the English bean,[380] maize,[618,619] and barley.[75,619]

Investigations of the transport of ethylene in rice showed that it is proportional to the fresh mass of the plant and does not depend on illumination nor on illumination-related changes in the transpiration stream,[727] which indicates that the exchange of gas between the plant and the atmosphere is not a factor limiting the total process of gas transport in the plant. This is supported by the investigations of Ando et al.[34] and Stelzer and Läuchli.[1121] The latter authors have found that oxygen enters the root system mainly at the stem base. It has also been found that oxygen can diffuse from roots to the surrounding medium in the case of rice,[42,43,1238,1239] bog-bean,[258] maize,[618] barley,[1228] and bog plants.[41] In the case of bog plants, internal gas transport suffices to cover the oxygen requirements of the whole root system; with mesophytes, it constitutes only a supplementary factor. Quantitative determination of the role of internal longitudinal diffusion is complicated by the closing of stomata under conditions of insufficient oxygenation.[604,1101,1102] However, it follows from the publications quoted previously[727,1121] that changes in the diffusive resistance in leaves should not affect the resultant oxygen transport in the roots to any significant degree.

The distribution of oxygen concentration at a depth in a straight vertical root of a constant diameter and length (L), under conditions of no radial transport to or from the surrounding medium, assuming that both the respiratory activity (q) and the oxygen diffusion coefficient (D) in the root are constant along the whole length of the root and that q is not a function of oxygen concentration, can be described in a manner analogous to soil[447] by the following equation:

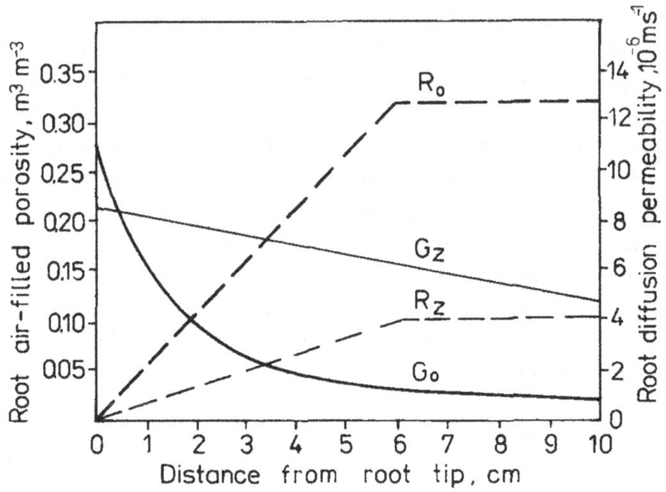

FIGURE 47. Air-filled porosity of soil (G) and radial diffusive permeability (R) of maize (z) and rice (o) as a function of the distance from the root tip. (From Gliński, J. and Stepniewski, W., *Soil Aeration and its Role for Plants*, CRC Press, Boca Raton, FL, 1985, 84.)

TABLE 11
Root Porosity

Plant	Porosity (m^3 m^{-3}): Abundance of oxygen in soil	Porosity (m^3 m^{-3}): Oxygen deficiency in soil	Ref.
Rice	0.05	0.25—0.30	75, 438
Maize	0.06—0.07	0.12—0.17	832
Festuca	0.02	0.06	832
Barley	0.03	0.04	1343
Sunflower	0.06	0.12	1343
Wheat	0.021—0.126	0.035—0.145	4, 761, 1100, 1102, 1241
Tomato	0.05	0.08	1343

$$C = C_o - \frac{q(2Lx - x^2)}{2D} \tag{52}$$

where C_0, in this case, is the concentration of oxygen at the beginning of the root, i.e. at $x = 0$.

Oxygen concentration at the tip of the root, i.e., at $x = L$, is

$$C_L = C_o - \frac{qL^2}{2D} \tag{53}$$

If we assume that the growth of roots stops when the critical concentration of oxygen, C', is reached at the root tip, then the limit of root penetration into a soil inundated with water, L', is

$$L' = \sqrt{\frac{2D(C_o - C')}{q}} \tag{54}$$

The root oxygenation depth, L_{an}, can be calculated by means of Equation 53 substituting values $C_L = 0$ and $L = L_{an}$. Then

$$L_{an} = \sqrt{\frac{2DC_o}{q}} \tag{55}$$

Replacing the atmospheric concentration of oxygen (C_o) and the diffusion coefficient in the root (D) with partial pressure of oxygen in the atmosphere (P_o) and the diffusion constant in the root (K), respectively,[447] we get:

$$L_{an} = \sqrt{\frac{2KP_o}{q}} \tag{56}$$

The length (L_{an}) can represent both the depth at which oxygen concentration in the roots drops to zero and the depth limit of root penetration in an anoxic medium, if the growth of the roots ceases at an oxygen concentration level of zero. A graphic interpretation of this relationship has been presented in Figure 48, where the value K/K_o, as long as the diffusion takes place in the gaseous phase, is equivalent to D/D_o and may be interpreted, in relation to root, as its effective porosity. We can see that when this porosity is below 0.05 $m^3\ m^{-3}$ at normal respiration values (approximately 100 mg $m^{-3}\ s^{-1}$), in the absence of radial diffusion from the soil, the oxygenated section of the roots is only about 7 cm, and therefore longitudinal transport is significant only for the part of the root system near the soil surface.

Literature on the subject does not quote any data concerning the longitudinal diffusion coefficient in the roots; hence it is difficult to use data from Figure 48 in practice. These difficulties are compounded by the simplifying assumptions adopted in the development of the equations in this part of the text. The assumptions referred to are no radial diffusion, constant values of the diffusion coefficient and of the respiratory activity of the root along its whole length, and q independent of oxygen concentration. All these assumptions are not met in reality, as q is greatest at the tip of the root[765] and decreases at oxygen concentrations below the critical value[447] and root porosity (and probably also the longitudinal diffusion coefficient) decreases in the direction of the growing tip of the root.

Calculations made by Luxmoore et al.[764-768] have shown that in the case of maize about 25% of the root requirement for oxygen is met by longitudinal internal transport. The investigations of Greenwood,[481] involving 12 vegetable plants, have shown that the internal aeration of their roots is significant for only a 5-cm section of root measured from the soil surface. A similar evaluation of the significance of internal transport in mesophytes has been arrived at by Yu et al.,[1343] investigating the penetration depth for the roots of five plants in a soil completely saturated with water. Under these conditions, the porosity of the roots of maize increased from 0.075 to 0.115 to 0.15 to 0.18 $m^3\ m^{-3}$, while the depth of root penetration into the soil was 17 cm. For sunflower the porosity increased from 0.06 to 0.11 $m^3\ m^{-3}$, and the penetration depth was 15 cm. The porosity of the roots of barley was about 0.035 $m^3\ m^{-3}$ and did not show any significant changes under conditions of inundation with water, while their penetration depth was 12 cm. In the case of wheat, the root porosity increased from 0.04 to 0.08 $m^3\ m^{-3}$, and the penetration depth was 10 cm. Another wheat variety, however, increased its root porosity from 0.06 to 0.15 $m^3\ m^{-3}$ with a root penetration depth of only 5 cm. Tomato roots increased their porosity from 0.04 to 0.08 $m^3\ m^{-3}$ with a penetration depth of 6 cm.

The difference in the penetration depth of roots with similar porosity may be the result of

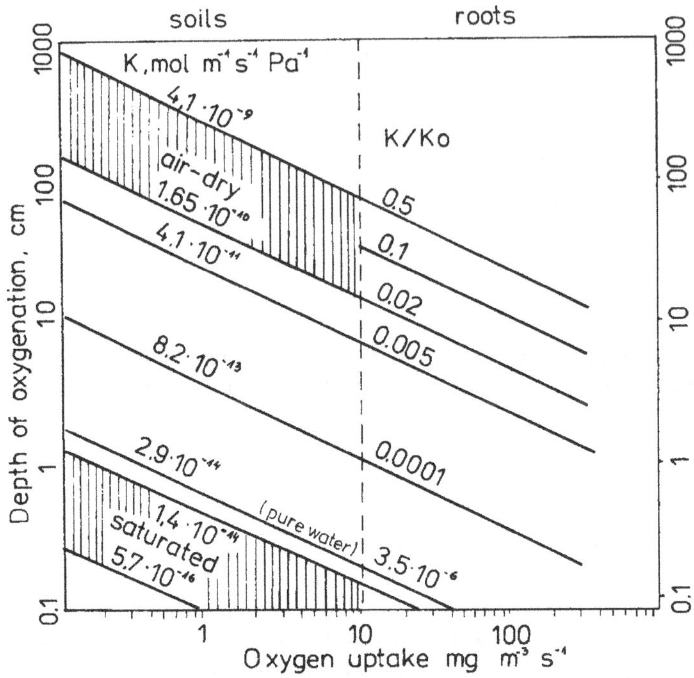

FIGURE 48. Soil oxygenation depth under steady-state conditions as a function of respiration rate of soils and roots, calculated from Equation 54: K_o, oxygen diffusion constant in air (equal to 8.26 10^{-9} mol m^{-1} s^{-1} Pa^{-1} at 20°C). (Adapted from Stepniewski, W. and Gliński, J., in *The Ecology and Management of Wetlands*, Hook, D. D., et al. Eds., Croom Helm, LTD, Beckenham, England, 1987, 269.)

differences in both their respiratory activity and in the values of their longitudinal diffusion coefficients. It seems, however, in view of the continuity of the intercellular pores, that the differences in the respiratory activity are probably more significant than in longitudinal diffusion. Some other aspects of gas exchange in the soil environment under flood conditions are treated elsewhere.[1124]

4.4.2 ROOT RESPIRATION AND FACTORS AFFECTING IT

Root respiration varies among plant species from that of several up to several hundreds and even up to 1600 cm^3 of O_2 or CO_2 per kilogram of root fresh weight per hour (Table 12). It is closely connected with root activity and especially with root growth (growth respiration), maintenance processes (maintenance respiration), and uptake and transport of ions (ion uptake respiration).[513,514,710-714,1246,1247]

According to Szaniawski,[1154-1156] the ratio of maintenance of respiratory activity of the whole shoot to that of the entire root system remains for a given plant constant irrespective of growth temperature (from 10 to 30°C) and shoot to root ratios (from 2 to 7). It was equal to about 1.1 for sunflower[1154] and to about 1.75 for cucumber.[1155]

Root growth respiration is the amount of oxygen used to produce 1 g of dry mass. The values of root growth respiration measured for several plant species by Lambers and Steingrover[711,712] have been found to be 390 to 1430 mg of O_2 per gram of dry weight. It was also found that for the same plant (*Senecio aquaticus*) grown in anoxic conditions, root growth respiration was three times lower than in the aerated culture.[711]

TABLE 12
Respiratory Activity of Root Fresh Weight in Cubic Centimeters Per Kilogram Per Hour

Plant	O_2 or CO_2	Respiration	Ref.
Bean	O_2	250—325	1154
Soybeans	O_2	330—470	900
Maize	O_2	360—580	900
Barley	O_2	100—650	204
Oats	CO_2	0—1500	392
Carrot	O_2	115—190	900
Onion	CO_2	300—1600	124
Potato tubers	O_2	2.3—3.2	1104
Alfalfa	CO_2	112	100
Trefoil	CO_2	114.2—136.6	100
Pears	CO_2	24.9—133.2	29
Sugar maple	O_2	134—245	207
Sugar maple	CO_2	95—225	207
Red maple	O_2	161—221	207
Red maple	CO_2	113—191	207
Bald cypress	O_2	168—226	207
Bald cypress	CO_2	119—202	207

Root activities mutually influence each other. For instance, a decrease in root respiration under low oxygen conditions has a negative effect on ion uptake, whereas a decrease in ion uptake by low ion concentrations has a negative influence on root respiration.

Veen[1247] examined, during 7 experimental days, the relation between root respiration and root activity of a maize root system. He found oxygen consumption per unit of root growth equal to 24.5 mg of O_2 per gram and per unit of ion uptake equal to 36.8 mg of O_2 per milliequivalent of NO_3^{-1} (NO_3^{-1} was used as a measure of the total ion uptake). Change of the light period (the amount of light supplied to the shoot) had a strong effect on ion uptake, but only a small effect on root respiration (Figure 49).

Studies by Hausen[513] revealed, however, that root respiration of *Lolium multiflorum* during 2- to 3-d periods of high illumination was three to four times higher than that of low illumination. After defoliation, root respiration decreased and was not affected by the illumination level until the regrowth of new leaves. Under normal conditions, ion uptake respiration is of considerable importance and amounts to 60% of the total root respiration which is shown in Table 13.

4.4.2.1 External Factors

4.4.2.1.1 Oxygen

Root respiration is the first process to be restricted by oxygen deficiency. Other disturbances of the vital functions of plants are the consequence of respiration disturbances. Andersen et al.,[29] after examining the effect of root anaerobiosis of six plant varieties on their respiration, found that respiration of feeder roots (<2-mm diameter) in air (0.21 m^3 m^{-3}) of *Pyrus betulaefolia, P. calleryana, P. comunis* and *Cydonia oblonga* was reduced by no more than 50% after 26 d of anaerobiosis. In contrast, root respiration of *Prunus persica* was reduced by 80% with anaerobiosis, whereas that of *Salix discolor* increased. Respiration rates of anaerobically or aerobically treated pear roots were inhibited by 25 to 50% when incubated in 0.5% O_2 compared to rates in air (Table 14).

The influence of oxygen concentration at different temperatures on the respiration rate of excised onion roots is shown in Figure 50. The curves in Figure 50 have two basic parts: the range of respiration limitation by oxygen concentration and a flat part where the respiration rate is not

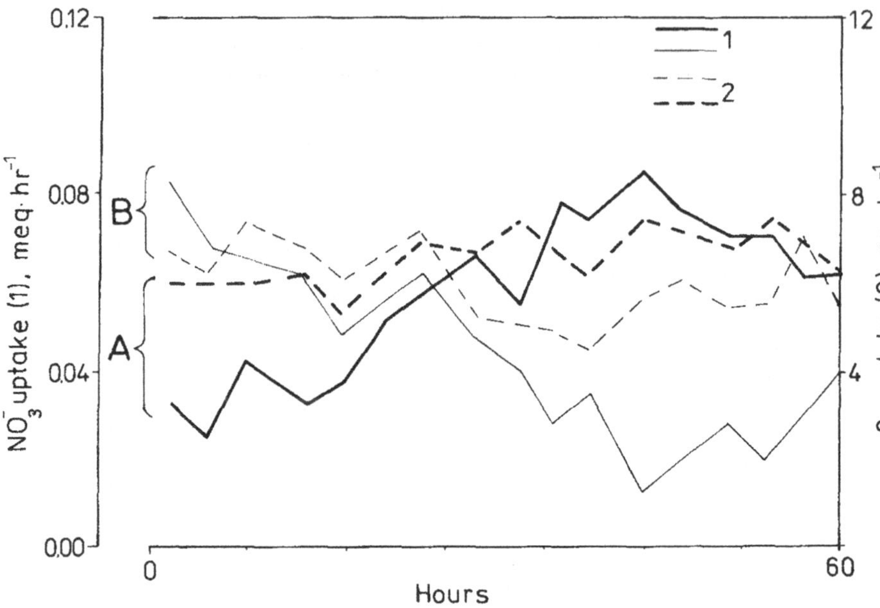

FIGURE 49. Effect of increasing (A) and decreasing (B) the light period on NO_3^- uptake (1) by maize roots and root respiration (2). (Adapted from Veen, B. W., *Plant Soil*, 63, 73, 1981.)

TABLE 13
Relative Energy Consumption of Uptake, Growth, and Maintenance Processes for Roots of Maize at Light Intensity 70 W m^{-2}

	Percent of total respiration used for		
No. days	Ion uptake	Growth	Maintenance
3	39	52	9
4	49	42	9
6	60	24	16

Modified from Veen, B. W., *Plant Soil*, 63, 73, 1981.

dependent on oxygen concentration. The respiration rate in this range of oxygen concentrations achieves its maximum value for a given temperature and will be called maximum respiration (q_{max}). The critical oxygen concentration (COC) values increase with temperature and are, for excised onion roots, equal from 0.10 to 0.15 m^3 m^{-3} at 15°C.[123,124] Equally high COC values for the respiration of excised roots were found for *Eriophorum angustifolium* (0.15 m^3 m^{-3}), pea (about 0.20 m^3 m^{-3}),[26] and *Senecio aquaticus* (0.15 m^3 m^{-3}).[710]

Atwell et al.,[53] examining COC for respiration of excised root segments of maize, found 0.117 and 0.065 mol of O_2 m^{-3} in the expanding and expanded zones of the roots, respectively. In contrast, the COC exceeded 0.20 mol of O_2 m^{-3} in the apex, suggesting that the O_2 supply for metabollic processes is most likely to be suboptimal in this zone. Their results show clearly that the adverse effects of low O_2 concentrations are unlikely to be a consequence of substrate shortage for either respiration or synthesis of macromolecules.

TABLE 14
The Effect of Treatment (Aerobic, Anaerobic), Incubation Conditions and Treatment Duration on Root Respiration on Three "Barlett" Grafted Pear Rootstocks

			Root respiration (cm^3 CO_2 kg^{-1} h^{-1} f. wt)		
Treatment	**Incubation (% O_2)**	**Duration (d)**	***P. betulaefolia***	***P. callergana***	***P. communis***
Aerobic	0.5	12	40.0	41.7	52.7
		26	60.0	106.2	64.5
	21	12	72.1	99.5	87.6
		26	84.4	133.2	74.0
Anaerobic	0.5	12	41.7	30.2	40.3
		26	45.6	48.0	26.6
	21	12	62.4	35.6	53.4
		26	97.7	66.1	24.9

Data from Andersen, P. C., Montano, J. M., and Lombard, P. B., *Hortic. Sci.*, 20, 248, 1985.

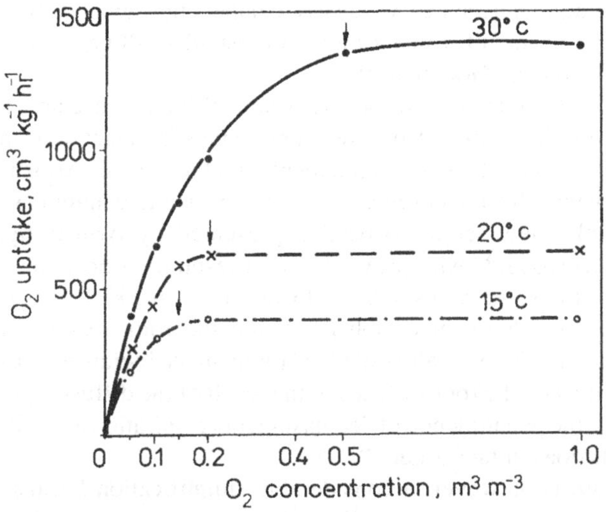

FIGURE 50. Influence of oxygen concentration, at different temperatures, on the respiration of excised segments of onion roots (0 to 5 mm from the tip). Arrows mark critical oxygen concentration, below which limitation of respiration by oxygen availability appears. (From Gliński, J. and Stepniewski, W., *Soil Aeration and Its Role for Plants*, CRC Press, Boca Raton, FL, 1985, 29.)

Because the activity of cytochrome oxidase decreases abruptly at very low oxygen concentrations (below 0.01 m^3 m^{-3}), one can assume that with a COC at the root surface, oxygen concentration at the root axis approaches zero. COC depends not only on the maximum root respiration (q_{max}, but also on other factors, shown by the equation:

$$COC = \frac{q_{max}R^2}{4D_i} \tag{57}$$

where R is the root radius and Di is the coefficient of radial diffusion of oxygen in root tissue. The thicker the root and the lower the oxygen diffusion coefficient within it, the higher the COC value.

In addition to COC, a half respiration concentration of oxygen (HRC), analogous to the Michaelis constant, is used. The HRC values for excised root segments are within 0.07 to 0.12 $m^3\ m^{-3}$ for maize, 0.05 to 0.25 $m^3\ m^{-3}$ for rice at 25°C,[765] and 0.025 $m^3\ m^{-3}$ for *S. aquaticus*.[710] The HRC values estimated from data of Berry and Norris[124] for excised onion roots are within 0.04 to 0.11 $m^3\ m^{-3}$.

There is an opinion[46,47] that those high COC and HRC values obtained *in vitro* for excised roots are an artifact of the measurement conditions due to the filling of the intercellular air spaces by liquid. For an intact root system, the oxygen diffusion coefficient should be much higher, implying lower COC values (compare with Equation 57). This point of view is supported by the results obtained by Armstrong and Gaynard,[47] who found, for roots of intact rice and *E. angustifolium* plants, COC values within 0.016 to 0.027 $m^3\ m^{-3}$ and HRC values below 0.001 $m^3\ m^{-3}$.

It can be seen that these values are almost one order of magnitude lower than those quoted above for excised roots. The results of Armstrong and Gaynard[47] are more convincing because the same authors found out that for excised roots of *E. angustifolium,* COC was equal to 0.15 $m^3\ m^{-3}$. Additionally, Lambers and Smakman[711] reported for intact root systems of six *Senecio* spp. grown in nutrient solution (but cut off from the tops) relatively low HRC values amounting to about 0.02 $m^3\ m^{-3}$, although their COC values are rather high (about 0.20 $m^3\ m^{-3}$). Similar HRC values (0.005 to 0.02 $m^3\ m^{-3}$) for intact roots of mustard seedlings were reported by Greenwood[482] and Greenwood and Goodman.[483]

The COC values obtained from roots grown in soil (Table 15) resemble those found for both excised and intact roots in solution. However, HRC values for intact roots in soil are apparently lower than for excised roots. This is in agreement with the view of Armstrong and Gaynard,[47] although the HRC values for intact roots are not so low as these authors suggest.

The most credible view seems to be that presented by Armstrong,[46] Armstrong and Gaynard,[47] and Greenwood,[480] who prove that root respiration does not decrease unless the oxygen partial pressure on the root surface is lower than 1 to 2 kPa. This is in agreement with physiological theory, as it has been established that cytochrome oxidase has a high affinity to oxygen. On this basis, the higher values of COC found in the literature must be the result of the presence of water films on the root surfaces or the result of the diffusive resistance of the plant tissue. Nevertheless, the phenomenon of decreasing root respiration at a soil air oxygen pressure of a few kilopascals does in fact occur.[1062,1158]

The above picture is, however, a considerable simplification because conditions of full oxygenation or of complete anoxia hardly ever, if ever, occur within plant roots. This is especially underlined by Crawford,[268] who draws attention to the fact that root meristems display some anoxic metabolism even in contact with atmospheric air. This is confirmed by the high respiratory quotients in meristems and the presence of ethanol in them.[268]

Because it is difficult to assess the thickness of the moisture films usually present on the root surface in soil conditions and their role in impeding the diffusion of oxygen, it is not easy to assign a value to the real oxygen concentration on root epidermis. This is the reason for the higher COC and HRC values observed in soil conditions compared with those predicted by biochemistry or found with the use of special experimental techniques.[47] The involvement of water films on root surfaces makes it difficult to relate root respiration and other plant responses to oxygen content in soil air directly.

When the oxygen concentration in the root environment decreases, the zone of anoxic metabolism localized in the meristem expands to the other parts of the root system, although

TABLE 15
Comparison of the Critical Oxygen Concentration (COC) and Half Respiration Concentration (HRC) Values for Plant Roots

Roots	Plants	COC ($m^3\ m^{-3}$)	HRC ($m^3\ m^{-3}$)	Ref.
Excised	Onion	0.10—0.50	0.04—0.11	123, 124
	Eriophorum angustifolium	0.15	0.03	46
	Peas	0.20		26
	Maize	0.05—0.16		53
	Maize		0.07—0.12	765
	Rice		0.05—0.25	765
Whole in solution	*Senecio*	0.15—0.20	0.02—0.025	711
	E. angustifolium and rice	0.016—0.027	0.001	46
	Mustard	0.20	0.005—0.10	482, 483
Intact	Maize	0.10	0.02	981
	Cotton	0.21	0.01	1159
	Dwarf peas	0.15	0.02—0.04	1062

rarely, if ever, occupying it entirely because its upper parts are supplied with oxygen through internal diffusion from the shoots. Thus, the physiological role of the decline in oxygen uptake by the root system as a whole, observed below the critical value of oxygen concentrations in the surroundings, may be of greater importance than is implied by the relative value of this decline as these anoxic areas, although limited in size, are located in the parts which are most important for root functioning. It should, however, be emphasized that a partial limiting of the root respiration rate does not necessarily mean a reduction in the root elongation rate.[483] It must be stressed that after a short anaerobic period which does not injure the roots permanently, an increased oxygen uptake by the roots appears (the so-called "respiration rebound"). This phenomenon has been observed both in excised roots[1348] and in intact roots.[82,282]

4.4.2.1.2 Carbon Dioxide

The effect of carbon dioxide on root respiration has been little studied. Harris and van Bavel[518] reported that 0.05 $m^3\ m^{-3}$ of carbon dioxide in soil air entirely inhibited root respiration of tobacco, maize, and cotton. However, Norris et al.[883] found that onion root respiration is less sensitive to carbon dioxide than is microbial respiration.

4.4.2.1.3 NPK Fertilization

Also the influence of mineral fertilization on root respiration has been little studied. Figliolia et al.[392] recorded that the respiration of oat roots in a nutrient solution was lowered by a deficiency of N and P, but was not influenced by K availability.

4.4.2.1.4 Ethylene

There are some papers[979,1104] indicating a stimulatory effect of ethylene on respiration of potato tubers, at concentrations which may occur in soil. A suppressing effect of 5 $cm^3\ m^{-3}$ of ethylene in soil air on respiration of grapevine roots was reported by Ishii et al.[596]

4.4.2.1.5 Temperature

Respiration of roots increases exponentially with temperature (Figure 51). Initially, this

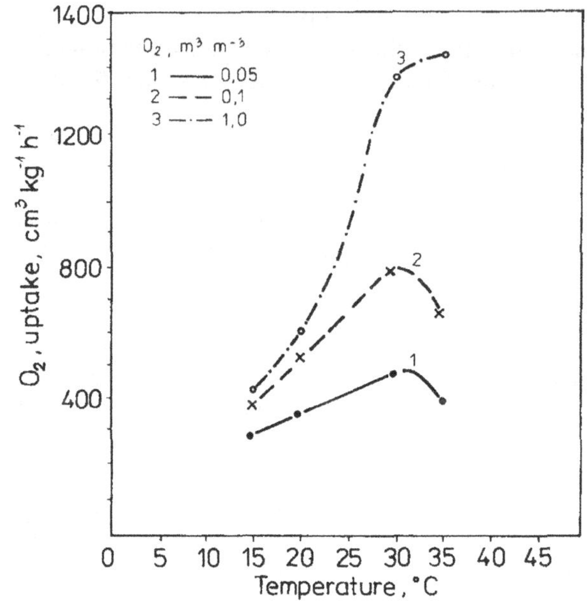

FIGURE 51. Respiration of excised segments of onion roots (0 to 5 mm from tip), at different O_2 contents, as related to temperature. (From Gliński, J. and Stepniewski, W., *Soil Aeration and Its Role for Plants*, CRC Press, Boca Raton, FL, 1985, 29.)

increase is rapid when the respiration rate is limited by chemical reactions. At higher temperatures, the transport of substrates and products of metabolism (sugars, oxygen, and carbon dioxide), based mainly on the diffusion process, becomes the limiting factor. At temperatures above 35°C, the protoplasm system starts to break down.

Limitation of respiration through the physical processes of diffusive transport may also occur at lower temperatures if the oxygen concentration is low, which is shown in Figure 51, by the curves corresponding to 0.05 and 0.10 $m^3\ m^{-3}$ of oxygen. These curves have been plotted on the basis of the data from Figure 50. Curves 1 and 2, corresponding to oxygen concentration below 0.1 $m^3\ m^{-3}$, illustrate the range of respiration limitation caused by oxygen, and curve 3, concerning pure oxygen, illustrates the limitation of respiration due to internal factors in the tissue.

In order to describe the influence of temperature on root respiration, the Arrhenius equation was applied:

$$\ln q = c + T \ln Q_{10}/10 \tag{58}$$

where q is the respiration rate at a temperature T; c is constant; and Q_{10} is the so-called Q-ten temperature coefficient, showing how many times the respiration intensity increases when the temperature increases by 10°C.

In most cases,[124,204,272,733] a change in the slope of the plot of lnq vs. 1/T was observed (Figure 52). This change of slope, which indicates a change of activation energy, denotes a change of the limiting factor with temperature. The values of activation energy for respiration of roots reported in the literature are very differentiated. Thus, for onion roots, Berry and Norris[124] found

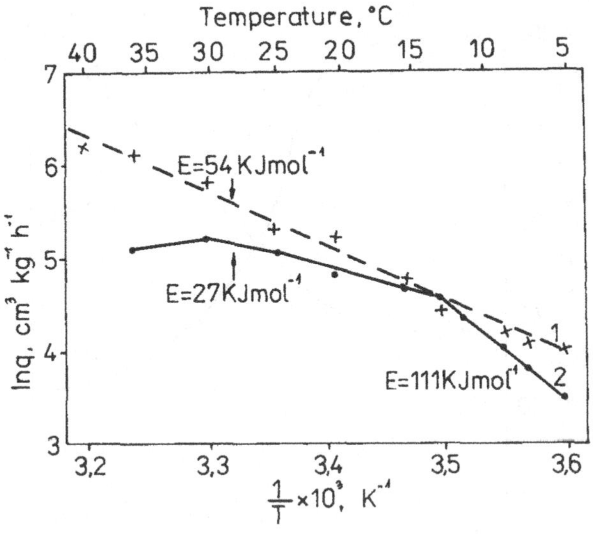

FIGURE 52. Arrhenius plot for respiration of excised roots of barley (1) and maize (2). (From Gliński, J. and Stepniewski, W., *Soil Aeration and Its Role for Plants*, CRC Press, Boca Raton, FL, 1985, 30.)

values within 7 to 59 kJ mol^{-1}, while Lemon and Wiegand[733] reported 45 to 75 kJ mol^{-1}. In turn, Carey and Berry[204] found 49 to 116 kJ mol^{-1} for maize and 54 kJ mol^{-1} for barley roots.

4.4.2.1.6 Soil Mechanical Impedance

The plant roots need more energy while penetrating a compact soil layer. Thus, their oxygen requirements in that situation should be higher. This has been proven in experiments on plants grown in nutrient solution with and without mechanical obstacles of glass beads, where oxygen uptake by bean roots has been studied,[1056] as well as in compacted soil in which carbon dioxide production by cotton roots was followed.[1159] It should be emphasized, in the latter case, that although a fivefold increase in carbon dioxide production per root mass unit, resulting from soil compaction, was observed, the mass of the entire root system and its total carbon dioxide production were negatively correlated with soil bulk density (Figure 53).

4.4.2.1.7 Soil Salinity

One of the external factors influencing root respiration is soil salinity, though its effect has not been studied extensively. It was found[636] that the respiration rate of excised root segments of *Tamarix tetragyna* decreased linearly with the increase of NaCl concentration in the solution.

4.4.2.1.8 Redox Potential and Soil Acidity

Anaerobic rice root respiration under controlled soil redox conditions and acidity was investigated by Tolley et al.[1194] Using the activity of inducible enzyme alcohol dehydrogenase (ADH) as an indicator of anaerobic root respiration, they found a decrease in redox potential resulted in an increase in root ADH (Figure 54). Soil acidity also seems to have had an effect since production of ADH at pH 5.5 was lower than at 6.5 and 7.5. This was apparently due to the relationship which exists between redox potential and pH.

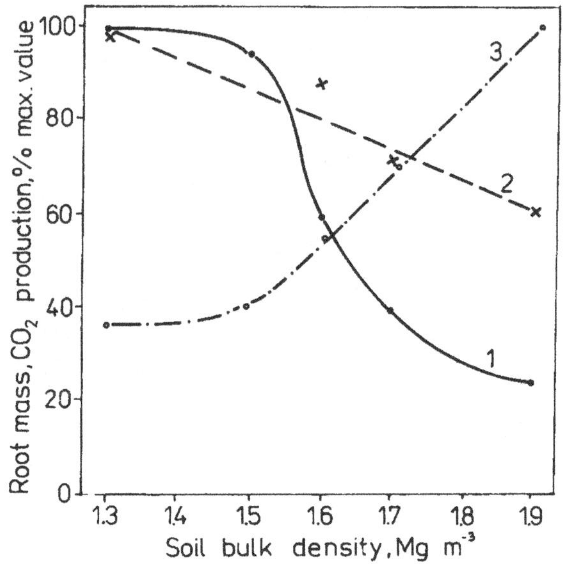

FIGURE 53. Influence of soil compaction on the respiration rate and weight of cotton roots in percent of the maximum values: (1) dry mass, (2) total CO_2 production by root system, and (3) CO_2 evolution per root dry mass unit. (From Gliński, J. and Stepniewski, W., *Soil Aeration and Its Role for Plants*, CRC Press, Boca Raton, FL, 1985, 30.)

4.4.2.2 Internal Factors

The limitation of root respiration by internal factors results from the properties of the respiring tissue (these being related to the genetic background and physiological age of the plant) as well as from the effect of the shoots. It was found respiration intensity is the highest in the apical part of the root and decreases with distance from the root tip,[123,124,765,778] as shown in Figure 55.

Root respiration during the vegetation period undergoes specific changes independently of the effect of temperature or other external factors. Thus, root respiration of the perennial desert shrub *Atriplex confertifolia*, measured at constant temperature, showed spring and autumn maxima.[560] According to the authors of that paper, this phenomenon may be connected with the reflection of environmental preconditioning, as well as with an overall genetically based program of plant development.

In the case of maize grown in controlled conditions,[35] maximum respiration of the entire root system was 800 cm^3 of CO_2 per plant per day during the appearance of panicles. Later the respiration rate decreased to about 400 cm^3 of CO_2 per day per plant and oscillated around this value until plant senescence.

Beside seasonal dynamics, plant roots also exhibit a diurnal variation in respiration rate. Huck et al.[580] pointed out that the root respiration of plants (maize, soybean, and *Derris elliptica*) grown at constant temperature was 25 to 50% higher during the day than at night. On the other hand, root segments excised from the plants at different times of the day and night showed no differentiation in respiratory activity. Also, no correlation was found between respiration rate and sugar content in the root, which eliminates the substrate level as a cause of the variation observed. The mechanism of the variation has not been fully explained, but it is undoubtedly connected with the action of light and the presence of shoots.[513,514]

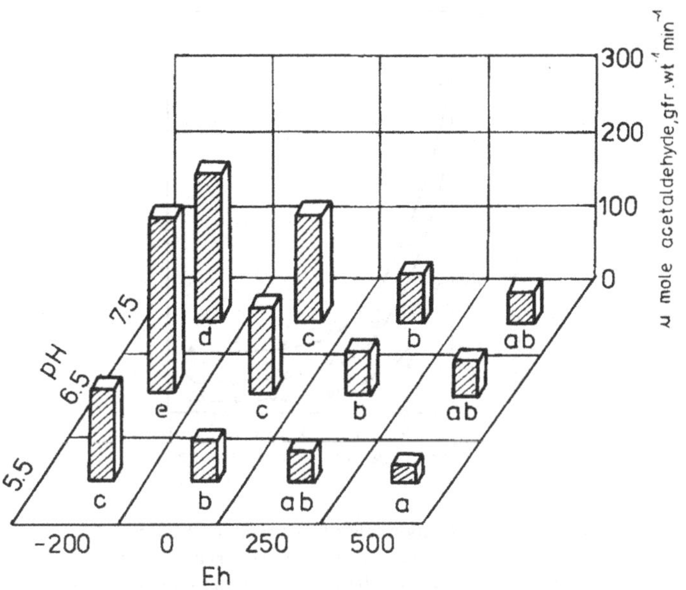

FIGURE 54. Root ADH activity as affected by soil pH and redox conditions (Eh). Values followed by the same letter are not statistically different at the 5% level. (Modified from Tolley, M. D., Delaune, R. R., and Patrick, W. H., Jr., *Plant Soil*, 93, 323, 1986.)

Studies by Hansen[513] revealed that the root respiration of *L. multiflorum* during 2- to 3–d periods of high illumination was three to four times higher than at low illumination. After defoliation, root respiration decreased and was not affected by illumination level until the regrowth of new leaves. Some differentiation in root respiration rate was also found among the plants of different climatic zones.[272] Roots of plants of cool regions showed higher respiration rates, at the same temperature, as compared to plants of warmer zones. According to Crawford and Palin,[272] high respiration rates at low temperatures can be an advantage for an earlier resumption of growth, but at high temperatures they can lead to an excessive consumption of carbohydrate reserves.

Lambers[710] reported that the root respiration rate of *S. aquaticus* (flood-tolerant species) was, at 20°C, lower than that of *S. jacobaea* (flood-intolerant). This would suggest that the low respiratory activity of roots may be an adaptive feature of increasing plant tolerance to flooding. The respiratory activities of flood pretreated roots of flood-tolerant and flood-intolerant trees (baldcypress, *Taxodium distichum;* red maple, *Acer rubrum;* and sugar maple, *A. saccharum*) were examined by Carpenter and Mitchell.[207,208] They found that flooding reduced the root respiration capacity in all species, affecting intolerant species the most and tolerant species the least (Figure 56).

Roots of all three species utilized O_2 with similar efficiency prior to flood stress. However, the respiratory capacity of sugar maple roots declined substantially during the first 8 d of flooding and declined more gradually from 8 to 22 d at 0.21 or 0.05 m^3 m^{-3} of O_2. Red maple roots declined gradually in respiratory capacity over the entire flooding period at 0.21, but not at 0.05 m^3 m^{-3} O_2. After an initial sharp decline, baldcypress roots gradually regained capacity to utilize O_2 from 8 to 22 d of flooding at all three levels of O_2. When tested at 0.21 m^3 m^{-3} of O_2, both red maple and baldcypress roots had two- to threefold higher respiratory capacities than had sugar maple roots after 22 d of flooding.

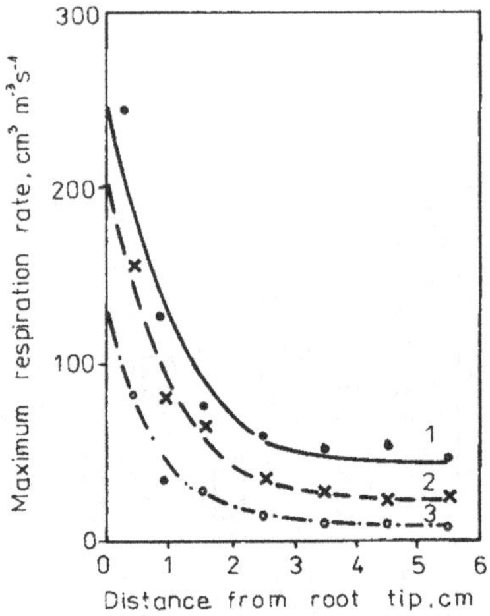

FIGURE 55. Respiration rate of rice roots in the atmosphere containing the following concentrations of oxygen: (1) 0.8 $m^3 m^{-3}$, (2) 0.208 $m^3 m^{-3}$, and (3) 0.044 $m^3 m^{-3}$. (From Gliński, J. and Stepniewski, W., *Soil Aeration and Its Role for Plants*, CRC Press, Boca Raton, FL, 1985, 31.)

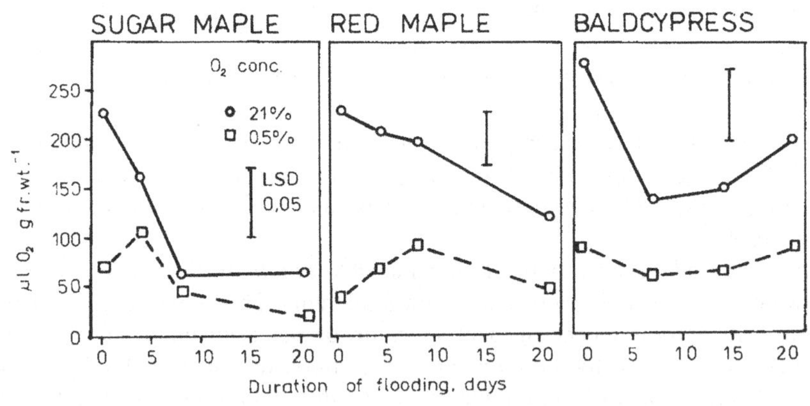

FIGURE 56. Oxygen consumption rates by excised tree roots at two O_2 levels after 0 to 22 d of flooding. (Modified from Carpenter, J. R. and Mitchell, C. A., *J. Am. Soc. Hortic. Sci.*, 105, 684, 1980.)

RQ values (the volumetric ratio of carbon dioxide evolved to oxygen consumed) were similar for all three species at 0.21 $m^3 m^{-3}$ and all decreased after flooding. At 0.05 $m^3 m^{-3}$, RQ values were higer for red maple and baldcypress than for sugar maple on both day 0 and day 8 (Table 16).

4.4.2.3 Respiration of Root Nodules

Pankhurst and Sprent[916] examined in a pot experiment the effects of temperature and oxygen tension on the respiratory activity of soybean and French bean nodules. They found an increase of the respiratory activity of turgid nodules with rising temperature over the range 10 to 30°C. However, below 10°C, the activity decreased rapidly.

Respiration by soybean nodules was not inhibited by high PO_2 partial pressure at low temperatures (Figure 57), but was saturated at a PO_2 of 40 kPa at 10°C and of 60 kPa at 15°C. At 20°C, respiration by the nodules showed a rather linear increase with increasing PO_2. Tjepkema[1193] noted that *Myrica gale* can transport O_2 to the submerged nitrogen-fixing nodules through aerenchymatous nodule roots which emerge above the water surface.

4.4.3 METABOLIC ACTIVITY AND ENERGY CHARGE OF ROOTS UNDER ANOXIA

Under anoxia, carbon dioxide, fermentative metabolites, such as ethanol, ethylene, and its precursor, 1-aminocyclopropane-1-carboxylic acid (ACC), and some organic acids may appear in roots. Also the available energy is limited, slowing absorption and translocation of water and nutrients in roots.

4.4.3.1 Carbon Dioxide Production

Aerobic and anaerobic CO_2 evolution was measured by Tripepi and Mitchell,[1198] who found a 66% decline in CO_2 production upon anaerobiosis by roots of the river birch and European birch. Barta[100] studied the same problem on waterlogging-tolerant bird's-foot trefoil and waterlogging-sensitive alfalfa roots placed in nutrient solution. He found that CO_2 evolution rate of waterlogged roots of both species showed significant declines relative to aerobic controls (similar to that reported by Tripepi and Mitchell[1198]). However, the percentage decline, relative to the aerobic control, was significantly less for trefoil roots.

This smaller percentage decline is consistent with a previous work of Barta,[99] showing that alcohol dehydrogenase (ADH) activity in waterlogged trefoil roots was stimulated to a level equal to or greater than that found in waterlogged alfalfa roots. There was no effect of waterlogging duration (1 to 4 d) on CO_2 evolution in either species which is a confirmation of previous results reported by Carpenter and Mitchell.[207,208]

4.4.3.2 Ethanol Synthesis and Loss from Roots

Exclusion of oxygen from metabolism causes a change of this metabolism toward fermentation. A few types of fermentation are possible which is shown in Figure 58. In the case of plants, the most common is alcohol fermentation.[265,268,270,572,661]

Ethanol is found in the root meristems, even if they are in contact with atmospheric air. In this case, its concentration rapidly decreases with increase in the distance from the root cap.[267] Ethanol formation in roots, when in anoxic conditions, is connected with the plant's capability of surviving periods of flooding. During flooding root ethanol production and alcohol dehydrogenase which is positively correlated with ethanol content in the plant[99,268,572] are stimulated. In fact, a high level of ADH activity and increased ethanol production has been used in the classification of flood-"tolerant" and "-intolerant" plants.

Barta,[99] in field experiments carried out on a fine, loamy soil with flooding of 2-year-old stands of alfalfa and bird's-foot trefoil for 12 d, found that root concentration of ethanol increased throughout the flooding regime in alfalfa roots from 0.12 to 0.22 μmol (g fr. wt.)$^{-1}$ and was significantly higher (0.91 μmol (g fr. wt.)$^{-1}$) in roots with shoots removed. This suggests that shoots either inhibited root ethanol production or facilitated its removal from the root.

It was confirmed by other authors, e.g., Soldatenkov and Chao,[1103] who suggested the positive

TABLE 16
Respiratory Quotient (RQ) for Excised Tree Roots at Two O_2 Levels (0.21 or 0.05 m^3m^{-3}) and After 0 or 8 d of Flooding

O_2 conc. (m^3 m^{-3})	Species	RQ Days flooded 0	8
0.21	Sugar maple	0.92	0.71
	Red maple	0.86	0.70
	Bald cypress	0.89	0.71
0.05	Sugar maple	1.45	1.37
	Red maple	2.68	1.79
	Bald cypress	1.69	1.70

Modified from Carpenter, J. R. and Mitchell, C. A., *J. Am. Soc. Hortic. Sci.*, 105, 684, 1980.

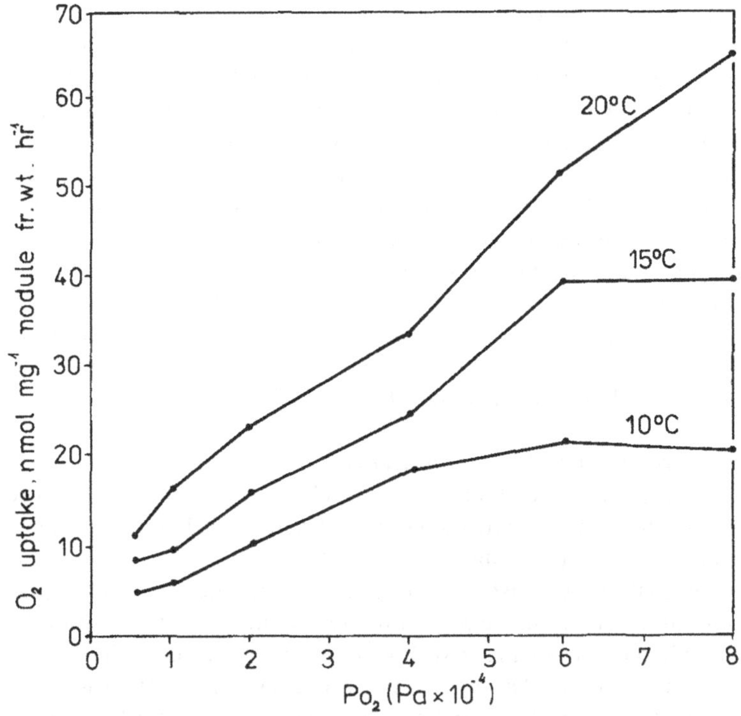

FIGURE 57. The effect of increasing PO_2 on the respiratory activity of turgid soybean root nodules at 10, 15, and 20°C. Each point is the mean of four replicates. (Adapted from Pankhurst, C. E. and Sprent, J. I., *J. Exp. Bot.*, 27, 96, 1976.)

effect of shoots on the inhibition of root ethanol production or its removal from the root reducing injury of flooded plants. No ethanol was detected in any trefoil (flood-tolerant species) root samples which is difficult to explain. It is possible that significant ethanol metabolism occurs

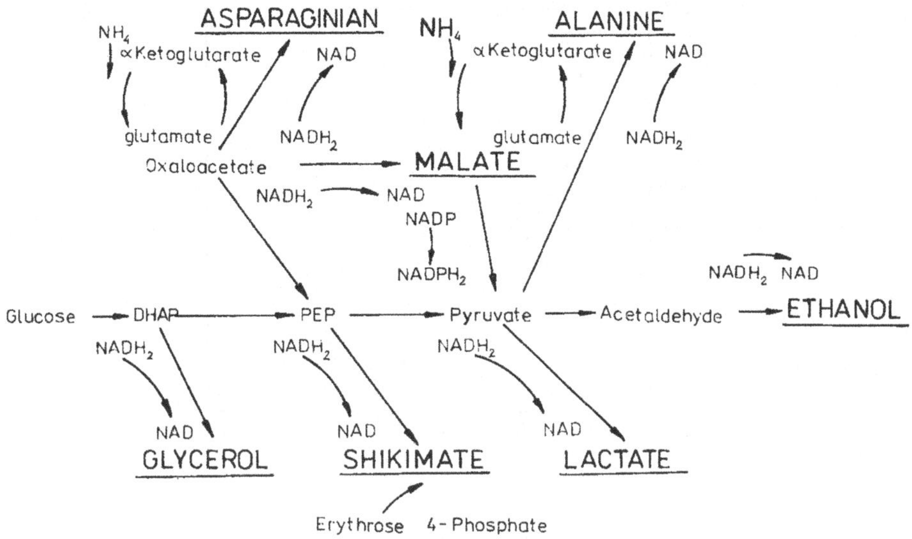

FIGURE 58. Diagram illustrating the various means of proton disposal and the range of end-products of glycolysis found in plants capable of enduring prolonged periods of partial or total anoxia. (From Gliński, J. and Stepniewski, W., *Soil Aeration and Its Role for Plants*, CRC Press, Boca Raton, FL, 1985, 142.)

in flooded trefoil roots and is an important mechanism in its removal from the root. This phenomenon was not found in pot experiments with the same plants.[98,99]

Atwell et al.,[53] examining the concentration of ethanol in maize roots grown in solution at two levels of O_2 concentration (0.25 and 0.04 mol m^{-3}), found a decrease of ethanol content in roots of 66 and 86% depending on the kind of nutrient solution.

Ethanol formation in plant roots is connected with environmental conditions. Fulton and Erickson[411] stated that ethanol accumulation in flooded tomato plants was closely correlated with ODR values in the soil which is shown in Figure 59. Further investigations of the relationship between ethanol accumulation in the plant and soil conditions showed that ethanol concentration is higher in plants subjected to sudden anoxic stress than in plants grown for a long time in conditions of insufficient oxygenation and, moreover, depends on the physiological age of the plants.[56]

Examining the ethanol balance in the tomato plant, Bolton and Erickson[149] and Fulton and Erickson[411] found the maximum ethanol concentration in the bottom of the stem and in the top roots. Its highest concentrations in the plant were 300 to 350 μg g^{-1}. The decrease of ethanol concentration in the lower part of the root is explained by its secretion to the environment.

The possibility of toxic effects of ethanol on plant tissues depends on its concentration, which is the result of the ethanol formation processes and of the processes of its removal by roots and shoots and also of the extent of its remetabolization within the plant tissues. Crawford[267] gives as an example rice and swamp tupelo trees (*Nyssa sylvatica*), which easily remove ethanol by adventitious roots and tolerate flooding well despite the fact that they show the Pasteur effect (the phenomenon of increased substrate use in anoxic conditions) and that ethanol is a main product of anoxic respiration.

Injury caused by ethanol consists of a deleterious effect on the cell membranes,[668] resulting in leakage of electrolytes, organic acids, amino acids, and sugars from the roots.[490,806] This cell injury caused by ethanol explains acceleration of the rate of glycolysis (Pasteur effect) which

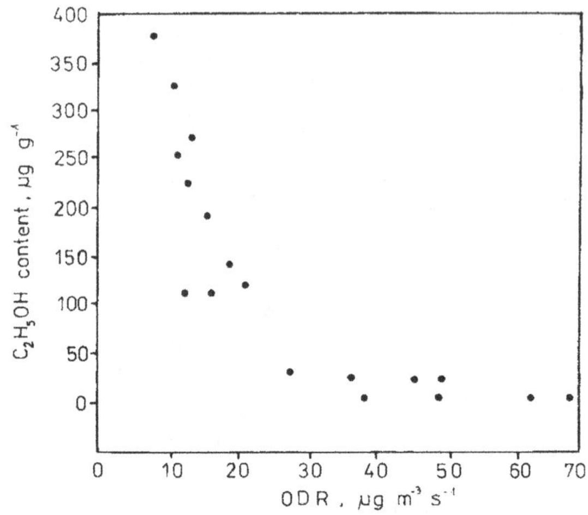

FIGURE 59. Ethanol content in xylem exudate of tomato plants vs. soil ODR. (From Gliński, J. and Stepniewski, W., *Soil Aeration and Its Role for Plants*, CRC Press, Boca Raton, FL, 1985, 144.)

is observed in flood-intolerant plants. This, in turn, results in a decreasing ATP level in the cells and, as a consequence, an increasing ethanol production. This autocatalytic process, according to Crawford,[267] explains the susceptibility of many plants to even oxygen stresses of short duration.[374,579]

According to Barta,[99] the ethanol toxicity in plant roots has been definitively proved. It has never been demonstrated that the level of ethanol accumulation in flooded roots is sufficiently high to be lethal to the tissue. Jackson et al.[606] showed that exposing pea roots to levels of ethanol up to 100-fold greater than that found in xylem sap of flooded roots had little toxic effect and failed to simulate flooding injury.

4.4.3.3 Ethylene

Ethylene is always present in plants and its increased concentration is caused by various stresses: thermal, water, mechanical, chemical, and disease.[3,497] Moreover, root oxygenation is a very important factor. An increase of ethylene concentration in roots and other parts of waterlogged plants was found in broadbean,[363] Callitriche,[863] chrysanthemum,[648] maize,[330] eucalyptus,[240] maple,[797] radish,[648] sunflower[648,649,1276] sweet potato,[925] tomato,[163,166,601,602,604,648,649] and willow.[224]

Ethylene concentration in waterlogged plants increases as the result of two processes: (1) anaerobic stimulation of ethylene production[163,166,601,604,651] and (2) the water jacket effect of water in preventing ethylene diffusion out of the roots.[330,648,650,863] Under flooded conditions, ethylene becomes a growth-promoting hormone for plants.[167,606,653]

Generally, it is known that ethylene is formed in plant tissue under oxic conditions.[746,795] This is true for roots, as the presence of ethylene was observed, in amounts of approximately 0.9 cm^3 m^{-3}, in the roots of intact maize plants grown in an aerated nutrient solution.[330,604] In addition, Konings and Jackson[675] observed the evolution of ethylene from undisturbed root systems of rice, white mustard, and three varieties of tomato cut off from the shoots and incubated in a humid atmosphere containing 0.21 m^3 m^{-3} of oxygen.

The production of ethylene in this case was positively correlated to the rate of root elongation;

thus rice roots, evolving ethylene in the amount of 1.5×10^{-3} mm^3 (g fr. wt.)$^{-1}$ h^{-1}, elongated at a rate of about 20 mm d^{-1}, while the roots of mustard, producing ethylene at a rate of 6.5×10^{-3} mm^3 (g fr. wt.)$^{-1}$ h^{-1}, elongated by about 65 mm d^{-1}.

On the other hand, other authors[330,603] observed that the ethylene concentration in the roots of plants grown in a nonaerated solution (up to 1.9 cm^3 m^{-3}) was twice as much as when grown in an aerated one (0.9 cm^3 m^{-3}). Also, Kawase[648] reported an increase in the ethylene concentration in the roots of garden plants (chrysanthemum, sunflower, tomato, and radish) under the effect of flooding the roots in the soil with water.

The examples that reported an increase in the ethylene concentration in roots due to the lack of aeration of the nutrient solution or flooding the soil with water cannot constitute proof of an increase in the production of ethylene by roots under anoxic conditions, as these conditions cannot be considered anoxic. In the nonaerated nutrient solution, oxygen concentration, as a result of internal oxygen transport through the plant, was equivalent to approximately 0.012 m^3 m^{-3} in air.[330,603] In the case of the waterlogging of garden plants mentioned,[648] there must have also been a certain amount of oxygen in the plant roots from internal transport. In turn, under anoxic conditions (in an atmosphere of nitrogen), no evolution of ethylene was observed during the incubation of excised roots.[330,604]

On the basis of the foregoing, it can be stated that ethylene in the roots is only produced in the presence of oxygen, although the required oxygen content is low. The lowest limit of oxygen concentration, as well as its optimum concentration for ethylene production, has not yet been determined.

Bradford et al.[164] and Bradford and Yang[165] found that under low oxygen conditions in the soil during flooding, tomato roots synthetize ACC, the precursor of ethylene, and transport it in the xylem to the shoot, where it is rapidly converted to ethylene. Adams and Yang[6] indentified this precursor in apple tissue. Inhibitors of the ethylene biosynthetic pathway are available for further testing of this ACC transport hypothesis: aminooxyacetic acid (AOA) or aminoethoxyvinylglycine (AVG) block the synthesis of ACC, whereas Co^{2+} prevents its conversion to ethylene.

It is worthwhile to mention that some authors, e.g., Chrominski,[229] found the effect of salts on ethylene production in *Virginia radiata* hypocotyl sections. (In their experiments, NaCl and KCl decreased ethylene production, when $CaCl_2$ increased.) Aerenchyma formation, and the growth of adventitious roots is favored by increases in the ethylene content of roots exposed to poorly aerated aqueous environments.[330,600]

4.4.3.4 Other Anoxic Metabolites

Besides metabolites discussed earlier, there are other substances which may appear in roots when the supply of oxygen is insufficient. Their formation and role are not yet fully explained.

Barta,[100,101] summarizing knowledge on the problem of metabolic responses of roots under anoxia, states that root anaerobiosis leads to rapid changes in membrane permeability, as evidenced by increased efflux of electrolytes and organic metabolites. Loss of O_2 uptake has also been attributed to irreversible injury of membrane-bound oxidase enzymes. He found, for bird's-foot trefoil and alfalfa, a significant lowering of glucose-6-phosphate in trefoil root exudation of K^+, sugars, and amino-N, especially during the first days of root anaerobiosis. Also alcohol dehydrogenase activity (ADH) increased severalfold with flooding of roots of these two species.[98] A significant increase in root ADH in the flood-tolerant *Spartina alterniflora* under hypoxic root environments was also found by Mendelssohn and McKee.[818] This enzyme which catalyzes a step of the fermentation process that yields ethanol CO_2 and energy from sugar has also been studied in plant roots under anoxic conditions by John and Greenway,[620] Chang et al.,[221] and Hook and Brown.[567] Crawford[264,268] and Stolzy and Flühler[1130] have reported that in anoxic conditions pyruvic, fumaric, acetic, aspartic, glutamic, propionic, butyric, malic, shikimic, glycolic, and α-aminobutyric acids are formed in roots.

Visser et al.[1253] observed that the amount of carbohydrates in apple roots decreased in the case of nonaerated nutrient solution. Atwell et al.[53] found no major carbohydrate depletion due to O_2 deficiency in the excised maize roots, but the total soluble sugars and amino acid concentrations in roots were generally greater at low (0.02 to 0.06 mol O_2 m^{-3}) rather than at high (0.16 to 0.25 mol O_2 m^{-3}) oxygen concentrations. This applied specifically to root apices (0 to 2 mm) and expanding (2 to 15 mm) tissue.

In anaerobic stress during waterlogging, the synthesis and translocation of growth regulators, such as cytokinins (CK) and gibberellins (GA), are suppressed in the roots. The rapid appearance of an anaerobic stress reduces root conductance of water flow, but generally in waterlogged plants, root conductance remains identical or often exceeds that of control roots.[167]

Cellulase activity is also increased by waterlogging in roots of bean, Brussel sprouts, celery, chrysanthemum, lettuce, mustard, pea, radish, and sunflower plants.[653] In tobacco, under anoxic conditions, a decrease of the nicotine concentration in roots was noted.[518] The results obtained by Blevis et al.[141] show an increase of particulate viologen utilizing nitrate reductase (EC 1.9.6.1) in barley roots situated in sterile, anoxic conditions.

4.4.3.5 Microbial Colonization and Activity

Trolldenier and Hecht-Buchholz,[1201] Trolldenier and von Rheibaben,[1202] and Stumpf,[1139] investigating microbe population accompanying wheat roots growing in aerated and unaerated nutrient solution, have revealed that reduced root growth due to the lack of aeration was accompanied by a higher root microbe population. The mucilaginous material (slime) surrounding the roots was absent in unaerated and abundant in aerated root tips. This missing or incomplete mucilaginous layer on root tips is associated with unfavorable growth factors, such as heavy metal toxicity[532] and attacks of pathogens.[1009]

4.4.3.6 Energy Charge

Adenylate energy charge (AEC) of plant tissues, defined as follows: AEC = (ATP + 0.5 ADP):(ATP + ADP + AMP), under aerobic conditions does not differ from that of animal tissues and microorganisms and is close to 0.9.[1030,1031] Saglio et al.[1030-1032] measured simultaneously energy production expressed as AEC, soluble sugars, and glycolytic activity (as estimated both by sugar consumption and buildup of terminal products such as ethanol and lactic acid) in excised maize roots tips after various periods of time of aging in the air. The transition of root tips from air to N_2 atmosphere induced a rapid drop of AEC from a high value close to 0.9 in the air to lower values ranging from 0.6 to 0.15. They concluded that, in anoxia, there is a quantative relationship between the AEC value and the level of metabolic activity via fermentation (Figure 60).

Similar results were obtained by Barta[100] for trefoil and alfalfa roots under anaerobic stress. The energy status of both species showed a sharp decline in ATP/ADP ratios amounting to 40 to 60% after 1 d of anaerobiosis. The significantly higher ATP/ADP ratios in trefoil in comparison to alfalfa support the earlier hypotheses[1198] that the decline of AEC during anaerobiosis is slower in waterlogging-tolerant species than in waterlogging-susceptible ones. It was found that waterlogging-tolerant rice is able to increase its AEC with prolonged flooding, while wheat cannot.[226]

Mendelssohn and McKee[818] found that the root energy status of *S. alterniflora* during hypoxia is maintained at levels similar to that during root normoxia. The dependence of the AEC or the ATP/ADP ratio on oxygen partial pressure[974,1033] was used to study internal transport in roots of maize[973,1033] and bulrush.[1120]

4.4.4 OXYGEN DEFICIT AND ROOT GROWTH AND WATER UPTAKE

4.4.4.1 Root Growth Rate

The reaction of roots to anoxic (hypoxic) conditions manifests itself in various ways.

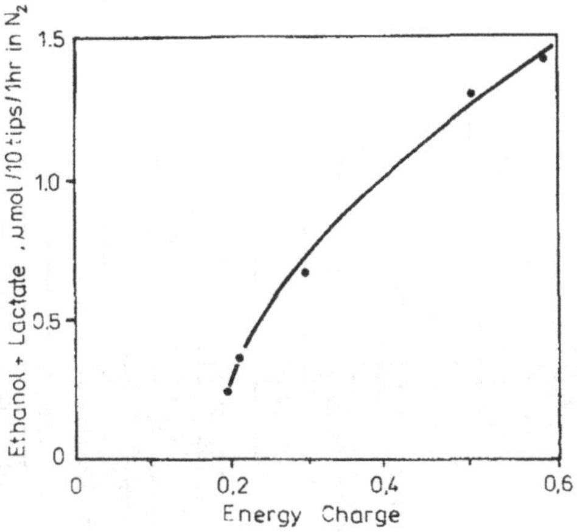

FIGURE 60. Relationship between energy charge after 30 min of anoxia and the rate of glycolysis (measured after 1 h of anoxia) in excised maize root tips. (Adapted from Saglio, P. H. and Pradet, A., *Plant Physiol.*, 66, 516, 1980.)

Completely anoxic conditions stop root growth entirely within 2 to 3 min. This has been observed with cotton and soybean.[579] When the stress did not last longer than 30 min, the growth rate subsequently returned to normal. Anoxic stress lasting for 3 h for cotton and 5 h for soybean killed the taproots.

Roots of plants which are very sensitive to anoxic conditions (e.g., tobacco and legumes) die after a few days of flooding,[537,682,742,1326,1327] but the first symptoms of damage by waterlogging may appear after 1 to 2 h and permanent damage may appear after 24 h, especially on warm sunny days.[686,742]

As has already been mentioned, plants which can survive an anoxic stress nevertheless decrease the rate of growth of their roots. This has been observed with tomato,[602] maize,[53,429,430,444,1215] barley,[428,429] wheat,[827,1072] pea,[241,428,429] oak,[1071] snapdragon,[1132] tulips,[1310] sugar beets,[1315] bean,[1057] and mustard.[483] As a result of these root growth disturbances (due to an insufficient supply of oxygen to the roots), the mass of the root system decreases. This was observed with sorghum,[571] barley,[334] oats,[198,1281] wheat,[1004,1204,1241,1281,1309] maize,[327,762] apple tree,[1253] sugar beets,[1303] lemon,[700,701,1131] orange,[699,1131,1273] jojoba,[982] avocado,[703] grasses,[737,1264] cotton,[926] legumes,[241] sugar cane,[927] bean,[1057] broad bean, cabbage, and potato.[1309]

Oxygen stresses (expressed by ODR and Eh values) influenced by waterlogging of soil in pot experiments[704] caused greater differentiation in dry root mass and the total yield to root mass ratio for smooth bromegrass and orchardgrass than in tall fescue and reed canarygrass (Figure 61).

Under anoxic conditions, the root mass of flood-intolerant plants decreases earlier and to a higher degree than the mass of the shoots. Hence the ratio of shoots to roots is higher in anoxic conditions than in the presence of oxygen. This was found in the case of wheat,[4,1203] lemon,[698] jojoba,[982] orange,[1273] and apple tree.[1253]

Roots growing in an environment with restricted oxygen supply are thicker, shorter, less branched, and of light color,[241,1264] have larger cells,[952] and in the case of legumes, possess fewer nodules.[241] Roots which are near the anoxic zone in the soil show an ability to avoid this zone,[1072] which results in formation of a shallow root system,[330,758,1158] and in some cases, the roots become aerotropic (negatively geotropic) and emerge from the soil.[602,1264,1308] Formation of the shallow

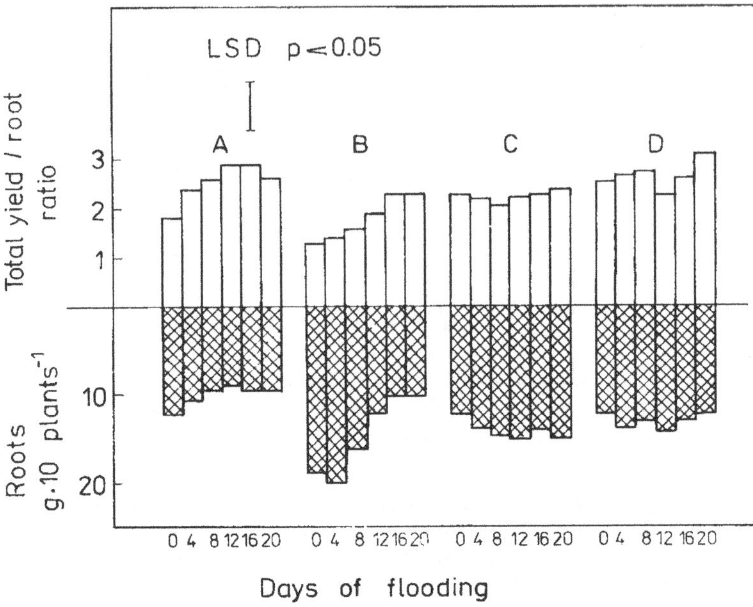

FIGURE 61. The mean dry mass of roots and the total yield to root mass ratio of smooth bromegrass (A), orchardgrass (B), tall fescue (C), and reed canarygrass (D) as influenced by periodic water flooding of soil. (Modified from Labuda, S., *Probl. Agrofizyki*, 52, 1988.)

root system results from stimulation of the growth of lateral roots following death of the taproot tip.[579] The effect of oxygen concentration on the growth rate of constrained and unconstrained roots is shown in Figure 62.

It is widely agreed that the best indicator of soil oxygen availability for plant roots is the ODR. The reaction of roots of different plants to ODR is shown in Figure 63. We can see that the number of roots in the soil (Figure 63A) begins to be reduced already at an ODR of 50 $\mu g\ m^{-2}\ s^{-1}$ and drops rapidly to less than one fifth of its original number over the ODR interval of 50 to 40 $\mu g\ m^{-2}\ s^{-1}$. The ODR value of 40 $\mu g\ m^{-2}\ s^{-1}$ may, accordingly, be accepted as critical, and 50 $\mu g\ m^{-2}\ s^{-1}$ may be accepted as limiting for wheat root growth. Similarly, the root growth rate of desert shrubs is restricted at an ODR of 50 $\mu g\ m^{-2}\ s^{-1}$ and is completely suppressed at an ODR of 30 to 35 $\mu g\ m^{-2}\ s^{-1}$ (Figure 63B).

Blackwell and Wells[137] found 15 $\mu g\ m^{-2}\ s^{-1}$ as a limiting value of ODR for the growth of winter oat. They also found a significant differentiation in root development which is shown in Table 17. A decrease in rooting depth of potato, tomato, and tulip, growing in soil affected by low oxygen supply, was indicated by Wiersum.[1311]

The critical ODR values for other plants are differentiated and (in $\mu g\ m^{-2}\ s^{-1}$) are equal to 25 for barley,[740] 42 and 20 for maize,[126,1230] 33 for sunflower,[738,740] 33 for snapdragon,[1132] and 33 for cotton.[738] Growth of grasses was inhibited at an ODR of 25 $\mu g\ m^{-2}\ s^{-1}$ for turfgrass, 17 $\mu g\ m^{-2}\ s^{-1}$ for ryegrass,[466,737] 8 to 15 $\mu g\ m^{-2}\ s^{-1}$ for Kentucky bluegrass, and 33 $\mu g\ m^{-2}\ s^{-1}$ for Newport Kentucky bluegrass.[739] Desert shrubs such as *Artemisia tridentata*, *Larrea tridentata*, and *Franseria dumosa*[760] have critical values of ODR in the range of 30 to 35 $\mu g\ m^{-2}\ s^{-1}$.

Root growth of plants adapted to prolonged flooding (e.g., rice) is better in such conditions than when not flooded. This causes a decrease of shoot-to-root ratio due to flooding of the soil.[626,762] These plants generally have more porous roots and sometimes, e.g., black mangroves (*Avicennia nitida*), form special aerial roots (pneumatophores) containing aerenchyma, which supply oxygen to normal roots.

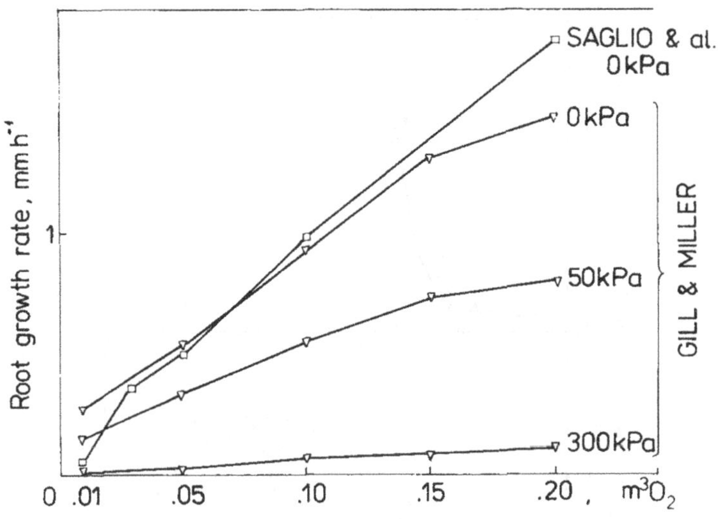

FIGURE 62. The effect of oxygen concentration on growth rate of constrained and unconstrained roots. (Data from Gill, W. R. and Miller, R. D., *Soil Sci. Soc. Am. Proc.*, 20, 254, 1956 and Saglio, P. H. and Pradet, A., *Plant Physiol.*, 66, 516, 1980.)

4.4.4.2 Root Growth Force and Pressure

The effect of oxygen on root growth pressure was studied by Eavis et al.[349] for peas and cotton by means of a dead-load technique. Average root growth pressure values did not change significantly for 3.8 and 21% O_2 treatments, and in the case of peas, they ranged from 1.1 to 1.4 MPa, while those for cotton nodules decreased from 0.95 to 1.1 MPa at 8 and 21% O_2 to 0.5 MPa at 3% O_2.

Souty and Stepniewski[1110] measured root growth force and pressure for young (2 to 4 d) maize seedlings using a strain gauge technique at oxygen concentrations ranging from 0 to 20%. Oxygen concentration was changed around the entire seedling or around its radicle only. Both root growth force and pressure as well as the growth rate decreased at oxygen concentrations below 3 to 5% (Figure 64). As can be seen, the plant response was more pronounced when the entire seedling was in the gas mixture of lowered oxygen concentration. When only the plant radicle was exposed to oxygen stress, the growth force (and thus pressure) was less affected, at least partly, due to possible oxygen diffusion through the plant tissue from the coleoptile to the radicle. The maximum mean root growth force of the radicles was 0.6 N before their deformation and 1.1 N after deformation of the impeded roots. This corresponds to the growth pressure 0.4 and 0.65 MPa for undeformed and deformed roots, respectively.

4.4.4.3 Water Uptake

It has been known that flooding the roots with water causes partial or complete inhibition of water uptake and transpiration.[4,28,192,559,602,604,742,1203,1322,1327] A positive linear correlation has been found between oxygen uptake by roots and water use by plants in undamaged root systems.[559]

The water-conducting ability of *Lycopersicon esculentum* Hill.,[167,604,682] *Nicotiana tabacum,*[686] and several *Pyrus* spp.[28] was found to decline in response to flooding. Several authors suggested the explanation that the decrease in the water permeability of the roots due to anoxia leads to a decrease in the water potential of leaves and, in consequence, the closure of stomata observed in numerous herbaceous and woody plant species.[259,604,936,1101]

It is true that some authors[363,441,682,686] have observed the intensification of water stress after

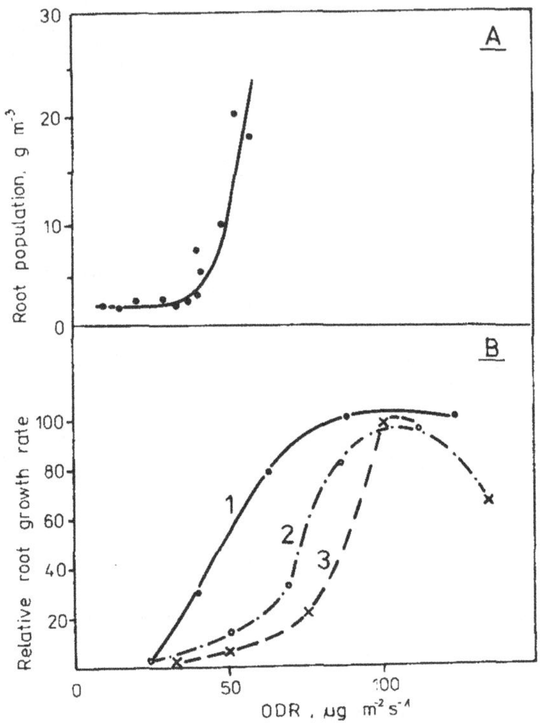

FIGURE 63. Root response to ODR. (A) Population of wheat roots in soil under field conditions; (B) relative root growth rate of three desert shrubs: *Franseria dumosa* (1), *Larrea tridentata* (2), and *Artemisia tridentata* (3). (From Gliński, J. and Stepniewski, W., *Soil Aeration and Its Role for Plants*, CRC Press, Boca Raton, FL, 1985, 151.)

flooding the roots. In many cases, however, this phenomenon was not confirmed, and even an opposite effect, i.e., an increase of water potential in leaves under conditions of root anoxia induced either by flooding[604,936,1203] or by introducing nitrogen to the soil,[1102] has been revealed.

This means that the decreased water permeability of the roots is not the only factor reducing transpiration, as the cause-and-effect relationship between root permeability and stomata closure has not been proved. Thus further work is required to elucidate the nature of signals responsible for stomatal closure with root anoxia and to evaluate the role of root permeability in decreased water uptake by plants.

4.4.5 NUTRIENT CONTENT AND UPTAKE BY ROOTS UNDER OXYGEN STRESS

Oxygen stress in the soil causes disturbances in the functioning of the root system and also in the nutrient uptake and content in the root. A review of papers[241,328,608,626,698,700-703,745,982,1082,1125,1131,1253,1273,1309] on the changes of nitrogen, phosphorus, potassium, calcium, magnesium, sodium, chloride, and some micronutrients (boron, zinc, copper, manganese, and iron) content and uptake in roots of various plants under the influence of oxygen stress is presented in Tables 18 to 21.

As can be seen, the direction of the changes is different in different plant roots. In all cases,

TABLE 17
Dry Weight, Numbers, Diameters, and Lengths of Winter Oat Roots After 8 d of Growing in Waterlogged and Freely Drained Soil

	Plants in	
Root features	**Waterlogged soil**	**Freely drained soil**
Average dry weight of roots (g/plant)	0.021	0.024
Average diameter of roots (cm)	0.066	0.039
Average number of lateral roots per plant	10	38
Different-sized seminal roots (cm)		
0—1	17.5	6
1—5	41	32.5
5—10	14.5	15
10—20	27	20
20+	0	26.5

Modified from Blackwell, P. S. and Wells, E. A., *Plant Soil*, 73, 129, 1983.

the uptake is decreased which is connected with a distinct decrease in the mass of the root exposed to oxygen deficiency.[447] In almost all cases, a decrease in potassium, magnesium, chloride,[447] and ammonium[1001] content in roots was found, as well. A tendency to increase manganese, iron, and zinc content in roots when oxygen concentration decreases is also visible.

In flooding conditions, changes of the pH of the medium may change the availability of some elements for roots. It has been recently revealed[325] that within 24 h pH of the nutrient medium of sunflower plants in the flooded treatments was by 1 pH unit higher than that of the nonflooded treatments and remained higher for several days.

4.4.6 ADAPTATION OF PLANT ROOTS TO ATMOSPHERIC STRESS IN THE SOIL

Atmospheric stress in the soil is referred to as oxygen deficiency, waterlogging, and anaerobiosis.[687] All involve a buildup of carbon dioxide, ethylene, and other potentially toxic gases and oxygen depletion which leads to a reduction of anaerobic respiration.

It has been proven that no plant can survive under complete anoxia for prolonged periods. Although few plants are subjected to complete anoxia, many are exposed to periodical oxygen deficiency caused by water-saturated, compacted, or surface-sealed and crusted soils. In the case of waterlogging, it should be noticed that it occurs not only when soil is inundated, but also when water fills a critical proportion of the soil air spaces.[653]

Anoxia-intolerant plant species survive oxygen-free conditions for periods ranging from hours to days or they eventually degenerate.[327] This period for cotton taproot is in the range of 0.5 to 3 h[579] and for rice seminal roots it is in the ramge of 96 to 120 h.[125] Anoxia-tolerant plants or organs survive much more extended periods. Crawford[269] found a number of wetland species that survived overwintering in anaerobic mud for a period of 2 months or more (*Scirpus maritima*) and for 1 month or more (*Schoenoplectus lacustris, S. tabernaemontani, Typha angustifolia, Phragmites australis,* and *Iris pseudocorus*).

Waterlogging of the soil rapidly alters both the physical and biological environment of plant roots and drastically limits the diffusion of oxygen from the soil to the root system. It inhibits root growth and root hair formation and increases susceptibility of roots to attacks by predators and pathogens. Under prolonged flooding, roots may blacken, die, and eventually rot.[687]

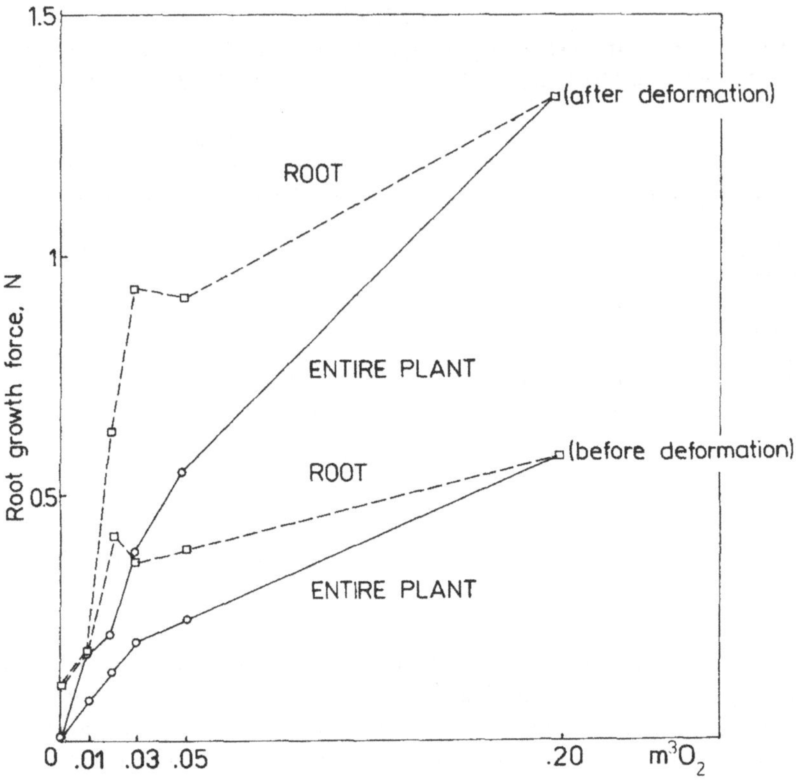

FIGURE 64. The dependence of growth force of maize seminal roots before and after their deformation due to facing the obstacle (each point is a mean value of four seedlings) on oxygen concentration in the gas mixture surrounding the entire plant and the radicle only. (Adapted from Souty, N. and Stepniewski, W., *Agronomie* 8, 295, 1988.)

Plants subjected to long-lasting atmospheric stresses in the soil form some adaptative mechanisms which allow them to grow, develop, and yield. Plant tolerance to root flooding is relative and depends on their variety and stage of development. There is, however, a controversy in the literature as to whether some phenomena, e.g., adventitious roots or lignification, are symptoms of flooding injury or a beneficial adaptation to waterlogging. Numerous authors identified three different kinds of plant adaptation to anoxia in the soil, namely, anatomical and morphological adaptation in internal transport, shallow rooting, and metabolic and physiological adaptation.

4.4.6.1 Anatomical Adaptation

This adaptation has been reviewed by many authors (see References10,17,37,38,44–46,260,304,326,328,334,441,569,570,644,762,1014). It is mainly evaluated by the capability of taking O_2 into the plants via the aerial parts, translocating O_2 by the process of longitudinal diffusion to the root systems, and partly evaluated by diffusing out of the root systems.

One of the most common anatomical adaptations is the formation of increased intercellular air spaces in the root cortex that provide canals parallel to the axis of the root to a distance of 2 to 3 mm from the root tip through which gases can diffuse longitudinally (Figure 65). They are found in the suberized root tissue known as aerenchyma.[304,328,330,653,654]

TABLE 18
Changes in N, P, and K Content and Uptake by Roots of Various Plants Under a Shortage of Oxygen in Soil

	Nitrogen			Phosphorus			Potassium		
Plant	**Cont.**	**Upt.**	**Ref.**	**Cont.**	**Upt.**	**Ref.**	**Cont.**	**Upt.**	**Ref.**
Lemon	+		702	0		702	0		702
	+	–	698	0		698	–		698
	+		701		–	701		–	701
Lemon and orange	–		1131	–		1131	–		1131
Apple tree	–		1253				–		1253
Jojoba	+		982	–		982	–		982
Avocado	+		703	0		703	–		703
	–		1082	0		1082	–		1082
Potato		–	1309		–	1309		–	1309
Barley	–	–	335				–		334
Rye		–	1125		–	1125		–	1125
Peas	–	–	241	–	–	241	–	–	
Soybean							–		1125

Note: Cont, content; Upt, uptake; Ref., references; –, decrease; +, increase; 0, insignificant changes.

TABLE 19
Changes in Ca, Mg, and Na Content and Uptake by Roots of Various Plants Under a Shortage of Oxygen in Soil

	Calcium			Magnesium			Sodium		
Plant	**Cont.**	**Upt.**	**Ref.**	**Cont.**	**Upt.**	**Ref.**	**Con.t**	**Up.t**	**Ref.**
Lemon	0		702	–		702	0		702
	–		698	0		698	+		698
		–	701		–	701		–	701
Lemon and orange	–		1131	0		1131			
Jojoba	0		982	–		982	–		982
Avocado	+		703	–		703	–		703
				–		1082	+		1082
Potato		–	1309						
Broad bean		–	1309						
Cabbage		–	1309						
Wheat		–	1309						
Peas		–	241	0	–	241			
Rye		–	1125				0	–	1125
Apple tree	–		1253						
Soybean							–		1125

Note: Cont, content; Upt, uptake; Ref., references; –, decrease; +, increase; 0, insignificant changes.

The porosity of mesophyte roots is slight (of the order of several percent of the root volume). However, under conditions of soil inundation with water, even these with plants it is quite common to find a development of adventitious roots containing aerenchyma[334] and having a porosity mudch greater than in the primary roots (Figure 66).[762] This results in an increase on root porosity under conditions of insufficient oxygen supply from the soil which was reported by many authors.[4,13,75,210,595,761,832,1073,1100,1102,1241,1343]

TABLE 20
Changes in Cl, B, and Zn Content and Uptake by Roots of Various Plants Under a Shortage of Oxygen in Soil

	Chloride			Boron			Zinc		
Plant	**Cont.**	**Upt.**	**Ref.**	**Cont.**	**Upt.**	**Ref.**	**Cont.**	**Upt.**	**Ref.**
Lemon	–		702	0		700	+		700
	–		698	0		702	0		702
		–	701	+		698	+		698
					–	701		–	701
Lemon and orange	–		1131	0		1131	0		1131
Avocado	–		703				+		703
	–		1082						
Barley	–		608						
Jojoba				+		982	0		982

Note: Cont, content; Upt, uptake; Ref., references; –, decrease; +, increase; 0, insignificant changes.

TABLE 21
Changes in Cu, Mn, and Fe Content and Uptake by Roots of Various Plants Under a Shortage of Oxygen in Soil

	Copper			Manganese			Iron		
Plant	**Cont.**	**Upt.**	**Ref.**	**Cont.**	**Upt.**	**Ref.**	**Cont.**	**Upt.**	**Ref.**
Lemon	+		700	+		700	+		700
	0		702	0		702	0		702
	–		698	+		698	+		698
		–	701		–	701		–	701
Orange and lemon	+		1131	+		1131	+		1131
Jojoba	0		982	+		982	+		982
Avocado	0		1082	0		1082	0		1082

Note: Cont, content; Upt, uptake; Ref., references; –, decrease; +, increase; 0, insignificant changes.

Two kinds of aerenchyma have been described: lysigenous and schizogenous.[379] The first is formed when intercellular space arises through the dissolution of entire cells. The second is connected with a separation of cell walls from each other along more or less extended areas of their contact.

McPherson[812] was the first who clearly defined lysigenous aerenchyma formation. He observed that aerenchyma formation in maize roots is preceded by the disappearance of protoplasm from the cell, bulging cells, death of cells, and by collapse of the cell walls. He observed that:

1. Aerenchyma formation in maize roots is prevented by O_2 application.
2. Increased respiration by raising temperature invariably increases aerenchyma formation.
3. Aerenchyma is never produced in the stelar tissue, but in the cortex, which is distal to the conducting tissue of the food.

Based on these findings, his stipulation is that death of the cells is caused by starvation since

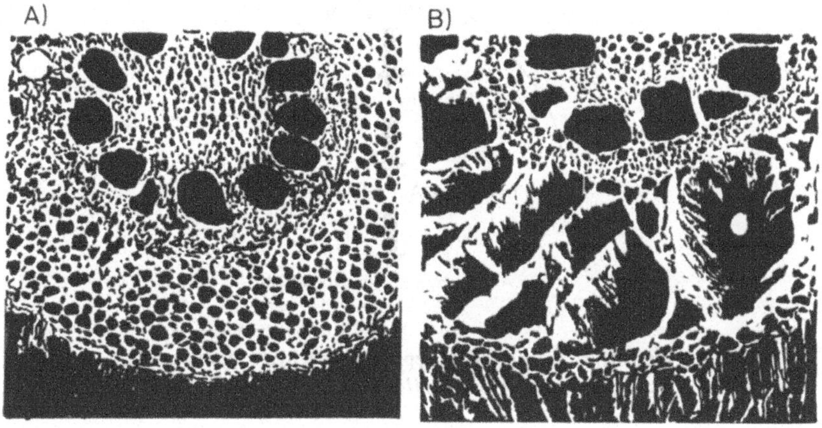

FIGURE 65. Scanning electron microscope of transverse sections of second whorl adventitious roots grown in aerated (A) and nonaerated (B) medium. (Adapted from Drew, M. C., Jackson, M. B., and Giffard, S., *Planta,* 147, 83, 1979.)

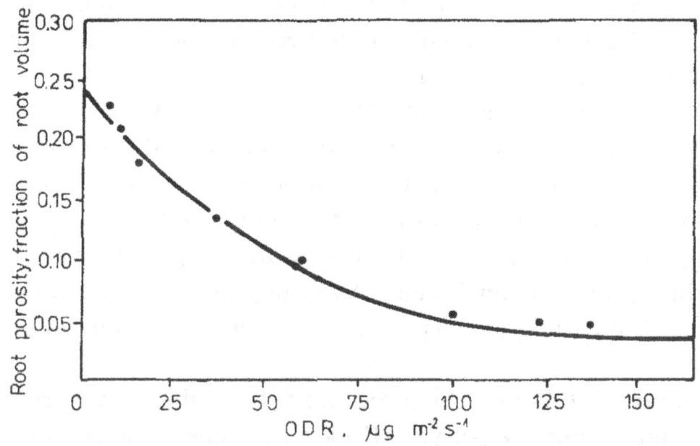

FIGURE 66. Rice root porosity in active tillering phase as affected by soil air. (Adapted from Gliński, J. and Stepniewski, W., *Soil Aeration and Its Role for Plants,* CRC Press, Boca Raton, FL, 1985, 150.)

anaerobic respiration is insufficient to maintain cellula integrity. The results described earlier suggested[652] proposal of a chain of events in which waterlogging leads to aerenchyma development through elevation of ethylene concentration, which in turn increases cellulase activity (Figure 67).

Beckel,[105] examining *Bouteloua gracilis,* in which aerenchyma develops into root cortex under normal conditions, without waterlogging, found that this plant has aerenchyma in the basal region of the root except for a short distance behind the growing point. Also, there is no aerenchyma near the tips of the adventitious roots formed in the water by waterlogged sunflower plants.6[54] It seems that normal adventitious roots develop in sunflower plants even in water. If they are kept continuously in water, aerenchyma develops gradually as the roots increase in age. Philipson and Coutts[949] found that O_2 diffuses at the bases and tips of lodgepole pine roots grown

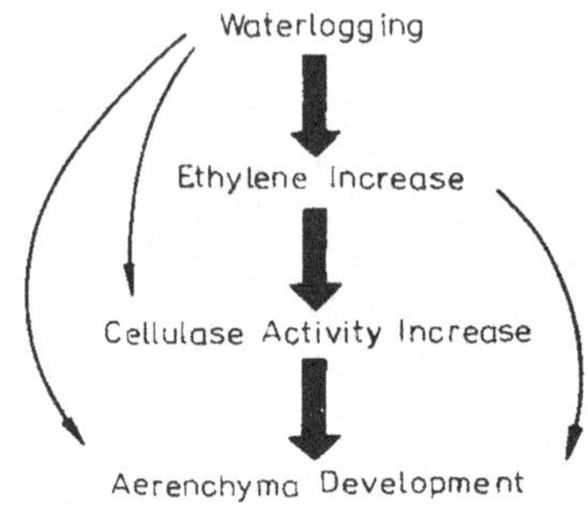

FIGURE 67. Sequential steps in aerenchyma development. (Adapted from Kawase, M., *HortScience*, 16, 31, 1981.)

anaerobically, while it diffuses only at the bases of the roots grown aerobically. It shows that metabolically active root tips get a better O_2 supply with this kind of aerenchyma development.

Williams and Barber[1324] drew attention to another role of aerenchyma. They stated that aerenchyma provides the greatest strength with the least tissue and that minimal tissue requires the minimum amount of O_2 under anaerobic conditions. In other words, there is a decreased demand for oxygen arising from the smaller number of cells present in aerenchymatous roots. It was shown that aerenchymatous roots show unimpeded nutrient uptake rates, even if only one-fifth of the midcortex remains unchanged.[329,333]

After investigating two types of aerenchyma (lysigenous and schizogenous) in waterlogged roots, Yamasaki[1337] observed two types of radial cortical cell arrangement: columnar and oblique (Figure 68). The columnar type had much more air space per unit area of cortex than the oblique type, provided the cell sizes are equal. He found the columnar type (intensified by the development of lysigenous aerenchyma) to be more waterlogging-tolerant than the oblique type (connected with schizogenous aerenchyma). As examples, *Fagopyrum esculentum* (with the oblique type of cortical cell arrangement) and *Polygonum thunbergii* (with the columnar type) have been presented.

It is a general opinion that plants which root do not readily form aerenchyma (e.g., peas[326] belong to the group which is most sensitive to flooding). Some plant species develop specialized knee roots and pneumatophores, constituting a diffusion path for gases.[441,1193] In the mangrove, a swamp plant, roots possess a specific mechanism for the entry of O_2 into it, e.g. lenticels.[568,797,949]

According to Yamasaki,[1337] also the lignification of the cortex and epidermis which occurs more strongly in waterlogging-tolerant species than in nontolerant species may tell about plant adaptation to oxygen deficit. It may increase the mechanical strength of the plant and increase disease resistance. He suggests the classification of waterlogging tolerance based on the limitation, cortical cell arrangement, and aerenchyma development in the roots (I, the weakest, to VIII, the strongest):

I. No lignification, oblique, less aerenchyma — Welsh onion, carrot, and *Astragalus sinicus*
II. Lignified epidermis, oblique, — potato, pepper, tomato, cosmos, and buckwheat
III. Lignified epidermis and cortex, oblique, aerenchyma development — barley and wheat
IV. Lignified stele — rape, Japanese horseradish, and sweet potato
V. Tannin-like substance in root, oblique — *Saxifraga stolonifera*, *Equisetum arvense*, and taro
VI. Aerenchymatous epidermis, rarely lignified epidermis and cortex — *P. thunbergii*,

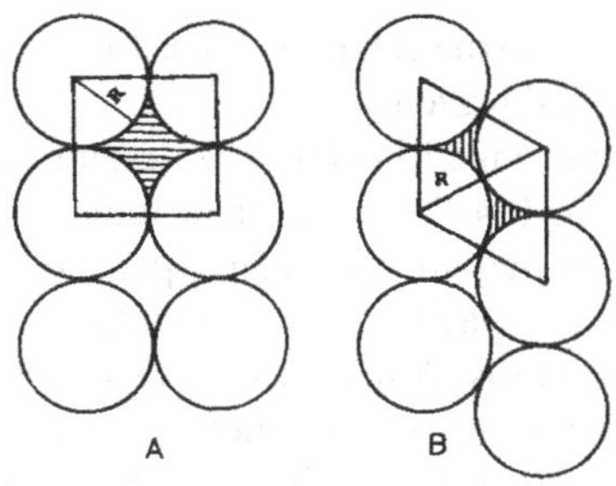

FIGURE 68. Schematic illustration of cross section of root cortex describing two types of cortical cell arrangements. (A) Columnar; (B) oblique. R and shaded area indicate cell radius and intercellular space, respectively. Vertical line of each drawing is tangential to the root cross section. (Adapted from Kawase, M., *HortScience*. 16, 31, 1981.)

Acorus calamus, Cyperus microiria, C. malaccensis, Eleocharis tuberosa, Sagittaria trifolia, Nalumbo nucifera, and *Oenanthe stolonifera*

VII. Strongly lignified epidermis, large aerenchyma — rice, *Zizania latifolia, Beckmania erucaeformis,* maize, *Setaria italica, Coix Ma-yuen,* and *Commelina communis*

VIII. Lignified epidermis, cortex, and root hairs; aerenchyma in cortex — *Juncus effusus, J. alatus,* and *I. laevigata*

Krizek[687] has reported that fruit tree species have a wide-range of tolerance to long periods in water-saturated soils. The roots of apple, pear, and quince are relatively tolerant to flooding, while those of peach, apricot, and almond are generally considered to be highly sensitive. "Myrobalan" plum is intermediate in tolerance.[1015] Good and Patrick[452] show differences in the ability of two species, water oak and green ash, to maintain rhizosphere oxidation under prolonged periods of flooding and to prevent the accumulation of potentially phytotoxic reduced compounds.

While discussing the anatomical adaptation of plant roots to atmospheric stress, it is worthwhile to mention the studies carried out by Trolldenier and Hecht-Buhcholz[1201] on the ultrastructure of wheat roots grown in aerated and unaerated nutrient solutions. They found that the ultrastructure of the root cap meristem cells of unaerated roots was similar to that of the aerated ones. The same was true for the central root cap cells of the root cap. The cytoplasm of these cells appeared to contain less organelles in the unaerated roots than in the aerated ones.

A curious example of adaptation to flooding is the floating of deep water rice grown in the floodplains of southern Asia where the water can rise up to 6 m during the rainy season.[971] The distinguishing characteristic of this rice is its ability to elongate with rising waters, and growth rates of 20 to 25 cm d^{-1} have been recorded, with the total plant height reaching up to 7 m. Its survival depends on its ability to keep the tip of its foliage above the water surface, and completely submerged plants cease to elongate and eventually die. Raskin and Kendle[971] found that in partially flooded deep water rice, continuous air layers occurred between the hydrophobic, compacted surface of the leaf blades and the surrounding water, constituting the major path of gas transport. The total volume of the air layers on both sides of a leaf blade is about 45% of its volume. The conduction of gases through the internal air spaces of the leaf is a negligible component compared to the conduction through the external air layers. Gases move through the air layers both by diffusion and mass flow. The air layers are vital for the survival of the partially submerged rice plants. Plants without air layers (e.g., washed with a surfactant solution) do not grow in response to submergence and the submerged parts of the plant deteriorate, as is evident by the rapid loss of chlorophyll and protein.[971]

Another mechanism of gas exchange found in water plants is pressurized ventillation. This phenomenon was reported first in 1841[963] in leaves of lotus (*Nelumbium*). Merget[822] demonstrated the physical nature of this phenomenon in an experiment with dead lotus leaves which, after remoistening, exhibited their capacity for pressurization. Recently this phenomenon was studied in detail in the yellow water lily (*Nuphar luteun*) by Dacey.[281,282] This plant, growing in lake sediments, has a rhizome measuring several meters in length and up to 10 cm in diameter and several rosettes of leaves that rise up through the water on petioles which may be 2 m long. The plant has an extensive system of continuous gas spaces or lacunae occupying about 60% of the petiole volume and up to 40% of the volume of the roots and rhizome.[282] It was found that young leaves of the water lily show a small overpressure of up to 0.2 kPa. Due to this pressure several liter of air per day move to 50 cm min^{-1} and then move up the petioles of the older leaves to the atmosphere. The overpressure in young leaves is created by purely physical processes of thermal transpiration and hygrometric pressure induced by the temperature difference between the leaf and the atmospheric air. It occurs only during the day when solar energy creates the temperature differences. This mechanism obviously not only supplies oxygen, but also removes methane and carbon dioxide from plant tissues.[283,284] It is not known how widespread this pressurized ventillation is among aquatic plants, and further studies on this topic are necessary.

Due to the anatomical adaptation to internal gas, transfer roots are able to oxidize the surrounding soil, leading to formation of oxidized coatings on the root surface, the amount of which is thought generally to be a function of the content of reduced Fe and Mn[225,328,479,741,1168] and the oxidizing capacity of the roots.[81,637,819,1200]

As an example of a plant of high-oxidizing power of its root, rice is very often cited.[637,1200] Trolldenier,[1200] growing rice seedlings in an agar and sand media with ferrous iron leucomethylene blue, observed oxidation of ferrous iron cations and leucomethylene blue as well as solubilization of ferrous sulfide. Reddish-brown coloration due to the formation of ferric oxide/hydroxide was observed around the roots and on the root surface during 1 d of incubation. Within a few hours, leucomethylene blue was oxidized to methylene blue on and near the roots. Iron was also deposited on spherically shaped structures in the rhizosphere and near the agar surface as well as in the slimy layers appearing on the root surface (being bacterial colonies). It is accepted that the oxidizing power of the roots protects the plant against excessive ferrous ion uptake and harmful reducing substances like sulfides.[206] Good et al.[451] examined the composition of the coating on green ash (*Fraxinus pennsylvanica* Marsh.) roots (Al, As, Ca, Fe, K, Mg, Ni, P, and Zn) as the effect of soil wetness. They proposed to first use these root coating constituents as an indicator of rhizosphere oxidation.

4.4.6.2 Morphological Adaptation

The basic method of morphological adaptation of herbaceous wetland species to flooding tolerance is the formation of adventitious roots containing aerenchyma that emerge from the base of the shoot (at and above the root collar).[46,239,653] Clemens et al.[239] examining root systems of three young seedlings of *Eucalyptus* sp. during 7 to 10 weeks of waterlogging, found that more than 90% of their root systems was replaced with new adventitious roots. They concluded that the greater the adventitious root formation, the greater the waterlogging tolerance of plants.

Adventitious roots were also found in 17 woody species by Hosner and Boyce,[575] e.g., sycamore and green ash, in water tupelo by Hook and Brown,[567] in red maple and sugar maple by Maronek and Wott,[797] in sunflower and tomato by Kramer[683] and Jackson,[607] and in cucumber and tomato by Hozumi.[576] There is a controversy as to the origin of adventitious roots — do they appear as the result of an anaerobic effect causing the accumulation of auxin or by the "water effect" of waterlogging?[683,1275,1276]

4.4.6.3 Shallow Rooting

Shallow rooting is the third way of plant adaptation to oxygen-deficient media.[447] The roots remain in the surface layer of the soil which is more abundant in oxygen than the deeper horizons. Moreover, the path for internal diffusion in such roots, which are shorter than in plants grown in well-oxygenated media, is usually shorter. According to Huck,[579] shallow horizontal rooting is explained by the death of the top roots after reaching the anoxic layer followed by stimulation of lateral root growth. Another mechanism responsible for shallow rooting is aerotropic root growth,[1264,1308] i.e., upward growth to better oxygenated soil layers. Due to this, roots sometimes grow just at or even above the soil surface.[603]

4.4.6.4 Metabolic and Physiological Adaptation

Metabolic and physiological adaptation to withstand anoxia has been discussed recently by a number of authors.[63,81,208,221,225,265,326,327,451,567,620,809,818,819,1168,1303] The mechanism of such adaptation, which seems to be important, at least during short-term oxygen deficiency, is rather obscure. Such adaptation involves:

1. A decrease of the Pasteur effect
2. Ethanol removal from the roots
3. An alternative metabolic path leading not to ethanol, but to less toxic products, such as malate, lactate, shikimate, glutamate, amino acids, or glycerol

The results of the study carried out by Carpenter and Mitchell[208] with flood-tolerant and flood-intolerant roots of four tree species indicate that the roots of all these species possess CN-resistant as well as CN-sensitive O_2 consumption mechanisms. The use of specific respiratory inhibitors, cyanide (KCN) and salicylhydroxamic acid (SHAM), also indicated that different species possess different proportions of the two pathways (oxic and anoxic) and that these proportions shift in response to flooding. The final statement is that cumulative flooding causes damage to the aerobic respiratory mechanism in roots of flood-intolerant species.

Lambers[710] reported that the root respiration rate of *Senecio aquaticus* (a flood-tolerant species) was lower than that of *S. jacobaea* (a flood-intolerant species). This was confirmed by Carpenter and Mitchell[207,208] with reference to roots of some tree species (*Acer saccharum, A. rubrum, Taxodium districhum,* and *Taxus cuspidata*). This would suggest that the low respiratory activity of roots may be an adaptive feature, increasing root tolerance to flooding.

Several authors[260,265,266] try to explain why certain tree species are resistant to damage by waterlogged soils. One possibility is that enough O_2 diffuses through lenticels and intercellular spaces to sustain aerobic respiration in submerged roots of tolerant species.[260] Alternatively, flood-tolerance in stress,[265,266] as in certain herbaceous species,[811] may depend on the ability of the roots to cope with products of anaerobic respiration. Such tolerance mechanisms need not be mutually exclusive, since flooded roots of *F. pennsylvanica* and *Nyssa aquatica* not only receive O_2 internally from the shoot, but also accumulate ethanol.[567]

Spartina alterniflora (the dominant intertidal angiosperm in the salt marshes along the Atlantic and Gulf coasts of the U.S. is a good example of adaptation to hypoxia[817,818] by stimulating a high rate of alcoholic fermentation which provides the high root energy status needed to support plant metabolism. A similar metabolic adaptation was found for the black mangrove, *Avicennia germinous.*[809] When the root system of the seedlings of this plant were switched from an aerobic to a hypoxic environment, oxygen concentration in the roots decreased from 16 to less than 2% within 1 h and the capacity for alcoholic fermentation, as measured by ADH activity, increased threefold.

Plant hormones have been shown to also play a role in the adaptation of roots to flooding. In general, increased resistance to this stress is associated with a rise in endogenous abscisic acid

(ABA) and a drop in gibberellins[687] which cause an increase in the root permeability to water through enhancement in hydraulic conductivity of the root. Recently, attention was drawn to the possibility that the composition of membrane lipids (especially in relation to the synthesis of unsaturated fatty acids which is dependent on molecular oxygen[1139] could play a key role in anoxia tolerance.[228,547,1244]

Summarizing, these results indicate that soil oxygen plays a fundamental role for plant root functioning, influencing it both directly and indirectly. Its role is particularly important in root respiration. Adaptative responses of roots of some plants (e.g., aerenchymatous root formation, shallow rooting, and aerotropic root growth) to anoxic conditions are well documented. Much attention has been devoted to a better understanding of the transportation of oxygen and other gases in the root zone. Numerous authors attempted to describe these processes by means of mathematical models. The effect of oxygen is closely related to other soil physical factors. This is discussed in a more detailed way in Chapter 4.7.

4.5 TEMPERATURE

4.5.1 ROOT GROWTH

Plants are very sensitive to temperature. Generally speaking, the plant roots have evolved with a temperature optimum somewhat lower than the shoots and they are more sensitive to sudden fluctuations in temperature.[877] It influences ion absorption, hydraulic conductivity, processes at the cellular level, such as osmotic potential, hydration of ions, Gibbs free energy available for work, membrane permeability, solute solubility, diffusion and enzymatic activity, and development of symbiotic organisms. The above-mentioned processes or reactions have their own characteristic response to temperature expressed as the Q_{10} value (see Chapter 4.4.2.1.5).

For plant tissue, the Q_{10} value lies in the interval of 1.5 to 3.0 and at temperatures below 23°C.[447] Most plant species need an optimum temperature for maximum root growth.[252,879] Table 22 shows the optimum temperatures for root growth of various species. The optimum temperature for root growth differs not only among the genera of plants, but also among species, genus, within a genus,[538,1153] and even among cultivars within a species.[190,409,622,861,1113]

The optimum soil temperature gets lower as the plants become older.[1328] The increase of temperature from 10 to 30°C usually results in decrease of root diameter and increase in color darkness, indicating advanced maturity or suberization of the tissue.[252] An example of the influence of soil temperature on root growth of white oak is shown in Figure 69.

Macduff et al.[777] studied the effect of increasing culture solution temperature on parameters of root growth of oilseed rape and barley. They found that temperature increased root extension and density of root hairs and decreased root radii. Similar results were reported by Agamuthu and Broughton[11] for oil palm roots. The temperature rise also resulted in a higher force exerted by roots.[472,473]

Cutforth et al.,[279] studying the effect of soil temperature and water content on root growth of maize, found a positive effect of both factors on root growth rate (Figure 70). The roots were thinner as temperature increased. Significant increase of root length of maize grown in soil resulted from the combined effect of temperature and phosphate level increase (Figure 71). High root temperature during flooding had a severe effect on root weight of alfalfa.[190] It also enhanced flooding damage of alfalfa and sainfoin roots, but did not influence bird's-foot trefoil roots.[537] Allmaras and Nelson,[25] investigating maize root configuration in row and interrow soil, showed the importance of soil temperature management in the row and water management in the interrow.

The combined effect of root temperature and NO_3^- NH_4^+ ratio on strawberry roots resulted

TABLE 22
Optimum Temperature for Root Growth

Plant	Direction of change	Optimum temperature (°C)	Ref.
Sunflower (*Helianthus* sp.)	Root elongation rate	20	418
Loblolly pine (*Pinus taeda*)	Root elongation rate	20	92
Pecan (*Carya illinoensis*)	Root elongation rate	30	1335
Tomato (*Lycopersicon esculentum* Mill.)	Root elongation rate	30	1293
Maize (*Zea mays*)	Root elongation rate	30	32 135
Cotton (*Gossypium hirsutum*)	Root elongation rate	33	48
Cotton	Taproot extension rate	25	1134
Oats (*Avena sativa*)	Root mass	5 (at maturity)	878
Kentucky bluegrass (*Poa pratensis*)	Root mass	15	179
Strawberry (*Fragaria* sp.)	Root mass	10—20	177
Rape (*Brassica napus*)	Root extension	About 23	849
Oil palms (*Elaeis guineensis* Jacq.)	Root growth	30—35	11
White oak (seedlings) (*Quercus alba*)	Root growth	24	720
White oak (*Quercus alba*)	Root elongation rate	About 17	1187
Lolium perenne	Dry weight	17	237
Rice (*Oryza sativa*)	Root growth	25—37	639
Maize (*Zea mays*)	Root mass	26	1272
Agave deserti	Root elongation	30	624

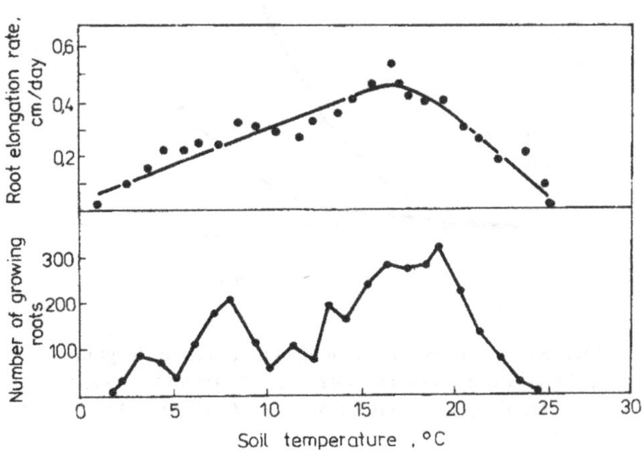

FIGURE 69. Effect of soil temperature of white oak. (Modified from Teskey, R. O. and Hinckley, T. M., *Physiol. Plant.*, 52, 363, 1981.)

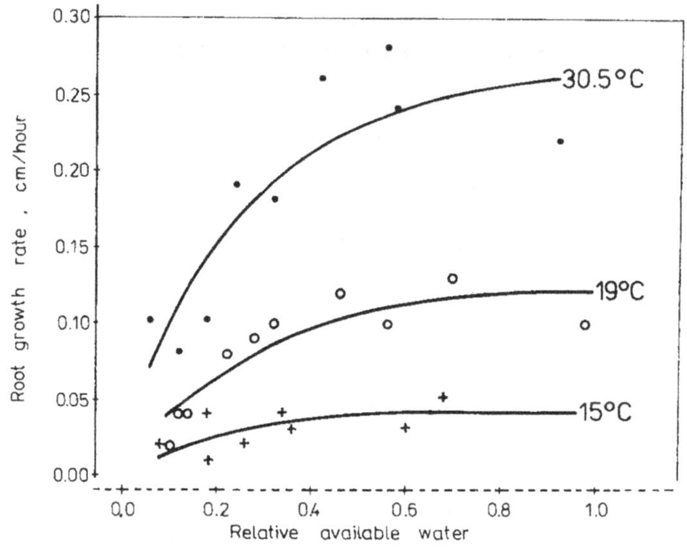

FIGURE 70. Root growth rate of maize as a function of temperature and relative available water. (Modified from Cutforth, H. W., Shaykewich, C. F., and Cho, C. M., *Can. J. Soil Sci.*, 66, 51, 1986.)

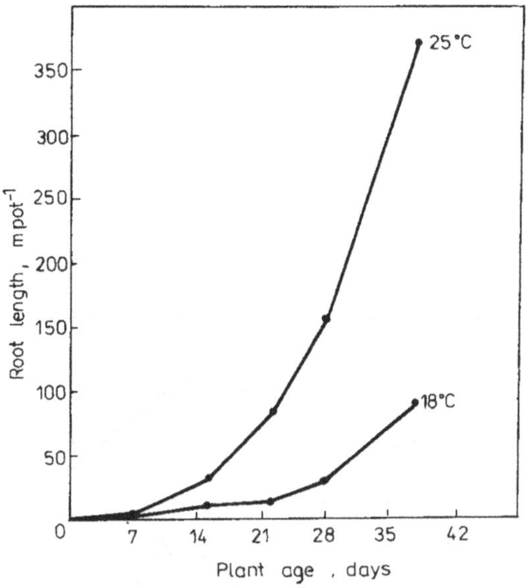

FIGURE 71. Effect of soil temperature on root length of maize grown in growth chamber. (Modified from Mackay, A. D. and Barber, S. A., *Soil Sci. Soc. Am. J.*, 48, 818, 1984.)

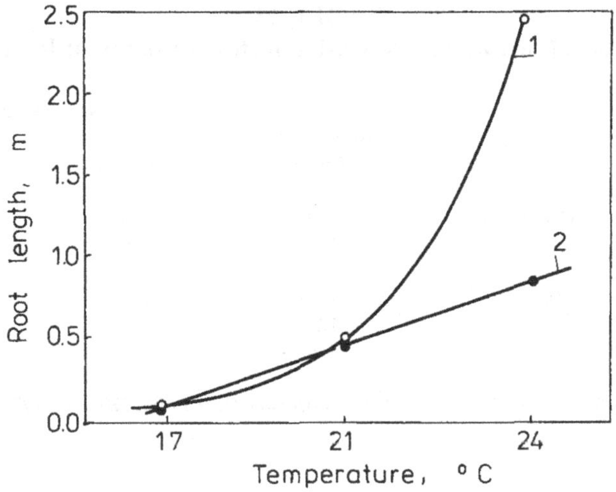

FIGURE 72. Root length of 6-d-old seedlings as influenced by temperature: (1) total and (2) axes. (Modified from Logsdon, S. D., Reneau, R. B., Jr., and Parker, J. C., *Agron. J.*, 79, 221, 1987b.)

in a general decline in root growth with increasing root temperature and increasing NH^+_4 fraction in the nutrient solution. This decline was followed by a decrease in the sugar concentration of the roots.[420]

The number of growing roots in relation to temperature did not change uniformly. The increase of soil temperature from 5 to 20°C positively affected the initiation and elongation of new roots on transplanted *Pinus radiata* D.[865] Low temperature had the most dramatic effect on maize seedling root elongation and root initiation compared to the influence of differentiated bulk density, moisture content, and ODR levels. After 6 d the total root length at 21°C was only about 20% of that at 25°C, whereas root length at 17°C was only about 5% of that at 25°C (Figure 72). Some desert succulents, e.g., *Agave deserti*, and *Ferocactus acanthodes* can grow in a wide range of lethal temperatures (+56 and –7°C).[624]

Root elongation rates for warm-season crops are reduced when these crops are grown during a cooler season and vice versa. Table 23 shows responses of wheat grown in winter and maize grown during the summer months in simulated conditions.

In general, warmer soil and longer periods of higher light intensities increased root elongation rates in these test conditions. Roots of crops evolved in tropical climates often elongate at their peak rates under warmer conditions than do crops characteristic of northern Europe which reach their peak rates at somewhat lower temperatures.

4.5.2 WATER UPTAKE

The most important root functions, namely, water and nutrient absorption, are temperature mediated. Temperature affects both passive and active water absorption by increasing permeability of the cell membrane and metabolic activity. As temperature increased from 13 to 28°C, the water absorption by rice roots was doubled.[1160] The increase of temperature from 5 to 22°C was accompanied by a sevenfold increase of water uptake by the pea.[178]

Frossard[409] found a significant increase of water content (Figure 73) in maize roots and the

TABLE 23
Effect of Season on Elongation Rates of Dominant Roots

Crop	Soil	Bulk density (Mg/m^3)	Root elongation Autumn	Root elongation Summer
Maize	Clay loam	1.10	3.9	6.3
		1.28	1.6	2.7
		1.44	0.6	0.7
Wheat	Clay	1.10	3.4	2.7
		1.28	1.5	1.3
		1.44	0.7	0.6

Modified from Trouse, A. C., Jr., *Zesz. Probl. Postepow Nauk Roln.*, 220, 349, 1979

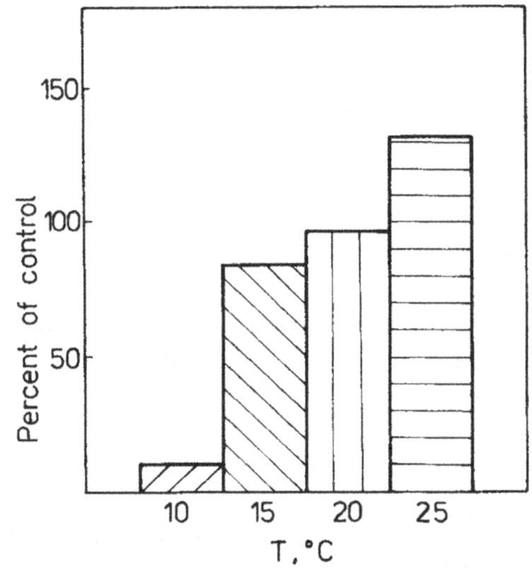

FIGURE 73. Water content (percent of control) in root of maize at different root temperatures. (Modified from Frossard, J. S., *Agronomie*, 5, 719, 1985.)

level of root respiration (Figure 74) when their temperature increased from 10 to 25°C. In the study of Borisova,[157] the increase of maize root temperature decreased hydraulic resistance and increased water uptake and exudation and extrusion of osmotically acting substances into xylem vessels. The root temperature also modified the water uptake by two tomato genotypes.[256]

4.5.3 NUTRIENT UPTAKE

Soil temperature significantly influences root absorption of available nutrients and assimilation of the nutrients into growth metabolites in the plant. Most studies on nutrient uptake by roots concern the uptake of nitrogen[237,238,256,421,422,515,726,1021,1274] and phosphorus.[237,461,464,849,928,1068] Few papers report the influence of root temperature on absorption of other ions, such as K, Mg, Ca, Fe, Mn, and Cd.[255,421,461,464,503]

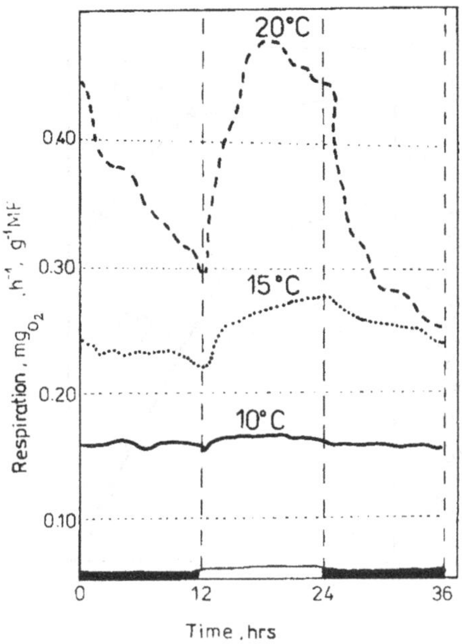

FIGURE 74. Root respiration vs. time at different temperatures. (Modified from Frossard, J. S., *Agronomie*, 5, 719, 1985.)

4.5.3.1 Nitrogen

It is a general tendency for nitrogen absorption to increase with temperature,[256] but this relation depends on the form in which the nitrogen is applied. Study of the influence of temperature on the uptake of NO_3^- and NH_4^+ suggests that NH_4^+ is absorbed more readily than NO_3^- at low temperature by roots of Italian ryegrass (Figure 75),[238] lettuce,[410] perennial ryegrass,[770] tomato,[419] strawberry,[421] barley,[776] wheat,[175] and sugar beets.[174] Ganmore-Neuman and Kafkafi[421] showed that this preference is not valid for the whole growing period.

4.5.3.2 Potassium

An increase of K content in plants with an increase in root temperature was found for tomato,[230,461,804] pepper,[464] ryegrass,[237] and peas,[178] but K content decreased in strawberry.[421]

4.5.3.3 Phosphorus

There is a general opinion that more P is required for growth when soil temperatures are below the optimum level.[213,470,879,1075,1076,1148] Moorby and Nye[849] found a decrease in phosphate inflow to rape roots and Clarkson et al.[237] found a decrease in phosphate inflow to ryegrass root at low (5°C) temperature as compared to higher temperatures. A similar response was found by Cornillon and Maisonneuve and Gosselin and Trudel for tomato[256,461] and for pepper.[464] Concentration of P in maize was not affected significantly by decreasing the root zone temperature from 25 to 15°C.[928]

4.5.3.4 Magnesium and Calcium

Root zone temperature did not affect Ca concentration in tomato leaves, but caused Mg increase with increase of the temperature from 12 to 24°C[254,461] and decrease at 30 to 36°C.[461]

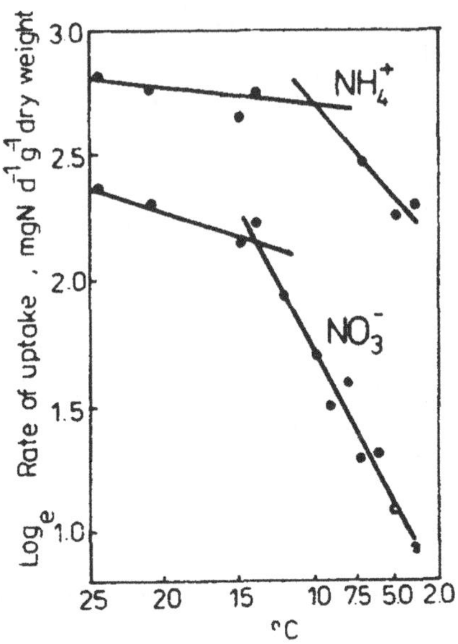

FIGURE 75. Relationship between rate of uptake of NH^+_4 or NO^-_3 by Italian ryegrass and temperature. (Modified from Clarkson, D. T. and Warner, A., *Plant Physiol.*, 64, 557, 1979.)

An increase of root zone temperature from 12 to 36°C reduced the concentration of Mg and Ca in pepper leaves[464] and reduced Mg in strawberry plants.[421]

4.5.4 CHEMICAL COMPOSITION OF ROOTS

The influence of temperature on the chemical composition of roots was studied by Cornillon.[254] The study concerned the content of fatty acids and free amino acids of tomato roots grown in nutrient solution at various temperatures (12 to 22°C). Crawford and Huxter[271] examined the influence of low temperatures: 2 to 14°C on the concentration of carbohydrates (fructose, glucose, and sucrose) in maize and pea roots.

4.5.5 ROOT NODULATION AND NITROGEN FIXATION

Root temperature has a marked effect on both the development and nitrogen-fixing activity of leguminous nodules.[286,339,414,515,726,748,861,1007] Figures 76 and 77 show the nodule fresh weight and the number of nodules as influenced by soil temperature. Maximum nitrogen fixation was at 24°C for *Medicago tribuloides* Desr. and *Vicia atropurpurea* Desf.,[922,923] at 27°C for *Vigna sinensis*,[287] at 22 to 26°C for *Trifolium subterraneum* L.,[439,440] and at 18 to 24°C for bird's-foot trefoil.[697]

The root temperature of 30°C appeared to decrease nitrogen fixation by nodulated *T. subterraneum*.[697,914] The effect of low root temperature on the efficiency of energy use in N_2 fixation of nodulated soybean roots was studied by Layzell et al.[725,726] and by Schubert and Evans.[1054] The low temperature delayed *Rhizobium* infection in *T. subterraneum*.[1007]

Some studies have been carried out on the effect of root temperature on the short-term variation in nodule activity (C_2H_2reduction).[352,915,1061] The relative efficiency [RE = 1 – (H_2 evolution/C_2H_2reduction)] of N_2 fixation varies widely with temperature, but there is a general tendency toward higher RE with lower root temperature.[726,915,967]

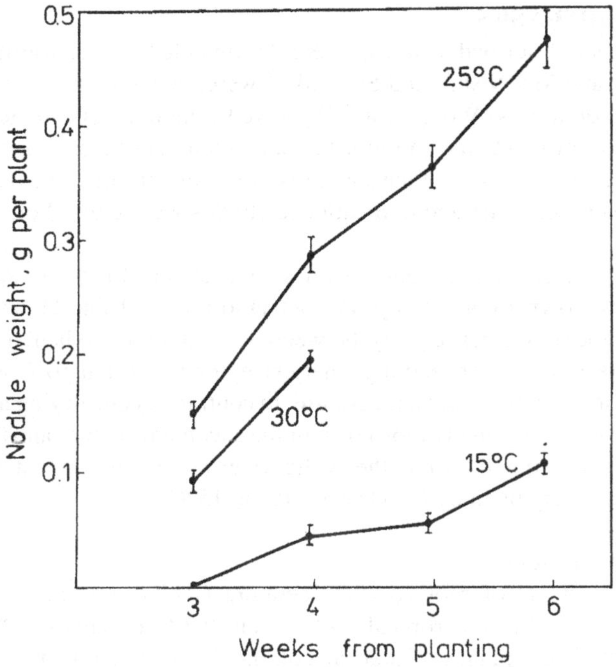

FIGURE 76. Total nodule fresh weight as influenced by soybean root temperature and stage of growth. (Modified from Lindemann, W. C. and Ham, G. E., *Soil Sci. Soc. Am. J.*, 43, 1134, 1979.)

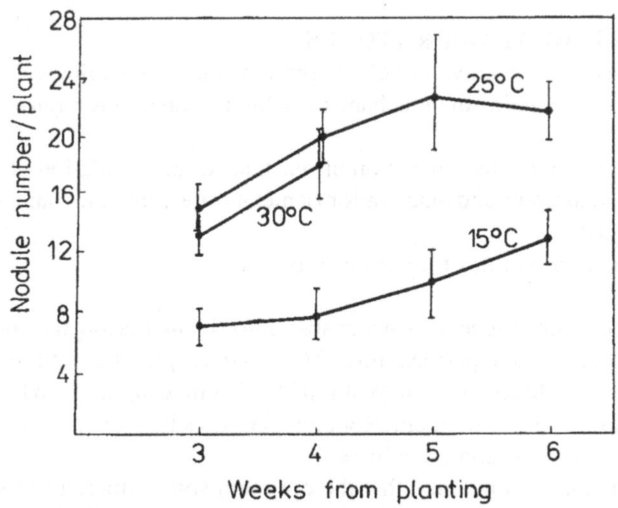

FIGURE 77. Number of nodules as influenced by soybean root temperature and stage of growth. (Modified from Lindemann, W. C. and Ham, G. E., *Soil Sci. Soc. Am. J.*, 43, 1134, 1979.)

4.5.6 ROOT ROT DISEASES

Root rots have been described as most severe during cold[659,1136] or warm[80,784] periods, and Kellock,[659] Barbetti and MacNish,[80] and Stovold[1136] were unable to associate root rot severity with soil moisture conditions. Wong et al.[1333] showed a number of interactions among 204 combinations of root rot incidents, soil moistures, and temperatures. Generally, root rot was most severe in cool soil, regardless of the soil moisture level, but some incidents caused severe root rot in warm, wet soils. The above-mentioned studies were carried out in pasteurized or artificial media.

Smiley et al.[1086] indicated that root rot on subterranean clover (*T. subterraneum* c.v. Woogenellup) plants grown in continually moist intact soil cores (Table 24) became more severe with increasing temperature, particularly between 10 and 15°C. When soils were dried to subwilting water stress soon after seedling emergence, the root rot at 10°C was higher than in the continually moist soil at the same temperature. In contrast, cyclic drying and wetting of soil led to less disease than in continually moist soil for the clay loam at 15°C and for the sandy loam at 20°C. In the clay loam, the disease in the cyclic water regime was less at 15°C than at 10 to 20°C. Maximum root weights in both soils occurred at 15°C.

4.5.7 ROOT ZONE HEATING

The comparative effects of root zone temperature (RZT) and shoot temperature were examined for growing, yield, and crop value of pearl millet,[485] tomatoes,[462,591,790,854,908] cucumbers,[463,669] peppers,[464,1028] eggplants,[1029] and calceolaria.[1292] Most results indicate that root zone heating was more beneficial than shoot heating. The economic data suggest that lower night air temperatures than those commonly used and higher root temperatures, related to growth phase, would increase the crop glasshouse production.[591]

Energy conservation in glasshouse production is nowadays of great importance. It is especially important for crops such as early tomatoes and cucumbers which require high fuel consumption. One approach to energy saving is to reduce air temperature and to increase the RZT because it is cheaper to heat the roots than the shoots.[1042]

4.5.8. HIGH AND LOW TEMPERATURES

Soil temperatures that are above or below optimum cause physiological stresses and injury for roots. Langridge[717] explains the mechanism of high-temperature injuries by:

1. Curtailed mass action due to exhaustion of reactants or accumulation of reaction products
2. Accelerated breakdown and inactivation of metabolites, enzymes, and other growth-regulating substances
3. Imbalance of reaction rates for various processes

A chilling soil temperature reduces root water absorption by increasing the viscosity of the water and decreasing cell membrane permeability.[274,683] An example of plant tolerance to high root temperature is given by Munevar and Wollum,[861] who investigated soybean genotypes and *Rhizobium* strains associated with them. Root temperatures had no significant effect on shoot-to-root ratios of lemon and orange seedlings.[1153]

Summarizing, it can be concluded that the optimum soil temperature for root growth and function is closely related to plant species and stage of development. Soil temperature lower than the optimum reduces absorption of nutrients and water by increasing the viscosity of the water and decreasing cell membrane permeability. Absorption of N in the form NH_4^+ is greater than that of NO_3^- at low temperature. Higher temperature causes the reduction of mass action and a breakdown of metabolic activity. The effect of soil temperature is greater on root growth at an

TABLE 24
Root Rot Index and Root Weight as Related to Soil Temperature and Moisture (A) of Sandy and Clay Loams

	Sandy loam		Clay loam		
Temperature (°C)	Continually moist	Dried and wetted	Continually moist	Dried and wetted	LSD (0.05)
		Root Rot Index (B)			
10	1.2	2.9	1.6	3.1	0.3
15	3.1	2.8	3.3	2.5	0.4
20	4.0	3.2	3.7	3.6	0.5
		Root Weight (mg/plant)			
10	3.0	2.9	2.4	2.8	0.2
15	3.7	4.1	3.4	4.0	0.1
20	1.6	1.7	1.7	1.5	0.2

Note: A, in continually moist soil, the soil water potential was 0 to –30kPa or dried and wetted (–500 to –50 kPa) cyclically; B, index $\frac{\Sigma[(\text{number of plants in each rating class 0 to 5}) \times (\text{value of rating class})]}{\text{Total no. of plants evaluated}}$

Data from Smiley, R. W., Taylor, P. A., Clarke, R. G., Greenhalgh, F. C., and Trutmann, P., *Aust. J. Agric. Res.*, 37, 633, 1986.

early stage of development. Plant roots are more sensitive to fluctuations in temperature than shoots. The response of roots to temperature is closely interrelated with other soil physical factors. The above is discussed in Chapter 4.7.

4.6 LIGHT PENETRATION THROUGH SOIL

The problem of the influence of light penetrating the soil on the root system is discussed. Tester and Morris[1188] paid attention to the role of light penetrating through the soil on the light-stimulated geotropic responses of roots which can be sensitive to very low influences.

4.7 INTERACTIVE EFFECTS OF SOIL PHYSICAL PROPERTIES ON ROOTS

It is impossible to consider the single effects of a given soil physical property on the roots without referring to other soil parameters. In this chapter some compound interactive effects of soil physical properties on roots are discussed. Most results reported in the literature deal with a combined effect of mechanical impedance and other physical factors (soil texture, soil moisture and aeration, pore size distribution, aggregate-size distribution, and temperature) on root growth and function.

4.7.1 SOIL MECHANICAL IMPEDANCE AND TEXTURE

The maximum elongation rate of sugarcane (*Saccharum officinarium*) roots grown in soil with strong silt-sized aggregations, occurred at a bulk density of 0.91 Mg m^{-3} while in clay soil

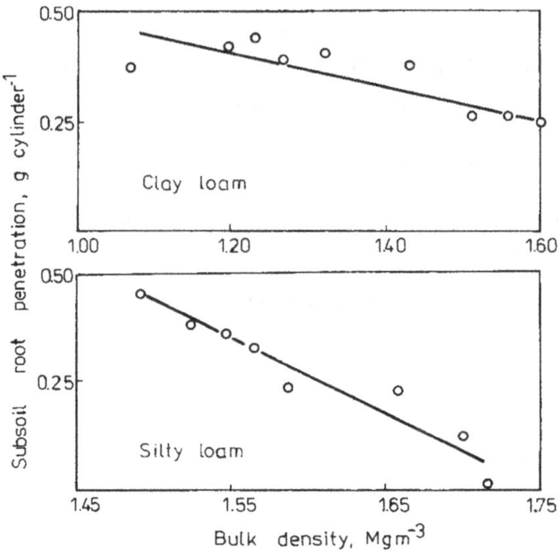

FIGURE 78. The relationship between bulk density of soil and subsoil root penetration. (Modified from Meredith, H. L. and Patrick, W. H., Jr., *Agron. J.*, 53, 163, 1961.)

it occurred at 1.1 Mg m^{-3}. Kar and Varade[639] reported that the maximum root growth of rice in sandy loam soil occurred at a bulk density of 1.6 mg m^{-3}, while in clay soil,[22] a drastic reduction occurred at 1.5 mg m^{-3}.

In coarse-textured soils, the significant mechanical impedance restricting root growth may result from the rough surface of the sand particles which resist particle displacement by slippage.[275] Such an effect on root growth of maize was reported by Chaudhary et al.[223] for loamy sand of relatively low bulk density. Another important feature of sandy soil for root penetration is the content of fine particles, organic matter, and cementing materials.[320,592,1128] Because of the low surface area of sandy material, a smaller percent of silt and clay is sufficient to cement sand grains than the same volume of finer particles. It was shown that the mechanical impedance of such mixtures was as high as that of silty clay loam with about 35% clay.[592]

Effects of compaction are usually less permanent in clay soils than in medium- and coarse-textured soils. The packing effect of clay soils, especially those containing the 2:1 type of clay minerals, can be alleviated due to swelling and shrinking forces which accompany hydration and dehydration. Montmorillonitic vertisols may alleviate previous compaction in a few days, while in micaceous clays, the swelling takes months.[536] Meredith and Patrick[821] studied the relationships between subsoil root penetration of Sudan grass and bulk density for two soils: swelling clay loam and nonswelling silty loam. It is interesting to note that the slope of the curves for this relationship was much steeper for the nonswelling soil (Figure 78). Therefore, compacted layers formed by heavy equipment most frequently occur in soils that do not swell and shrink upon wetting or drying. It is in agreement with the results of Kubota and Williams,[691] showing that adverse effects of soil compaction on root penetration were lower in heavy soil compared to light soil due to cracking in the former. The authors conclude that in clay soil, even though most of large pores may be destroyed by compaction, the soil is usually plastic enough for plant roots to penetrate readily if aeration is adequate.

The way in which the root growth of forest tree species[1279] upon soil compaction has been associated with the texture and origin of sandy soil is shown in Figure 79. Pine root growth was

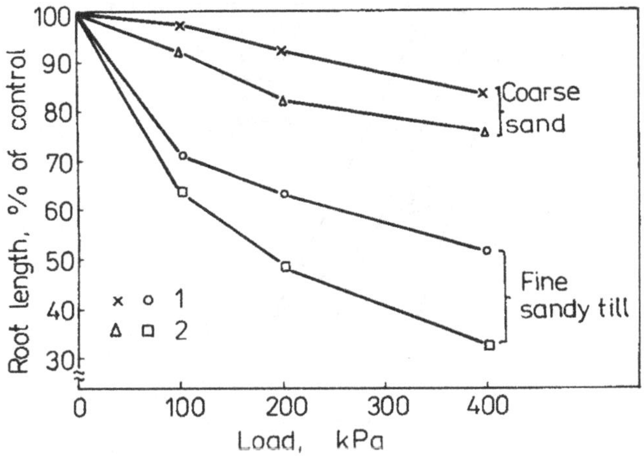

FIGURE 79. Root length of Scotch pine (1) and Norway spruce (2) seedlings at different levels of soil compaction. (Modified from Wästerlund, J., *For. Ecol. Manage.*, 11, 171, 1985.)

restricted at 1.4 and 1.6 Mg m^{-3} on silty loam and sandy loam,[349] respectively. To avoid an overestimation of critical bulk density for root growth in gravelly soil, the correction for pebbles and concretions should be made.[658] The influence of bulk density and other soil physical properties on boreal forest tree roots were studied by Strong and La Roi.[1137]

The literature on soil compaction effects on tree roots has been reviewed by Ruark et al.[1020] and Greacen and Sands.[477] Bowen[159] has pointed out that generally (with many exceptions) bulk densities of 1.55, 1.65, 1.80, and 1.85 Mg m^{-3} will severely impede root growth on clay loams, silt loams, fine sandy loams, and loamy fine sands, respectively. Heinonen[536] indicated that under Scandinavian conditions the "normal" bulk density appears to be a fair estimate of the optimum bulk density for cereal crops, but under more humid conditions, and by most dicotyledonous crops, the optimum bulk density would be lower. The "normal" bulk density can be determined by sampling the soils at field capacity after a moist period (to allow maximum swelling) from areas not tilled or heavily trafficked.

The above-mentioned results indicate that the optimum is shifting to a soil with a lower density when soil texture is fine instead of coarse. To eliminate the influence of soil texture and to make the comparison of the compaction effects easier between sites, use of indicators based on relative density of soil was suggested.[122,377,505] One such indicator was proposed by Eriksson et al.[377] and Håkansson[505] and it is called a "degree of compactness", i.e., a ratio of actual bulk density to that in a compacted, most dense state with a static load of 200 kPa (see Chapter 1). The method was tested over a 15-year period in an extensive series of field experiments.[505,990] The maximum crop yield was obtained at a similar degree of compactness irrespective of the soil type.

Bennie and van Antwerpen[122] proposed compaction classes to evaluate root-impeding characteristics of compacted soil layers. These classes, based on the degree of compaction being a ratio of bulk density minus minimum density and maximum minus minimum density, are as follows: <0.5, low degree of compaction; 0.5 to 0.6, medium degree of compaction; 0.6 to 0.7, high degree of compaction; and >0.7, very high degree of compaction.

Hemsath and Mazurak[540] showed that root elongation of sorghum in clay-sand mixtures with relatively low clay content was the highest, but as penetration resistance increased, root

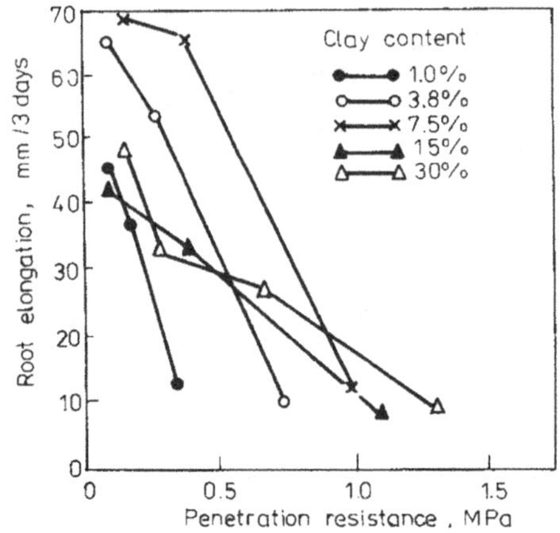

FIGURE 80. Root elongation of sorghum root in clay-sand mixtures for 33 kPa matric suction as related to penetration resistance (Data from Hemsath, D. L. and Mazurak, A. P., *Soil Sci. Soc. Am. Proc.*, 38, 387, 1974.)

elongation in these mixtures decreased more rapidly than in mixtures with higher clay contents (Figure 80). The ability of pea roots to penetrate a resistant clay has been studied by Champion and Barley[218] and Gerard et al.[434] The effect of gravels on maize root growth was reported by Babalola and Lal.[1152]

The critical soil strength, defined as the probe pressure at which root elongation stopped, was a function of percent of clay and ranged for cotton seedlings from 7 MPa in coarse-textured soils to 2.5 MPa in clay soils (Figure 81). It can be partly attributed to lower rigidity of pores in coarse-textured soils due to less cohesion. However, cotton taproot root penetration through medium- to coarse-textured soil materials was drastically reduced as soil strength increased to 2.5 MPa regardless of the soil texture.[1182]

Greacen et al.[465] pointed out that root penetration in coarsely textured or macroscopically structured soils cannot be predicted from values of probe pressure, q, measured with rigid probes even when mechanical forces are limiting. They ascertain that one of the main reasons why the apparent limiting values of q reported in the literature vary so enormously is the application of probe tests to soils that are too coarsely structured.

4.7.2 SOIL MECHANICAL IMPEDANCE, MOISTURE, AND AERATION

Usually when the soil moisture level decreases, there is often an increase in soil strength and reduced elongation.[477] Whether the observed decrease is due to the increase in soil strength, lowered moisture, or to combination of both is not clear. Some authors reported no effect of moisture per se on root elongation and have attributed reduced root growth to increasing strength,[85,360,476,1181] while in other studies, reduced elongation was related to greater moisture stress[102,834] or a combination of factors.[290,434,807]

Davis[290] studied the relative importance of soil moisture potential and soil strength on the root elongation of *P. radiata* D. Don in sandy loam soil and found that at higher levels of soil strength, the roots are more sensitive to soil moisture deficits (Figure 82). A similar relationship was found for root length of pea.[348] However, Muneer et al.[860] reported that at low bulk density of soil, soybean root length increased with moisture level, while at high-bulk-density soil, it decreased. These data agree with findings of Taylor and Ratliff,[1181] showing that at penetrometer resistance

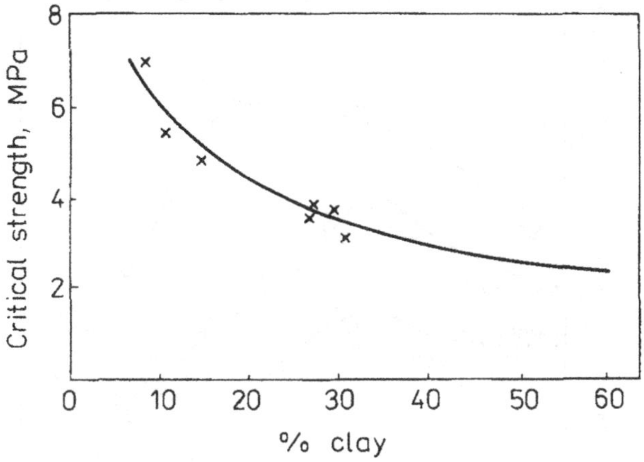

FIGURE 81. Relationship between soil strength which stopped root elongation and percent of clay. (Modified from Gerard, C. J., Sexton, P., and Shaw, G., *Agron. J.*, 74, 875, 1982.)

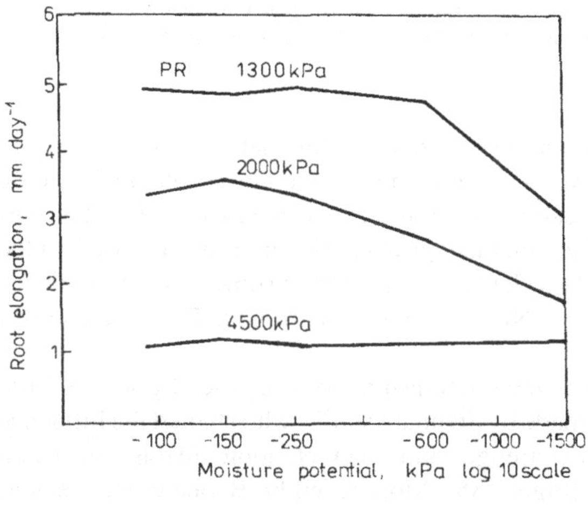

FIGURE 82. Pine seedling *radiata* root elongation with different levels of soil moisture potential and penetration resistance (PR). (Modified from Davis, G. R., in *Proc. Symp. on Site and Productivity of Fast Growing Plantations*, Vol. 2, South Africa, 1984, 871.)

as low as 5 kPa, root length of cotton was unaffected by water potentials between –17 and –70 kPa. The above-cited papers indicate that a change in moisture potential or water content has a much greater effect in a compacted than in a loose soil.

Phene and Beale[945] reported that the value of matric potential of sandy loam in the A_1 horizon at which soil strength limits root growth is –40 kPa and lower. A critical soil matric potential higher than –8 kPa would result in oxygen deficiency sufficient to retard plant growth in the soil. Eavis[347] considered the simultaneous effects of three factors: mechanical impedance, aeration, and water availability in restricting pea seedling root growth. The interactive effects are shown in Figure 83.

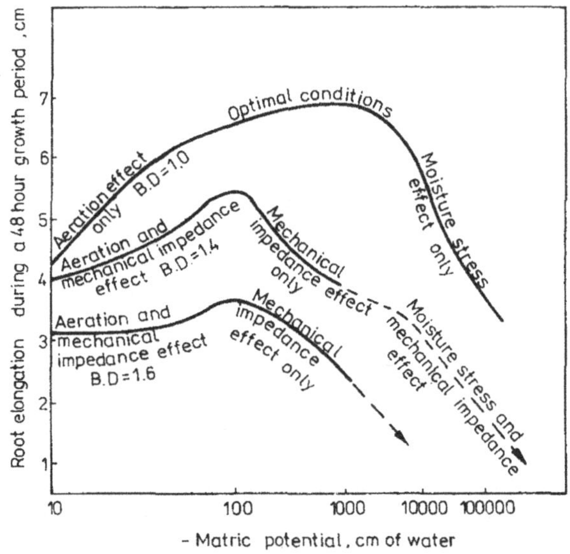

FIGURE 83. Role of mechanical impedance, aeration, and moisture stress on pea seedlings root elongation in a sandy loam soil held at different matric potentials and bulk densities. (Modified from Eavis, B. W., *Plant Soil,* 36, 613, 1972.)

Taking into consideration the relationship between soil physical properties and crop production, Letey[736] gives a modern approach to this problem (Figure 84). He also proposes a new indicator, the so-called "nonlimiting water range" (NLWR), which characterizes soil physical conditions affecting plant growth. This indicator is based on the data on soil aeration expressed as ODR and mechanical resistance in some soils. As bulk density increases and/or poor structure occurs, the NLWR becomes narrower. The principle of the above relations is shown in Figure 85.

A similar approach to the relationships between the physical soil properties important for plant growth has been made by Stepniewski,[1123] who studied the influence of the compaction of four Polish soils on ODR, strength, and water retention. He found critical values of the examined factors for these soils (Figure 86). More recently, Boone et al.[155] extended the approach and suggested critical limits of penetration resistance and aeration for root growth in relation to soil water potential. In this concept, root growth which allows optimal growth is supposed to be possible between the critical limits. The penetration resistance of 3 MPa is regarded as an upper critical mechanical limit (UCML) at which root growth is a function of the mechanical impedance or has completely stopped. As a lower critical mechanical limit (LCML), the penetration resistance of 1.5 MPa is proposed which causes a decrease in the root growth rate of 50%. Table 25 shows the values of matric potential corresponding to LCML and UCML for severely compacted sandy soils. The smallest difference in the potential between field capacity and UCML for fine sandy loam provides higher risks of suboptimal growth of roots and crop water supply.

These two trends confirm the data on the relationship between the specific root elongation rate and cotton and matric potential and between root elongation and soil impedance obtained in controlled conditions.[102] Recently, Busscher et al.[187] developed regression equations which relate changes of soil strength, bulk density, and soil water content between some measured

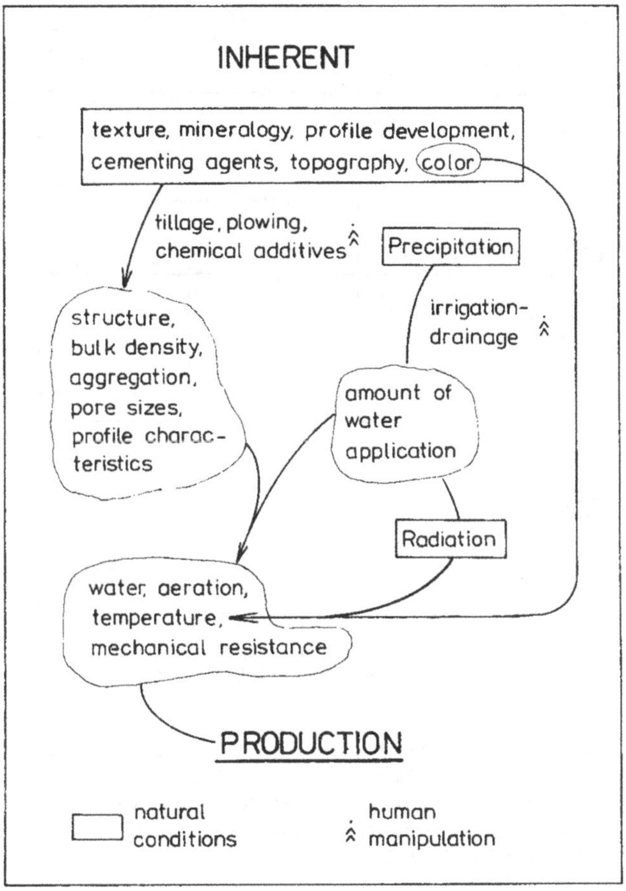

FIGURE 84. Relationship between natural conditions and human manipulation in affecting soil physical conditions directly related to crop production. (Modified from Letey, J., Ed., *Advances in Soil Science*, Vol. I, 1985, 277.)

value and the critical rooting conditions for soils at −100 kPa soil water potential. The equations can be used to estimate soil strength properties for critical rooting conditions.

Gerard et al.[434] conducted studies to determine the compound effect of soil moisture, density, clay content, voids, and strength on the root growth of cotton. They found that bulk density, voids, and clay content accounted for 76% of the variability in root growth. Barley and Greacen[86] indicate that when root growth is not restricted by soil physical properties, such as temperature and aeration, the main physical factor controlling root growth seems to be soil strength, which changes with bulk density and water content or matric potential.[295,835]

4.7.3 SOIL MECHANICAL IMPEDANCE, PORE SIZE DISTRIBUTION, AND AGGREGATE SIZE DISTRIBUTION

Influence of mechanical impedance on rooting depends to a high degree on the pore size distribution, especially of a diameter greater than that of the roots.[357,405,489,744,1141,1279] These pores are not detected by penetrometry measurements.[153,357,360,493,1024] Numerous studies indicated that

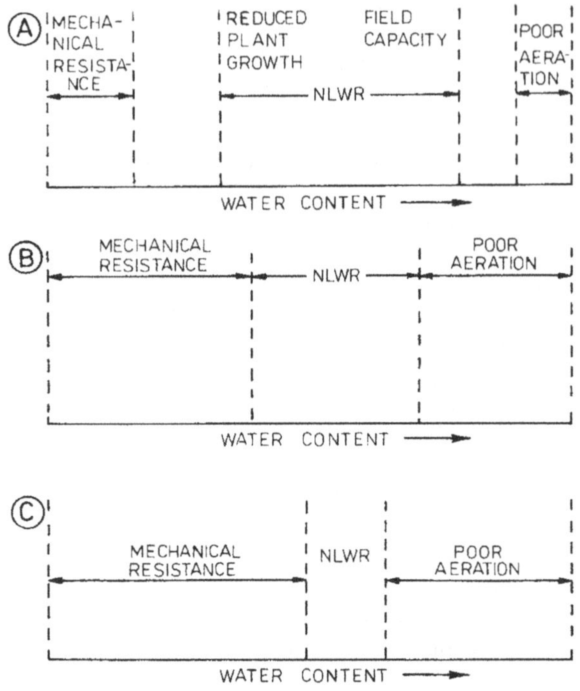

FIGURE 85. Generalized relationship between soil water content and restricting factors for plant growth in soils with increasing bulk density and decreasing aeration in going from case A to C: NLWR, non limiting water range. (Modified from Letey, J., Ed., *Advances Soil Science*, Vol. I, 1985, 277.)

a soil matrix with a larger pore size (structural cracks, macropores, and worm holes) will offer greater potential for undisturbed root growth because the roots can by pass the zones of high mechanical impedance, but on the other side, the weaker root-soil contact area in the large pores may cause a decrease of water absorption capacity of the roots[1166] unless the water supply into the roots is satisfactory.[526,753] The presence of pores equal to or larger than the growing root tips increased the limiting value of mechanical impedance for root growth.[153,154,357,360,368,489,1140]

Stypa et al.[1140] reported that corn root growth in the field was not reduced by subsoil bulk density of 1.5 Mg m^{-3} compared with an artificial medium that provided little mechanical impedance. It is suggested that natural structural cracks and biopores permitted root development with minimum restriction. The preferential root penetration into the zones of weakness in forest soil was demonstrated for *radiata* pine[1041] and for spruce.[1148] The increase of volume of voids[434] and macropores (pores of diameter of >50 μm)[457] enhanced root growth and activity, but they were impaired when their continuity was reduced and they became water-filled. The soil with connected voids which are oriented parallel to compaction forces in the soil profile would show better gas diffusion values besides forming channels through which roots can penetrate to the subsurface compacted layer better.[368,753,1105,1205]

In the study of Voorhees,[1259] the root elongation rate of pea seedlings in compacted soil increased with increasing air-filled porosity for 0.1 and 0.5 MPa of the normal point resistance. It is suggested that it may result from a gaseous concentration gradient across the compacted soil

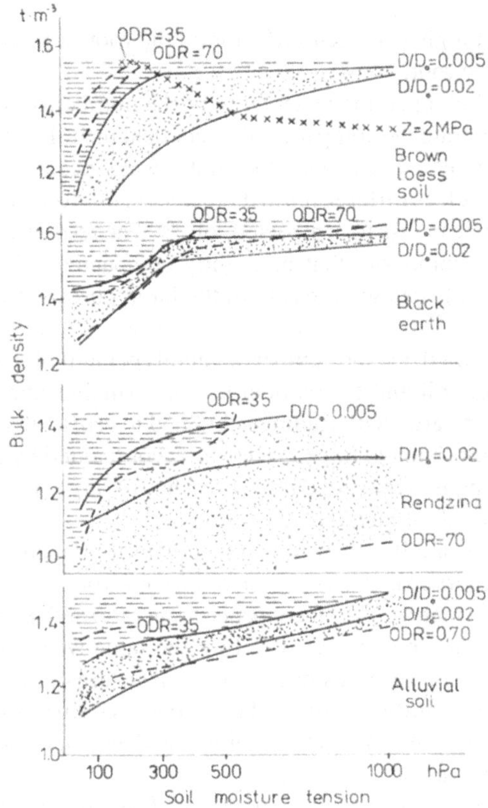

FIGURE 86. Ranges of soil bulk density and moisture tension corresponding to the boundary values of ODR, D/D_o, and resistance to penetration (Z) found in four soils. (Modified from Stepniewski, W., *Polish J. Soil Sci.*, 14, 89, 1981.)

TABLE 25
Matric Water Pontential ($-\psi_m$) at a Cone Resistance of 1.5 (LCML) and 3.0 MPa (UCML) and Corresponding Decrease in Volumetric Fraction of Water (Dq) Between Field Capacity (ψ_m = –80 hPa) and This Matric Water Potential, Respectively, For Severely Compacted Soils

	LCML		UCML	
Soil	**$-\psi_m$ (hPa)**	**$\Delta\theta$ ($m^3\ m^{-3}$)**	**$-\psi_m$ (hPa)**	**$\Delta\theta$ ($m^3\ m^{-3}$)**
Fine sand	160	0.029	1000	0.143
Loamy fine sand	160	0.008	1600	0.136
Fine sandy loam	130	0.015	500	0.073

Note: LCML, lower critical mechanical limit; UCML, upper critical mechanical limit.

Modified from Boone, F. R., van der Werf, H. M. G., Kroesbergen, B., and Boers, A., *Neth. J. Agric. Sci.*, 34, 155, 1986.

near the root surface and higher oxygen diffusion to the root or increased ethylene or carbon dioxide diffusion from the root.

As the direct effect of compaction is decrease of the number of large pores, the fine roots might be expected to penetrate the compact soil easier than the thicker ones. This was observed by Edwards et al.,[354] but such a response does not always occur[454,1069] and interpretation of various responses is difficult as differences in diameter are confounded with species, which for other reasons may vary in response to compacted soil.[1069] The response also depends on the level of compaction. Boone[151] points out that in slightly compacted soil, the root growth rate is relatively high because roots can grow easily in the large pores which still exist between the aggregates.

Boone and Veen[156] reported that the number of laterals per unit of main axis decreased as both mechanical impedance of soil and the number of pores smaller than the diameter of laterals increased. The data do not agree with findings obtained in artificial substrates (Ballotini glass beads) where increased mechanical impedance was accompanied by increased lateral branching.[83,455,1056] Different responses may reflect the presence of larger pores in artificial substrates, allowing unrestricted growth of laterals.

The influence of pores smaller than roots on rootability depends on soil wetness during growing season. It was observed that in compacted soil some roots were able to widen the narrow pore necks by plastic deformation.[136,153] Root penetration ratio and length declined linearly with decreasing air-filled porosity in compacted soil for the same water potential.[50] Root response was attributed to the decrease of pores having a diameter greater than the roots. The above-discussed papers indicate a significant role of the soil pore system in studies on the relationships between mechanical impedance and root penetrability and the importance of the factor increases as bulk density of soil increases.

Stypa et al.[1140] point out that these pores are usually formed in the field rather than in short-term laboratory experiments. Artificially made holes greater than the root diameter in the compact soil layer can also facilitate root penetration in laboratory conditions.[309,753,943] Generally speaking, the proportion of roots entering the holes in the compacted layer increased as their diameter and density increased.

However, the present methods for studying the pore system have not been well suited to evaluate the mechanical aspects of soil rootability.[151,155] Most limitations are related to pore size range or pore geometry. The highest matric potential applicable is –0.1 kPa and represents all pores with a diameter of >300 μm. Thin sections may provide more essential information on the system of these large pores, especially in compact soil with a limited number of these pores, but indirect physical methods which are often used for measuring the equivalent pore size distribution are not appropriate for the largest pores and results are highly affected by hysteresis.[155,1267] Real pore sizes, which are accurately measured with morphometric techniques, can only be estimated in sandy materials by means of physical techniques, while more important characteristics, such as continuity and direction of the pores and their spatial variability, however, are more cumbersome to measure quantitatively.[155,1027] None of these methods indicate directly any ease with which fine pores can be enlarged.[1027] The limitation is more important in compacted soil or in relatively undisturbed soil after cultivation or direct drilling.

The effect of the aggregate size (0- to 1-, 1- to 2-, 2- to 3-, and 3- to 6-mm diameter) of silt loam soil on maize root impedance was studied by Logsdon et al.[756] They developed an equation to describe root pressure necessary to avoid deflection around an aggregate. This pressure was predicted to increase with increasing aggregate size, decreasing diameter, and decreasing deflection angle of roots.

4.7.4 SOIL MECHANICAL IMPEDANCE, TEMPERATURE, AND MOISTURE

The temperature of the mechanically impeded root zone affects the root pressure and root

TABLE 26
The Influence of Bulk Density and Temperature of Soil on Dry Weight of Rice Roots

Bulk density (Mg m^{-3})	Dry weight of roots — Soil temperature regime (°C) 27—15	37—25	42—30	LSD (1%)
Clay				
1.0 (low)	0.79	1.54	0.52	
1.2 (medium)	0.87	1.18	0.60	0.09
1.4 (high)	0.50	0.78	0.36	
Sandy Loam				
1.6 (low)	0.62	1.44	0.45	
1.8 (medium)	0.72	0.87	0.56	0.10
2.0 (high)	0.37	0.50	0.28	

Modified from Kar, S. and Varade, S. B., *Agron. J.*, 68, 23, 1976.

size. Pearson et al.[932] pointed out that the response of root elongation to soil temperature decreased as the mechanical resistance of the soil increased. Greacen,[473] reviewing the results of maximum axial pressures, found that a 10°C rise in temperature from 20°C resulted in double the pressure exerted by pea roots.

Table 26 shows that the dry weight of rice roots was maximum at the submerged temperature regime of 37 to 25°C and was significantly lower at 42 to 30 and 27 to 15°C. The root growth at low (27 to 15°C) and high soil temperature regimes was significantly improved by compacting the soil to medium density (1.2 and 1.8 Mg m^{-3} in clay and sandy soil, respectively).

The effect of soil temperature and soil water potential on root growth of the white oak was examined by Teskey and Hinckley,[1187] who used a rhizotron. They found that the root elongation rate was linearly correlated with soil temperature. At soil temperatures lower than 17°C, temperature was the dominant factor effecting rate of growth, but at higher temperatures, soil water became an important factor.

4.7.5 MODELING OF INTERACTIVE EFFECTS

A computer simulation model of the growth of winter wheat root, based among others on soil temperature, soil moisture tension, and soil strength, was developed by Porter et al.[956] (Figure 87). The model of yield formation of storage organs, fibrous roots, and green organs of sugar beets based on climatic and soil conditions (water state, texture, and nitrogen content) was proposed by Hoffman.[558] Mettauer et al.[824] examined the influence of physical, mechanical, and nutritional soil factors on sugar beet morphology and root yield. They found that the growth in length was essentially conditioned by soil mechanical conditions (structural stability and sensitivity to compaction); the growth in thickness was much more related to nutritional factors and soil moisture (Figure 88).

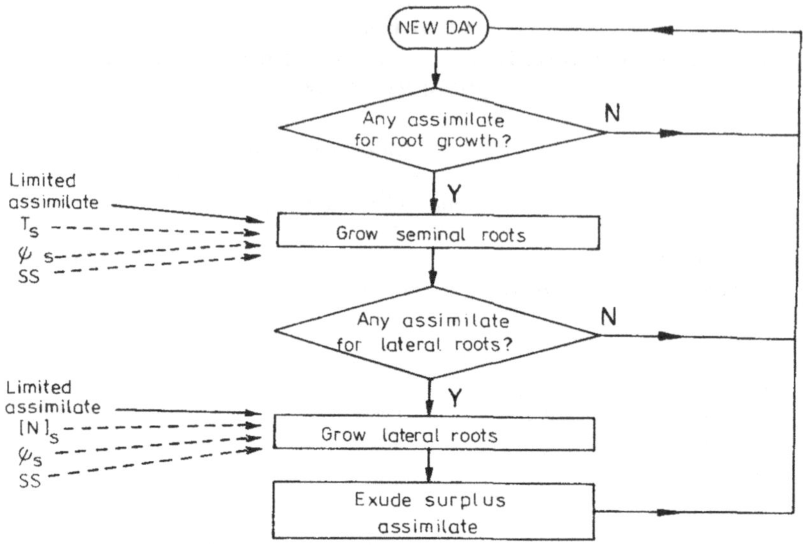

FIGURE 87. Flow diagram for the winter wheat root growth submodel: solid arrows indicate the principal driving variable; dashed arrows, modifiers; T_s, soil temperature; N_s, soil nitrogen concentration; ψ_s, soil moisture tension; and SS, soil strength. (Modified from Porter, J. R., Bragg, P. L., Rayner, J. H., Weir, H. H., and Landsberg, J. J., in *Opportunities for Manipulation of Cereal Productivity*, Monogr. British Plant Growth Regulator Group, 1981, 97.)

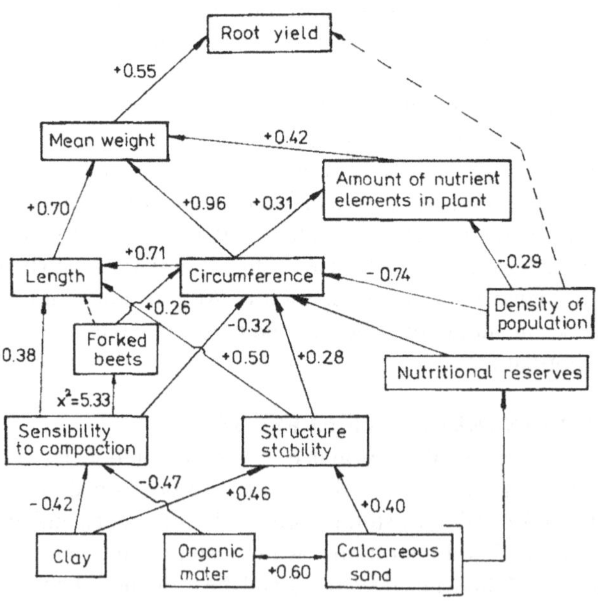

FIGURE 88. System for sugar beet root morphology and yield as related to soil conditions. (Modified from Mettauer, H., Tual, Y., Haeflinger, R., Conesa, A. P., and Trendel, R., *Ann. Agron.*, 29, 147, 1978.)

Chapter 5

INFLUENCE OF ROOT SYSTEMS ON SOIL CONDITIONS

5.0 INTRODUCTION

If much of the literature is concerned with the influence of soil physical properties on root growth and functions, very little research has been conducted to describe the changes in soil properties during the growing of crops. However, it is known that plant roots alter the chemical, physical, and biological nature of the soil adjacent to the root. They affect pH and the distribution of ions in the soil, both because of their physical presence and because of their absorption of nutrients and water. These phenomena introduce changes that occur in the rhizosphere zone which are described in detail by Barber[78] and Nye.[889] Plant roots influence such physical properties as structure stability, bulk density, porosity, soil mineralogical and mechanical composition, water and ion characteristics of soil, and also biological activity of the soil.

5.1 STRUCTURAL STABILITY

The role of plant roots in soil structure formation consists generally in soil aggregation. Roots and root hairs penetrating the soil produce lines of weakness along which the clod or soil mass may break into granules (primary elements). Also the pressure exerted by developing roots may induce aggregation. Root secretions may flocculate colloids and stabilize or cement aggregates. Moreover, the use of moisture by roots may cause dehydration of colloids, thus resulting in shrinkage and finally in cementation of soil aggregates.

The effect of living roots on soil structure stability, apart from the nature of the mineral substrates and climatic conditions, depends on the plant species. Lucerne and ryegrass[978,1192] improve structural stability of soil by increasing polysaccharides in the rhizosphere. Kobayashi and Otake,[673] studying the physical interaction of forage roots (timothy and upland rice) with grassland soil, found a remarkable increase of soil hardness and solid ratio. The increase in aggregate stability by maize, soybeans, and wheat roots observed by Monroe and Kladivko[848] was attributed to the physical entanglement of aggregates by roots and to the increased production of root exudates resulting from increased root growth. Maize roots, on the other hand, decrease structure stability of soils by chelating iron and aluminum, thus destroying chemical bonds with organic matter.

The effects of plant roots on the physical and physicochemical environment of calcareous soil are presented by Callot et al.[188] and Jaillard.[609] Root activity in such soils is connected with the formation of water flux toward the root and acidification of root environment. These phenomena form various types of soil structure (Figure 1) due to dissolution, movement, and reprecipitation of soil carbonates.

Extending roots are able to displace individual soil aggregates from their path in a loose seed bed. Whiteley and Dexter,[1299] based on the previously determined values for (1) the nonlinear elastic bending properties of plant roots,[1302] (2) the maximum buckling stresses of roots,[1301] and (3) the force displacement relationship for soil aggregates in loose beds of similar soil aggregates,[1297] presented a model for predicting the extent to which soil aggregates in a loose seed bed can be displaced by downward-growing seminal roots. They concluded that displacement of soil aggregates which lie between 20 and 100 mm from the surface of the aggregate bed is only likely to be important for roots with diameters of 0.5 mm or less when the diameter of the aggregates in the bed is less than 1 mm. However, for plant species which have relatively

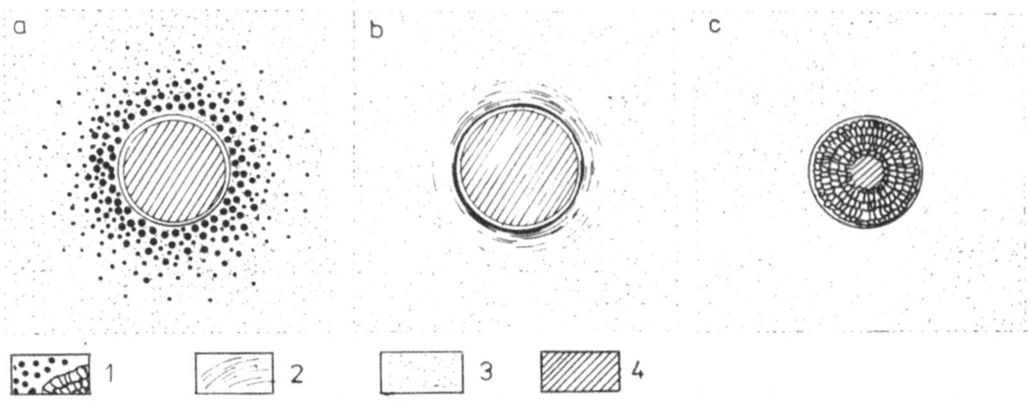

FIGURE 1. Different types of structure connected with root activity in calcareous medium. (a) Structure of accumulation near root surface; (b) structure of impoverishment near root surface; (c) structure of interrooting accumulation: (1) $CaCO_3$, (2) noncalcareous residue, (3) calcareous matrix, and (4) root. (Modified from Jaillard, B., *Zesz. Probl. Postepow Nauk Roln.*, 312, 229, 1986.)

large root diameters, such as the pea, significant displacement of aggregate of up to 4 mm in diameter may be possible. Roots of plants producing mucilages can form bridges between primary clay particles themselves and between these particles and the root surface, forming the so-called root soil aggregates[496,672,1191] during the process of soil desiccation and moistening (Figure 2).

5.2 BULK DENSITY AND SOIL STRENGTH

Roots growing in the soil occupy space that was previously occupied by soil pore space and soil particles. Since root diameters are usually larger than soil pores (root diameters of 0.1 to 3 mm and soil pore diameters of 0.002 to 0.2 mm), soil particles are pushed aside and the bulk density of the soil near the root increases.[78] The effects of the above depend on the the root diameter and the nature of the soil. Greacen et al.[474] showed increases in the bulk density of the soil next to the roots of 1.6 and 1.7 Mg m^{-3} from an initial level of 1.5 Mg m^{-3}.

Ruark et al.[1020] reported that the secondary growth of roots caused the increase of bulk density within the area of 1 to 2 mm. The area is characterized by a higher incidence of oriented clay and a higher percent of elongated grains oriented within 30° tangent to the root surface. Through the channels, water and air movement occur. Braunack and Freebairn[172] found that maize appears to compact the soil to higher densities than either wheat or pea roots (Figure 3). Under some circumstances, roots of certain plants increase soil strength. Waldron[1270] and Waldron and Dakessian[1271] showed that plants such as pine, oak, lucerne, and a range of grasses increase shearing resistance of soil. Willatt and Sulistyaningsih[1320] found that the root of rice increased both shearing resistance and bearing capacity of a loamy soil (Table 1). A higher proportion of clay and organic matter in the volume near the root, due to differential packing of soil particles, than at some distance from the root is found.

Barley[82] reported that maize roots compacted a synthetic sandy loam soil between the root channels and greatly reduced soil permeability. Also, local consolidation of soil by the pea was reported by Cockroft et al.[245] The influence of the stage of growth of a root crop (radish) on the bulk density of fine sandy loam and clay soils was examined by Gerard and Mehta.[431,432] They

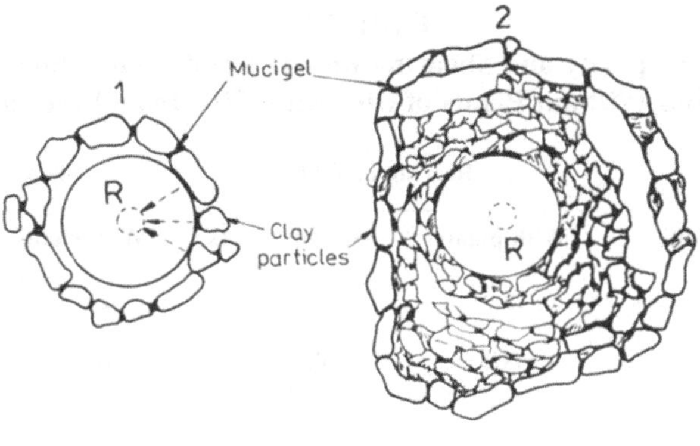

FIGURE 2. Model of aggregates formation at root surface: (1) junction of clay particles by mucigel bridges at the root (R) surface; (2) root aggregate. (Modified from Tinker, P. H. B., *Philos. Trans. R. Soc. London Ser. B*, 273, 445, 1976.)

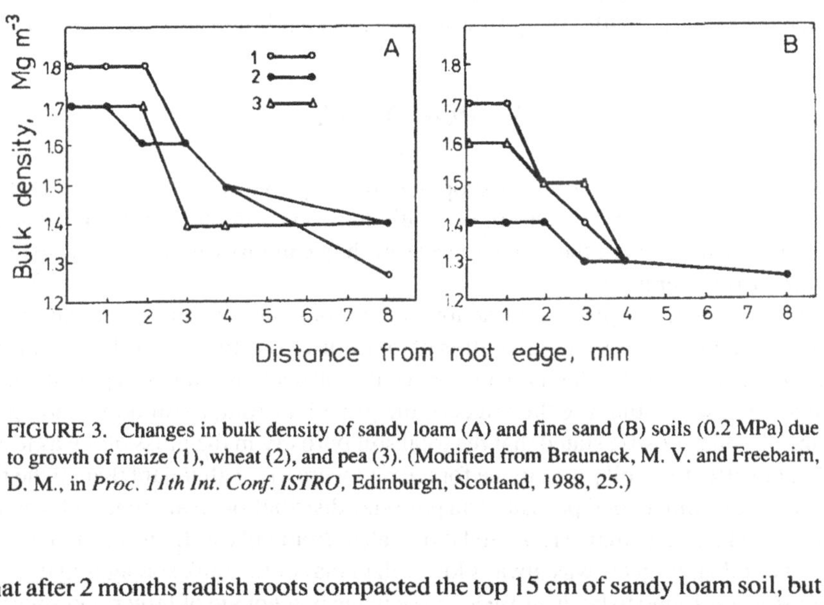

FIGURE 3. Changes in bulk density of sandy loam (A) and fine sand (B) soils (0.2 MPa) due to growth of maize (1), wheat (2), and pea (3). (Modified from Braunack, M. V. and Freebairn, D. M., in *Proc. 11th Int. Conf. ISTRO*, Edinburgh, Scotland, 1988, 25.)

found that after 2 months radish roots compacted the top 15 cm of sandy loam soil, but had little influence at the 15- to 39-cm soil depth. However, after 4 months, the enlarged radish roots decreased the density of the top 15 cm of soil, but compacted the soil at 15 to 30 cm of soil depth. Bulk density of clay soil decreased during the whole growing season and was higher at 15 to 30 cm than in the top 15 cm soil.

Several authors have pointed out that root penetration, especially by certain taprooted plants, e.g., Pigeon pea, *Leucaena, Festuca arundinacea, Pensacola bahiagrass,* robust wild plants, and *Stylosanthos guianeasis,*[364,389,536,632,989] can perforate compact layers and create easily accessible pathways for the roots of succeeding crops. On the above basis, proposals are made to screen suitable species and cultivars for their ability to penetrate and loosen compact semirigid soils[536,708,709] and therefore to become potential tillage tools.

TABLE 1
Bearing Capacity and Shearing Resistance of Loamy Soil (kPa) With and Without Plants of Rice After 70 d from Emergence

	Bearing capacity		Shearing resistance	
Irrigation treatments	With plants	Without plants	With plants	Without plants
Puddled soil irrigated each week with 30 mm of water	73.3	15.1	2.6	1.5
Nonpuddled soil irrigated each week with 30 mm of water	395.4	193.7	10.6	5.1

Modified from Willatt, S. T. and Sulistyaningsih, N., in *Proc. 11th Int. Conf. ISTRO*, Edinburgh, Scotland, 1988, 183.

A simplified model for soil compression around plant roots has been developed by Dexter.[311] The main assumptions of this model are that root volume is accommodated by the loss of porosity in the surrounding soil; that there is minimum soil porosity below which soil will not be compressed; and that the density decreases exponentially with distance from the root surface with an exponent which is a constant multiple of the root diameter. In addition, plant roots which are characterized by high tensile strength act against a landslide and thus increase soil resistance to erosion.[2,1047]

5.3 POROSITY

Plant roots penetrating soil leave macropores that improve water movement and gaseous diffusion. They contribute to the system of continuous pores in the soil. Roots of successive crops can either grow through these macropores or they can grow in the compacted soil itself because of improved aeration.

Due to the death of the appropriate fraction of the root system before plant maturity and the decrease of root diameter through degradation of their cortex due to invasion by microorganisms or dessication (see Chapter 3.1) their contact area with soil decreases and soil porosity increases. Kobayashi and Otake,[673] studying the effect of the growth of forage (timothy) and upland rice root systems on soil porosity, stated that the total soil porosity in the grass plot was less than in the rice plot, but effective soil pores in the forage plot were more abundant than in the rice plot. Guidi et al.[496] examined total porosity and pore size distribution in aggregates taken from the sandy loam soil adherent to maize roots and those taken from bulk soil. They found that the total porosity of the soil aggregates was always lower than that of the bulk soil aggregates. The loss of total porosity was found to be due to a decrease in the percentage of larger pores with respect to the total (Figure 4). Barley[82] and Sedgley and Barley[1063] found that grass roots growing through sandy loam affected pore size distribution of the soil without changing total porosity. Summing up, it is possible to conclude that some plant roots improve soil structure in the plow layer and some have the capacity to grow through compacted layers and improve soil below the plow layer.

5.4 MECHANICAL AND MINERALOGICAL COMPOSITION

The modifications of particle size distribution and composition of the clay fraction within the rhizosphere around French bean roots have been studied by Sakkar et al.[1035] The pot experiment

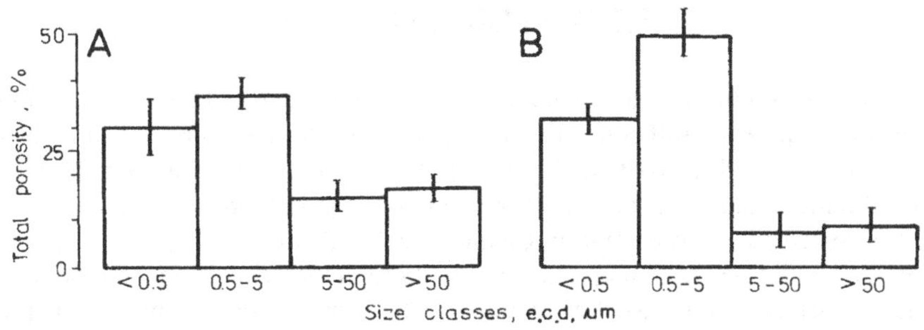

FIGURE 4. Pore-size distribution of bulk (A) and root (B) soil aggregates; e.c.d., equivalent cylindrical diameter. (Modified from Guidi, D., Poggio, G., and Petruzzelli, G., *Plant Soil*, 87, 311, 1985.)

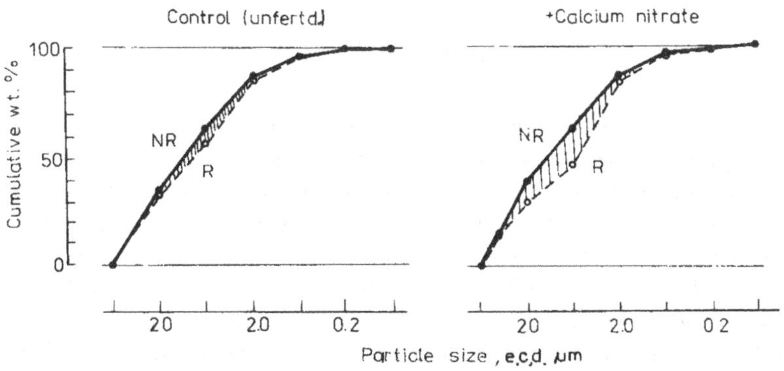

FIGURE 5. Mechanical analysis of rhizosphere (R) around French bean roots and nonrhizosphere (NR) silt loam samples (<63 µm) unfertilized and fertilized with calcium nitrate, after 6 weeks of experiment; e.s.d., equivalent size diameter. (Modified from Sakkar, A. N., Jenkins, D. A., and Wyn Jones, R. G., in *The Soil-Plant Interface*, Harley, J. L. and Russell, R. S., Eds., Academic Press, London, 1979, 125.)

was carried out with silty loam soil in which clay mineralogy was dominated by hydrous mica and chlorite, derived from the parental Ordovician shales, together with minor amounts of vermiculite and interstratified material. French beans were planted for 6 weeks, and after this time, "rhizosphere" samples were taken for mechanical and mineralogical analysis.

A consistent shift in the mechanical composition of the fraction <63 µm between the rhizosphere and nonrhizosphere soil samples was found. This involves a relative increase in the finer fractions (<6.3 µm) of 25 to 35% and of the fine silt (6.3 to 2.0 µm) in particular, at the expense of the medium and coarse silt fractions (Figure 5). It is explained by an accelerated disaggregation of compound shale fragments into their component silt/clay particles.

The mineralogical changes were manifested most clearly in the finest clay fractions (<0.0063 µm) and involved a decrease in both regularly and irregularly interstratified 0.0010- to 0.0014-µm material: "amorphous" (extractable) Al, Fe, and C accumulated in the rhizosphere, while Si was depleted. It was due to intensified weathering of soil materials in the vicinity of the plant root.

5.5 WATER AND IONS

The uptake behavior of roots changes water and ion movement and the accumulation profile in the rhizosphere differing salt concentration between the "rhizosphere soil" and the original soil. The proportion of dead roots and channels formed after root death results in a higher saturated hydraulic conductivity of soils.[941] Water uptake by roots and the water profile near the rhizoplane is described in more detail in Chapter 2.3.1 and 2.3.2.

The root system influences ion uptake directly through their flux in aqueous solution to the root surface by diffusion and mass flow induced by the transpiration stream (see Chapter 2.3.2) and also by modification of the rhizosphere pH (see Chapter 3.2). If water together with ions and other solutes arrive at the root faster than they are absorbed, their accumulation in the rhizosphere increases, increasing the osmotic potential. It causes the diffusion of ions away from the root until a steady state is reached when diffusion away just balances arrival by mass flow. Preferential uptake of either ions or water leads to depletion or accumulation profiles of ions. Examples of this are the depletion zones of phosphorus and potassium[629,1064] or the accumulation in the rhizosphere of calcium,[791] and the accumulation of sodium and chloride[1048] in saline soils.

5.6 BIOLOGICAL ACTIVITY

As described in Chapters 2.3 and 3.3, roots of numerous plant species are embedded in mucilaginous layers, released into the soil by outer root cap cells, derived from epidermal cells located behind the tip or produced by root-colonizing microorganisms. These compounds are supposed to display some properties of the pectin gels as well as to have hydrophylic characters. They affect soil aggregation (see Chapter 5.1) and protect plants against heavy metal toxicity (see Chapter 3.3). Hendrickson and Robinson[541] found that the products of the fine roots of forest trees, including exudation, are the major influence upon carbon and nitrogen mineralization rates in forest soil; a model of mineralization representing the available and humified carbon pools has been proposed.

Chapter 6

MODIFICATION OF ROOT DEVELOPMENT

6.0 INTRODUCTION

As was indicated in Chapter 4, plant root development is seemingly dependent on the soil physical characteristics of the root zone. These characteristics in turn are mostly influenced by tillage and soil and plant management practices. The influence of soil tillage on root development has been known for a long time.[450] Changes in physical properties and responses of roots due to tillage are markedly dependent on the soil type, plant species, stage of crop development, and weather and climate conditions.[65]

6.1 CONVENTIONAL AND REDUCED TILLAGE

Tillage is the most obvious way to modify soil structure deleterious to the root growth. Numerous studies have been conducted on root growth and distribution as affected by reduced tillage compared to conventional tillage. The trend toward zero or minimum tillage is increasing in many countries with concern for soil and energy conservation. The soils after direct drilling compared to plowed ones are characterized by a more compact top soil[370,950] and a greater number of longitudinally continuous earthworm channels (biopores),[64,90,361,458] resulting in greater water infiltration rates and saturated hydraulic conductivity,[356,458,707] better aeration,[322] wetter topsoil,[199,1024] lower nitrogen mineralization,[321] and greater concentrations of phosphorus and potassium in surface layers of the soil.[332] Aforementioned aspects are discussed in more detail by Cannell.[197]

Although bulk density of soil under minimum tillage or no-till soil was greater, root growth of barley and oats increased in the top layer.[212,360,370,423,516] Also, root length density[77,394,947] was greatest in the surface horizon of soil under minimum tillage, while total root size declined.[77] Root length under minimum tillage can be decreased in wet years.[24,368] Anderson[30] reported that increased root growth in the surface soil under minimum tillage allowed for the adequate uptake of nutrients from broadcast fertilizer without incorporation.

Experiments carried out on coarse-textured soils indicated that the growth of seminal axes was slower in a direct-drilled or shallow-cultivated soil compared to a plowed or deep-cultivated one[370,395] (Table 1), but on clay soil, root extension of winter wheat was slightly faster during winter after direct drilling than after plowing.[199]

In directly drilled sandy soil[68] and loess soil,[360] the root quantity in the 10- to 20-cm layer was smaller than on cultivated soil. Higher concentration of roots in that layer of cultivated soil was an effect of a traffic pan existing in the 25- to 30-cm soil layer which had higher bulk density and penetration resistance than any layer of untilled soil. In deeper layers, root quantity of cereals on loess soil[359] and clay soil[368] was consistently greater on direct-drilled than on plowed land. Accordingly, total water uptake from untilled soil in the 10- to 20-cm layer was lower and it was greater from the 20- to 60-cm layer compared to tilled soil.[359] More roots at depth on direct-drilled soil resulted in significantly higher water uptake by cereal roots in dry periods.

Ellis and Barnes[368] and Cascio and Venezia[212] reported that more roots of wheat at depth were consistently recorded on direct-drilled than on plowed land. Direct drilling also gave more uniform root distribution.[212] However, after prolonged wet soil conditions during winter, on a soil with a large clay fraction and a low hydraulic conductivity, root growth and penetration in spring, before the development of a soil water deficit, was more restricted on direct-drilled than on plowed land.

TABLE 1
Mean Length (in cm) of Seminal Roots of Winter Wheat During a Period 6 to 28 d After Planting as Affected by a Cultivation System

Cultivation treatment	Length of primary seminal root	Length of first pair of lateral seminal roots
Plowed	14.1	12.2
Direct-drilled	10.6	9.2
Shallow-cultivated	11.9	9.5
Deep-cultivated	15.1	12.0

Data from Finney, J. T. and Knight, B. A. G., *J. Agric. Sci. Cambridge*, 80, 435, 1973.

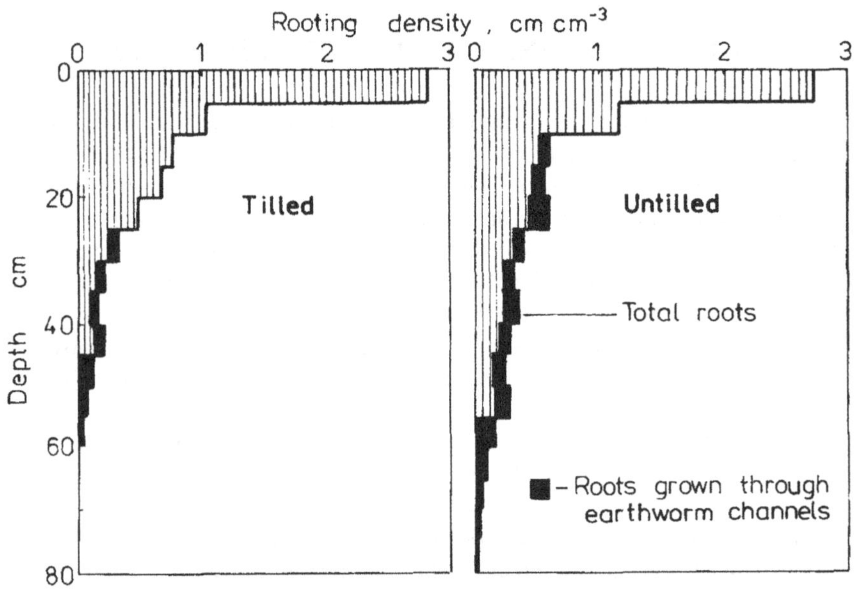

FIGURE 1. Total rooting density, rooting density in earthworm channels (black area), and rooting density of oats in bulk soil (lined area) as a function of depth in tilled and untilled loess soil. (Modified from Ehlers, W., Köpke, U., Hesse, F., and Böhm, W., *Soil Till. Res.*, 3, 261, 1983.)

In untilled, nonswelling loess soil, the biopores created by earthworms were utilized by roots as passages of comparatively low soil strength.[360] Rooting density growing in bulk soil and through earthworm channels within the soil profile of both tilled and untilled soil is shown in Figure 1. Opara-Nadi and Lal[907] reported the combined effect of tillage (tilled and untilled) and land clearing (tree pusher and manual) methods on maize root growth.

Goss[455] and Goss et al.[457] reported that root growth and activity were related to the volume of macropores (>50 μm) — also called transit pores.[1026] They were impaired when the continuity was reduced or in undrained soil which became water filled. The pores are also important for water drainage and air transport. Tillage implements such as the moldboard plow create macropores in the topsoil, but simultaneously reduce deeper root penetration.3[68]

Barber[77] observed finer root growth in tilled than in untilled soil, but Anderson[30] reported no effect of minimum tillage on root diameter. Tardieu[1161-1164] and Tardieu and Manichon[1167] studied the effect of three typical plowed-layer structures composed of: (1) fine earth and centimeter-sized crumbs; (2) compacted decimeter-sized blocks separated by cavities; and (3) uniformly compacted soil on the maize root system in the field. They found that plowed layer structures markedly affected the horizontal and vertical spatial distribution of roots over the whole soil profile. The changes were reflected in water uptake which was reduced by nearly half in compacted plots.[1161]

Zach[1345] reported that a rotavator (tillage to a depth 12 cm) is not suitable as a primary tillage implement for sugar beets. Mihalic et al.[829] indicated that on loess soil of high fertility, the effect of plowing to a depth of 30 to 40 cm and plowing at 30 cm, with subsoiling to 40 cm, on the root yield of sugar beet was stronger than that of mineral fertilization, whereas on soil of lower fertility and on soil with more expressed B-horizon, fertilizing was more decisive.

Effects of various tillage practices (no tillage, chisel plow, and moldboard plow) and residue management on rooting with 7 continuous years of maize were reported by Dowdy et al.[323] The maximum root length was associated with chisel plow treatment. The residues stimulated root accumulation in the surface soil.

As was shown in Chapter 4.4 root growth is strongly dependent on oxygen supply. By means of an electrical analog experiment, Kirkham[667] found that the conductance of oxygen to the root zone was more affected by the between rows width of tillage (the amount of soil surface that is not sealed for the flow of oxygen) than by the depth of tillage. However, the effect of the tillage depth increased as the amount of the tilled soil surface became larger.

The area and volume of compacted soil can be considerably reduced by confining vehicular traffic to narrow interrow paths.[224,1206,1209] Interrow compaction reduced horizontal root growth and increased deep root penetration.[224] There were also some favorable effects of interrow traffic.[23,1254] Voorhees[1254] reported that wheel traffic on both sides of soybean rows increased the yield up to 25% in seasons of less precipitation compared to rows having no wheel traffic. This effect seems to be related to better use of water and phosphorus by roots growing in compacted soil. Moderate compaction may also increase root branching which allows for better exploration of soil for nutrients. The results obtained by Gerik et al.[435] suggested that the favorable effect of controlled traffic may not necessarily occur in all soils, especially in soils with 40% or more of 2:1 clay minerals. Bauder et al.[103] evaluated the effects of various tillage systems with confined wheel traffic patterns on corn root growth and found (Figure 2) that although vehicular traffic affected the extent of root development and distribution, the type of tillage used had a much greater effect on overall root length and root distribution. Possibilities for reduction and alleviation of the negative effects of heavy traffic were considered by several authors.[506,536,1280]

6.2 DEEP LOOSENING

Soil layers that may restrict root growth are plowpans, fragipans, duripans, fine-textured B–horizons, claypans, and high clay horizons. The depth and thickness of the pans depend on site conditions. Restricted root growth can be a consequence of too low porosity accompanied by insufficient oxygen supply,[55,1311,1312] excessive mechanical impedance which is affected by soil moisture content,[215,295,360,489,594,1249] and the absence of pores of diameters greater than root tips.[155]

One of the diagnostic means of a root growth problem is a cone index. Because it is a moisture-dependent characteristic, it should be determined in a year of near average rainfall for the site before a subsoiling recommendation is made.[1249] It was indicated that when good fertilization and management practices are used on the plow layer, economic response to profile modification is doubtful.[351] In recent studies,[1278] the reversibility of compaction effects by disturbance or ripping was also questioned. The question results from the fact that ripping may produce large

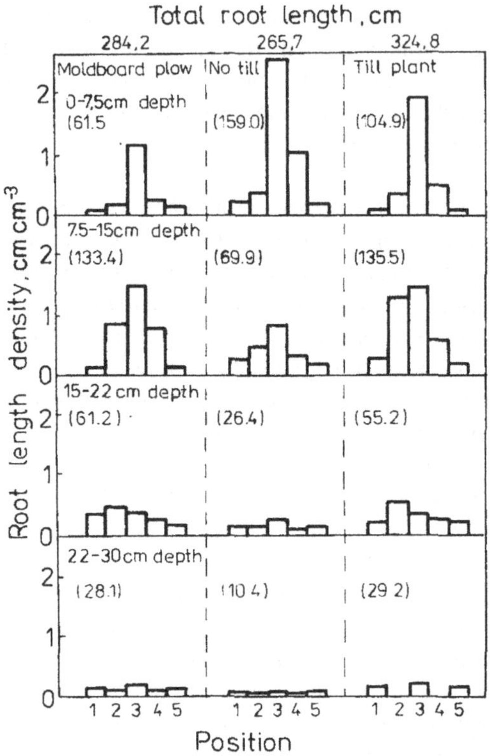

FIGURE 2. Total root length and root length density by position normal to the maize row and soil depth. Total within each tillage by depth combination is the total root length of all sampled depths and positions for each tillage treatment: position (1) nontrafficked interrow; (3) center of crop row; (5) trafficked interrow; positions (2) and (4) are intermediate between positions (1) and (3) and between positions (3) and (5), respectively. (Modified from Bauder, J. W., Randall, G. W., and Schuler, R. T., *J. Soil Water Conserv.*, 40, 383, 1985.)

voids between fairly large soil structure units (>2 mm), but compaction may alter smaller units (<1 mm). The importance of profile modification was discussed in detail by Eck and Unger,[351] Cassel,[214] Spoor,[1112] Swain,[1151] Unger,[1214] Wind,[1330] and Heinonen.[536] The pans restricted root growth of bean,[184,185] tobacco,[1249] potato,[1003] and grape.[1055]

Figure 3 shows that deeper-cultivated maize was characterized by higher root growth and higher rooting density in the subsoil compared to conventionally tilled fine sandy soil (25 cm deep).[120] During the period of vegetative growth, the root growth rate was 20.0 and 11.8 mm d^{-1} for deep- and conventionally tilled soil. Deep plowing (45 cm) of chernozem soil in comparison to the shallow one (25 cm) caused better growth of the wheat root system and its more even distribution in the soil.[232] The importance of profile modification for grapevine rooting was summarized by van Huyssteen.[1231]

The review of subsoiling effects in the U.K. by Russell[1022] indicated that only in 40% of cases were responses positive. Generally speaking, the coarse-textured soils were less responsive to

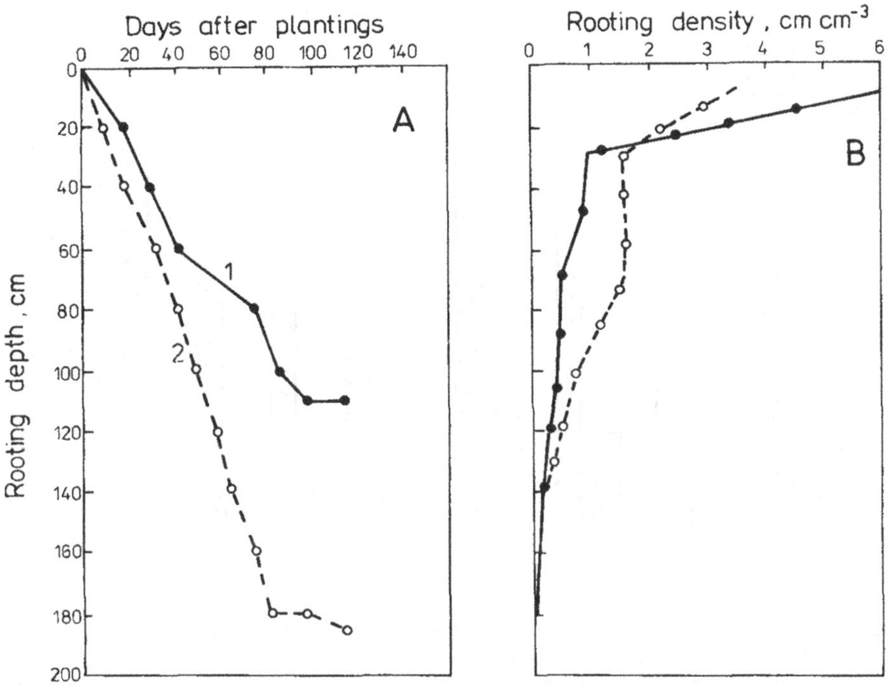

FIGURE 3. Maize rooting depth with time (A) and rooting density with depth (B) (99 d from planting) for two treatments: plowing 25 cm deep with uncontrolled wheel traffic (1) and plowing 40 cm deep and controlled wheel traffic (2). (Modified from Bennie, A. T. P. and Botha, F. J. P., *Soil Till. Res.*, 7, 85, 1986.)

subsoiling than the fine-textured soils. It can be related to the shorter duration effect in coarse-textured soils because of the low content of finer soil materials or organic matter to stabilize the pore system.[535]

In coarse-textured soil, where root growth was likely to be impeded by hindered displacement of sand particles due to their surface roughness, subsoiling induced deeper and greater rooting of maize, although the decrease in bulk density was only of the order of 0.1 Mg m^{-3}.[223] An important factor, which can limit root growth in coarse-textured soils, is a low plant-available water holding capacity. Subsoiling which loosens the subsoil without inverting it promoted deeper root penetration,[351,1151,1208,1214,1249] but crop yield responses varied with the plant species, cultural practices, soil, and environmental conditions during growth period.[223,351]

Ross,[1003] Acharya and Bhagat,[5] Ide et al.,[594] and Chaudhary et al.[223] showed that subsoiling of loamy soils resulted in deeper root penetration and greater profile water use and increased yield under drought conditions. The effect of subsoiling was more positive in those soils, where there was a definite layer which impeded root growth and water movement to deeper horizons.[194,319,638,1151]

Because of the high cost of profile modification, subsoiling only under the crop row, called "precision tillage" or "under-the-row subsoiling", is recommended.[511,1208] The cost of operation can be markedly diminished with crops grown in wide rows. Recently a pneumatic apparatus, requiring less energy, was successfully used for deep loosening of soils in vineyards.[1055]

The restricted root growth in fragipan horizons often results from both high bulk density or

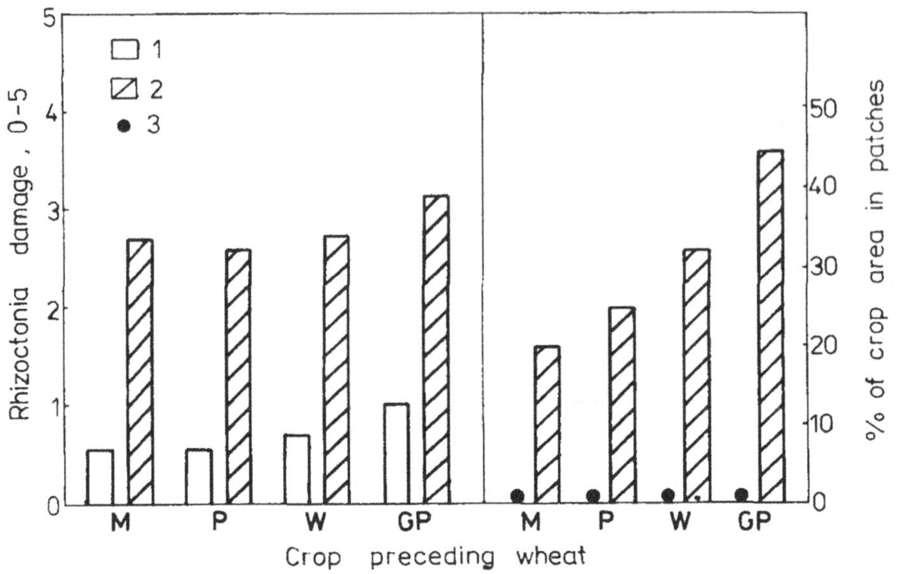

FIGURE 4. Effect of tillage and rotation on the damage of wheat roots by *Rhizoctonia* sp. (root rot rating scale was 0 to 5, with 5 being maximum disease on the area of bare patches in fields); M, *Medicago* pasture; P, peas; W, wheat; GP, grass (*Medicago* annual pasture); 1, cultivated; 2, direct drilled; 3, cultivated — no patches. (Modified from Rovira, A. D., *Phytopathology*, 76, 669, 1986.)

acidity.[138,162,1098,1099,1331] Blanchar et al.[138] reported that increasing the pH in the range of 4.2 to 6.5, especially from 4 to 5, resulted in a greater root length of pea in both B2 and fragipan samples. Acid soils may restrict root growth through a deficiency of Mo and Ca or by H, Mn and Al toxicity.[7]

A significant increase of rooting depth in hard and acid subsoils can be increased by deep tillage plus liming,[162] but incorporation of lime with the subsoil is technically difficult.[340,931] Adams[7] suggested in his review that the most plausible means for alleviating subsoil acidity lies in the judicious choice of surface-soil amendments.

6.3 TILLAGE AND ROOT DISEASES

Tillage practices, through their direct influence on physical properties of the soil, may in turn affect the viability of root pathogens and susceptibility or resistance of the host. One of the major root diseases of cereals is *Rhizoctonia* bare patch caused by *Rhizoctonia solani* Kuhn. The disease has spread recently in Australia[613,614,781,1008] and has also been reported in the U.K.,[110,862] U.S.,[962,1288] and Canada.[1186] Numerous researchers have found that the incidence of this disease is greater in direct-drillled crops than in those sown into cultivated soil.[110,782,783,962,1008,1013] Rovira[1008] assessed the damage to wheat by the severity of the areas of diseased patches of stunted plants in cultivated and direct-drilled fields (Figure 4). This figure also illustrates that there was significantly less damage to the crop following peas and medicago than following pasture.

The frequency of disease occurrence was inversely related to the amount of soil disturbance during seed–bed preparation.[783,962,1008] Pumphrey[962] reported that plants of winter wheat within patch-affected areas produced 52% less grain compared to unaffected plants. Neither time nor rate of nitrogen application had a significant effect on the occurrence of bare patch.

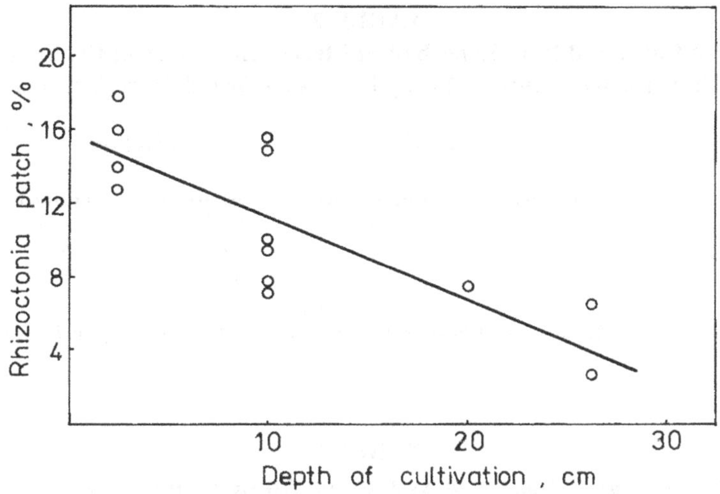

FIGURE 5. Relationship between depth of cultivation (D) and *Rhizoctonia* patch (RP = 15.9 to 0.45 D); (r^2 = 0.62). (Modified from Jarvis, R. J. and Brennan, R. F., *Aust. J. Exp. Agric.*, 26, 703, 1986.)

The effects of minimum tillage on *Rhizoctonia* root disease are attributed to significantly higher concentration of propagules of *Rhizoctonia solani* in the top layer of the soil profile,[1142] and during cultivation there is a fragmentation of the residues which reduces propagule size and may affect the pathogenicity of hyphae that have ramified through the soil. Rovira[1008] observed that *Rhizoctonia* patch in direct-drilled soil occurred in districts (South Australia and Victoria) which are subjected to hot, dry summers during which little or no decomposition occurs.

The incidence of *Rhizoctonia* root patch of wheat was reduced by the deeper cultivation on a fine sand (Figure 5).The cultivation probably disrupted the subsoil, allowing more rapid penetration to depth. This suggestion agrees with that of Hamblin and Tennant,[510] indicating that deeper working results in faster early root development and plant vigor which compensate for the root damage due to infection with *R. solani*.

Sumner et al.[1143] reported that deep turning removes the fungal propagules from the top layer and buries them deep. As R. *solani* is sensitive to high levels of CO_2, there is an incidence of pathogen decline in buried debris. The recent experiments already discussed[614,962,1008,1288] reveal that the occurrence of this disease may become a serious problem with continued use of conservation tillage practices (direct drilling). Populations of *R. solani* in topsoil are probably influenced more by tillage practices than any other soil-borne pathogen because the fungus survives in colonized plant debris in the soil.1 1[43]

The disturbance of sandy and clay loams (into units of 2 mm or less) largely reduced root incidence of subterranean clover compared to intact cores (Table 2). In clay loam, the roots were shorter in the intact cores than in the disturbed soil, but in sandy loam, root length was unaffected by these treatments.

Subsoil of sandy loam soil did not reduce infection of bean roots by *Fusarium solani*, but it allowed for the development of a deeper, more vigorous root system and for the increase of yields compared to moldboard-plowed plots.[185] In another paper, it was shown that propagules of *F. solani* were found mostly in the plow layer (0 to 30 cm) and that vertical distribution of the pathogen coincided with bean root distribution which is largely confined to the plowed layer by a compact plow sole.[184] Tillage tended to distribute the fungus throughout the plowed layer.

TABLE 2
Root Rot Index and Root Length of Subterranean Cover in Disturbed and Undisturbed Sandy and Clay Loams Collected from Pastures

	Sandy loam		Clay loam		
	Disturbed	**Undisturbed**	**Disturbed**	**Undisturbed**	**LSD (0.05)**
Root rot index[a]	1.7	3.1	1.2	3.3	0.7
Root length (cm/plant)	9	8	10	6	3

Modified from Smiley, R. W., Taylor, P. A., Clarke, R. G., Greenhalgh, F. C., and Trutmann, P., *Aust. J. Agric. Res.*, 37, 633, 1986.

TABLE 3
Effect of Soil Management Systems on Apple Root Characteristics

	Grass		Herbicide		Cultivation	
Length (cm)	**Length (cm)**	**Specific root density (cm g^{-1})**	**Length (cm)**	**Specific root density (cm g^{-1})**	**Length (cm)**	**Specific root density (cm g^{-1})**
0—10	71	125	135	63	3	50
10—20	86	68	108	42	10	62
20—60	100	33	98	43	116	31

Modified from Haynes, R. J., *Soil Till. Res.*, 1, 269, 1981.

Tu[1211] reviewed the influence of several tillage practices on pea root disease in Canada and indicated that fall chisel plow plus spring flat seedbed preparation and fall plow plus spring raised seedbed preparation significantly reduced root rot severity, compared with conventional fall plow plus spring flat seedbed preparation. The most severe root rot occurred with fall plow plus compaction.

6.4 SOIL MANAGEMENT

The effects of soil management on rooting have received much less attention than the effects of soil tillage. Haynes[529] investigated the influence of three forms of soil management in an orchard: maintenance of a grass sword and cultivation or use of herbicides on rooting of trees. He found that in comparison with grass, the herbicide treatment encouraged root growth of the apple tree, while in contrast, cultivation to a depth of 20 cm resulted in a very much decreased surface soil root activity (Table 3). In addition, the competitive effect of the grass sword promoted relatively thinner (as indicated by specific root density) and more ramified tree roots than did the herbicide treatment. The author concluded that root growth in the top 20 cm was practically annihilated by physical disruption due to cultivation, but was positively influenced by herbicide treatment.

6.5 SOWING DATE

Barraclough and Leigh[94] reported that in England, September-sown winter wheat crops had more roots throughout the growing season than October-sown wheat. The differences were dependent on accumulated thermal time. Coale and Grove[243] reviewing the results on soybean root production in relation to management practices, concluded that the effect of planting date on development of the roots is primarily a combination of the effects of temperature, moisture, and photoperiod.

6.6 SOIL MOISTURE REGULATORY PRACTICES

Root growth conditions can be considerably improved by drainage to avoid aeration problems in wet soil or by irrigation to avoid critical values of penetration resistance in dry soil.[144,151,199,539] Spiers,[1111] comparing the influence of the two soil-mulching moisture-supplementary practices and irrigation on root system development in *Vaccinuum ashei*, found the following ranking: mulch > incorporated peatmoss > irrigation. Mulching resulted in uniform root distribution from the plant crown outward and in concentration of the roots in the upper 15 cm of the soil.

6.7 MULCHING

The response of roots to surface mulching depends on soil moisture and placement of the mulch.[23] In the season, with moderate soil moisture, surface plant residues over the interrow stimulated maize root development, but retarded root development when placed over the row. In the wet year, both surface placements retarded all root activity because of both reduced soil temperature and excess moisture. Mulching practices may also influence root growth via changes in the surface soil temperature, especially when plants are young.[25,1255]

6.8 CHEMICAL TREATMENTS AND PLANT HORMONES

In some cases, e.g., when excessive mechanical impedance and insufficient aeration of soil occur, chemical treatments are used to increase rootability.[20-22,456,1034,1315-1317] Incorporation of 3,5-diioodo-4-hydroxy-benzoic acid (DIHB) significantly enhanced root growth of peas[1316] and alfalfa[1034] grown in compacted soil and barley grown in inadequate aeration (5% oxygen).[456] DIHB acts by modifying the cell wall extensibility factor. It was also suggested that DIHB is antagonistic to the action of endogenous ethylene.[456,719,993] The results obtained by Goss et al.[456] suggest that DIHB is of practical significance in promoting rapid root penetration through soil layers where mechanical and aeration constraints to growth occur simultaneously.

Root growth-promotion properties in compacted soil were also demonstrated by 3,5-dichloro-4-hydroxybenzoic acid (DCHB).[21] The effect increased as concentration of the chemical in soil solution increased (Figure 6). It is suggested that the influence of DCHB is associated with either its beneficial action on osmoregulation or its conversion into a compound possessing auxin activity and a synergistic action with ethylene-induced endogenous auxin.

Another compound promoting root growth in compacted soil is 2,4-dichlorophenoxyacetic acid (2,4-D). Inclusion of the chemical in compact soil led to a significant increase in the root growth of rice seedlings and nitrogen uptake (Table 4). This compound alleviated compaction effects on barley seedling growth[20] as well. The mode of action of 2,4-D is not clear.

Root growth and their functions are strongly reduced when subjected to salinity.[1221,1265,1266,1284] Recently, Waisel[1266] found that NaCl concentrations which inhibit the growth of the whole plant stimulate the growth of single roots when applied to them individually. Such a phenomenon has ecological significance in nonuniform environments.

Abscisic acid (AbA) is a plant hormone, production of which is enhanced by water stress.[112,496,521,987] The results of Markhart et al.[796] and Fiscus[399] showed that AbA decreased

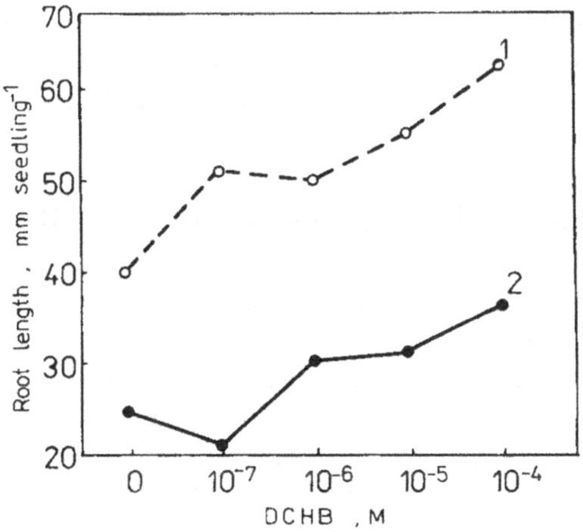

FIGURE 6. Effect of 3.5-dichloro-4-hydroxy-benzoic acid (DCHB) on length of longest root of seedlings of two rice varieties (1, IR-56; 2, *Azucena*) grown for 7 d in compact soil. (Modified from Alejar, A. A. and Erasga, A. B., *Kalikasan Philipp. J. Biol.*, 13, 295, 1983.)

TABLE 4
Effect of 2,4-D on Length, Fresh and Dry Weight and Nitrogen Content of Rice Root Seedlings Cultivar IR-56 Grown for 7 d in Compact and Loose Soil

Treatment	Length (mm)	Fresh wt (mg)	Dry wt (mg)	Total N (%)
		Compact soil		
Water	14.12	8.08	1.69	1.67
10^{-6} *M* 2,4-D	21.69	13.44	2.58	1.76
		Loose soil		
Water	56.29	14.00	2.60	1.46
10^{-6} *M* 2,4-D	67.75	15.22	2.72	1.50

Modified from Alejar, A. A. and Erasga, A. B., *Philos. Agric.*, 67, 441. 1984.

hydraulic conductance of roots and the degree of decrease was strongly dependent on the concentration of AbA applied. Volume and/or ion fluxes through plant roots have been reported to increase,[247,378,399,643] to remain the same,[378] or to decrease[953] under the influence of AbA.

Fiscus[399] found that the exact response of the *Phaseolus* root system to applied AbA was highly dependent on the magnitude of the pre-AbA volume flux density, the relative contributions of osmotic and pressure-induced flow, and the applied dosage. Another root hormone, indol-3–ylacetic acid (IAA), promoted chestnut rooting at the juvenile phase.[1245] Dawkins et al.[293] suggested that the presence of ethylene in low concentration (less than 1.0 μl^{-1}) can mediate a compaction response in the root zone by the promotion of cell expansion at the expense of elongation.

6.9 PLANT BREEDING AND SELECTION

Moss et al.[853] listed factors controlling water use by roots which could be regulated genetically. These are water absorption, water transport, hairs, ability to grow in dry soil, aeration (internal), penetration, size (diameter and length), branching, respiration, and reaction to temperature.

The efficiency of the roots depends on the supply of water and nutrients in the environment. de Willigen and van Noordwijk[303] indicate that when the larger root system is not accompanied by increased supply of water and/or nutrients, plant production may decrease instead of increasing. Plant breeders make efforts to select cultivars for extensive and deep root systems to improve the efficiency of water and nutrients use[680] and to reduce leaching of nitrate. Ghaderi et al.[437] present different approaches in plant selection in relation to soil stress conditions. There are contradictory opinions whether the selection should be made under optimal or stress conditions. The relative importance of plant breeding or soil management in alleviation of stress conditions is also considered.

Under drought conditions, the extensive root systems result in high yields.[298,589,590,836] It was confirmed by Nour and Weibel[884] in the study of ten sorghum cultivars known to vary in drought resistance.

Bragg[168] reported that greater root lengths and many seminal roots produced by barley cultivars would be more likely to encounter cracks in compacted field soil and thus be able to grow and proliferate sufficiently to meet the needs of the developing plant. Adaptation for drought tolerance of rice is of heritable and quantitative nature.[465] Differences in drought tolerance were found in a cross between wild rice and a Japonica rice.[851] O'Toole and Soemaratano[909] and Ekanayake et al.[362] proposed the root pulling resistance measurement (the vertical force needed to pull a plant from the soil) to evaluate the differences in drought tolerance of various cultivars and crosses. The same technique was used to find the genetic differences among maize inbreds.[616,934]

Heen[533] pointed out the importance of the seminal roots of some Norwegian barley, wheat, and oat varieties for water and nutrient uptake. The number of these roots is genetically determined in contrast to the crown root development which varied according to water supply.

The possibility of increasing root hair length of a white clover population by selection was indicated by Caradus.[202] This increase resulted in an 11% increase in the volume of soil explored by root hairs.

Increasing the hydraulic resistance of the root system through breeding of the existing cultivars for smaller xylem vessels would increase grain yield. As a result, the rate of extraction of water from the soil is decreased without decreasing the amount of available water. Meyer and Alston[825] verified this idea. So and Jayasekara[1093] found in split root systems that the sorghum cultivar with low total and high root axial resistance to water flow was less tolerant to drought than cultivars with high total and low axial resistance. The latter resulted in conservative water use with a better ability to extract water from the subsoil. When crops are irrigated, extensive root systems using a large quantity of photosynthates are unnecessary. Therefore different genotypes are required for irrigated and unirrigated conditions or other farming practices.

There are also proposals to select plant species and cultivars for their greater ability to penetrate and loosen compact soils,[536,708,709] to produce higher air-filled porosity[1343] and thus lower requirements for external aeration, and for their ability to decrease the susceptibility to plant lodging.[59] Selection of maize genotypes with high root densities, particularly in the subsoil, can be used to reduce winter leaching of nitrate, consequently decreasing groundwater contamination.[1313]

The possibilities of increasing porosity by selection and maintaining the penetration ability of the roots are greater for plants grown in soilless cultures than in soil.[303] As the time needed for selection or breeding of new varieties, more resistant to stress conditions, is rather long, certain proper soil management practices can be more profitable.[40]

6.10 FERTILIZER APPLICATION

Root system development can be altered by the individual nutrient ion level and its availability as well as by overall fertility. These factors can be modified by correct fertilizer application.[79,1144] The response depends on plant species,[158,1144,1279] phase of crop growth,[244] and method of fertilization.[16,1144] Generally speaking, roots grow faster in soils of greater fertility and concentrate more in undepleted areas.[1287] Among nutrients, nitrogen has the greatest effect, followed by phosphate and potassium. Roots of several plant species have grown better when NH^+_4 compared to the NO^-_3 form was used.[263,980,1043]

The effect of NPK fertilization on root growth increased with increasing soil moisture level.[968] An example of the combined effect of NPK fertilization on root size (compare to Figure 1 in Chapter 4) shows the increase of root weight with the increase of NPK level. This increase takes place in certain amounts of NPK.

The modification of root growth through fertilizer application in relation to soil factors, such as moisture, density, oxygen level, temperature, microbial population, and others, was discussed in more detail by Sumner and Boswell[1144] and Barber.[73] Attempts are undertaken to increase the efficiency of fertilization by means of a foliar application of plant nutrients. The results obtained are not always positive.[205] In acid soils, root growth can be enhanced by liming through its effect on availability of plant nutrients and decreasing aluminium toxicity,[7] decreasing soil salinity,[556] or decreasing soil compaction level.[377]

6.11 OPTIMUM ROOT SIZE

For most crops, root growth per se is of no economic interest, and if the crops are not harvested, it could waste nutrients and photosynthetic potential. The depth of effective roots is probably more important for plant growth than the maximum penetration.[300,787] Optimum root size is not always the largest one and root growth in excess of that needed to meet water and nutrient requirements can lead to dissipation of the products of photosynthesis.[1025] In favorable conditions with a good supply of water and nutrients, some restriction of root quantity can be beneficial.[251,753] However, under restricted supply, greater root systems may absorb more nutrients and/or water. In the field under drought conditions, restricted root systems and smaller water absorption before anthesis may lead to a greater water supply in the latter and possibly to an increase of crop yields.[919,921]

de Willigen and van Noordwijk,[303] based on the assumption that a plant is dealing with diminishing returns of extra water uptake when investing in additional root growth in a zone already containing roots, defined root length density for which the marginal water yield just equals the cost to the plant. From their analysis, results that the marginal root water use efficiency of 0.02 cm^3 water can be used to estimate optimum root length density. In respect to water per centimeter of root length uptake, the optimum root length density is of 2 to 6 cm cm^{-3}, depending on the soil type and internal moisture tension in a plant. Barraclough and Weir[95] reported that a relatively small root length density of 1 cm cm^{-3} in winter wheat was capable of extracting most of the available water (at matric potential of –10 to –1500 kPa). For sorghum, as Jordan and Miller[625a] reported in a reviewed paper, this root length density was larger than 2 cm cm^{-3}. Such a definition of optimum root length in respect to nutrient uptake is not possible because the relation between nutrient uptake and photosynthesis is less clear than that between transpiration and photosynthesis.[303]

Chapter 7

METHODS FOR STUDYING ROOTS

7.0 INTRODUCTION

A wide review of methods to study the root system was given by Böhm.[146] In this Chapter, recent methods which can be used to evaluate the response of the roots to soil physical conditions are discussed.

7.1 ROOT QUANTITY AND DISTRIBUTION

7.1.1 ROOT OBSERVATION TECHNIQUES

Recently, much attention has been devoted to further testing and improvements of non-destructive root observation techniques through transparent materials.[145,169,209,407,484,597,611,662,826,833,1038,1119,1173,1216-1219,1261] This interest results mainly from the fact that the observation techniques seem to be rapid and less destructive and allow study of continuous root development at the same place.

Observation of the roots can be made with the aid of tubes inserted into the soil and with a glass wall in the soil container and a rhizotron. The root length can be measured on the basis of the "line intersection method" developed by Newman.[868] As these methods make it possible to measure only a fraction of the total root growth occurring in the soil volume, they were compared with more commonly used techniques.

It was reported that root length densities obtained from vertical root observation tubes (minirhizotrons and microrhizotrons) are underestimated in upper soil layers and overestimated at depth.[169,484] Higher root concentration near the observation window was also reported by Taylor and Böhm.[1173] However, root elongation rates along a soil–container interface were significantly lower than those measured within the bulk soil. These differences were attributed to higher soil strength at the interface or to an attracting electrical charge on the container. In rhizotron studies,[1176,1177] it was found that root growth observed through rhizotron–viewing panels was highly correlated with root growth inside the soil compartment. The results from observation tubes placed at an angle to the surface rather than vertically give better correlations with measured root length densities.[169] Meyer and Barrs[826] reported that the r^2 for the relationship between root intercept counts on horizontally inserted tubes and destructively determined values of root length density, varied from 0.78 to 0.96 and were higher for a tube of 38.5 mm than for a 25-mm-diameter tube, but the greater difficulty of both installing and observing within the larger tubes exceeded this advantage. The use of horizontal root observation tubes and a flexible fiberscope in large soil cores was found to be a practical means of obtaining a semicontinuous record.

Itoh[597] hypothesized that the volume of the tube as an obstacle is directly related to the rooting density in the bulk soil by changing environmental conditions. To verify this hypothesis, he used different radii (6, 9, 12, and 15 mm) acrylic tubes which were installed upright in potting soil. He found that rooting density increased and approached the actual rooting density as the radius of the tube decreased (Figure 1).

Vos and Groenevelt[1261] reported that the amount of wheat roots viewed by an endoscope in horizontally installed square tubes (cm cm^{-2} equaled 0.29 times the root length density (cm cm^{-3}) in adjacent soil, determined by core sampling. However, the results for potato plants with lower root density could not be interpreted as unequivocally as those for wheat. The authors indicate

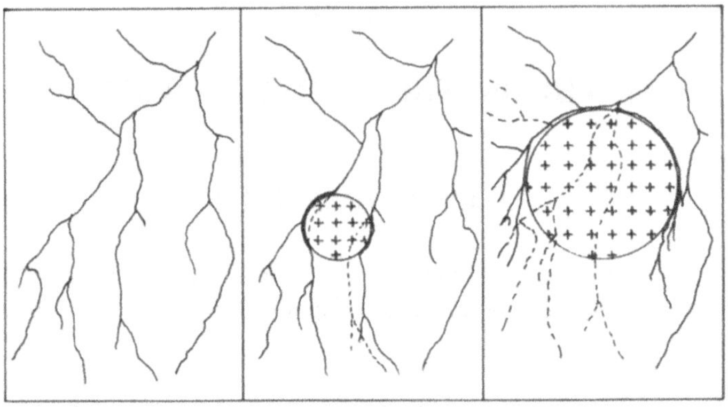

FIGURE 1. Diagram of root growth and root distribution around an obstacle in the bulk soil. (A) Without obstacle; (B) small obstacle; (C) large obstacle. The solid line shows the root distribution. The dotted line indicates the root distribution expected when the obstacle is replaced by the bulk soil. (Modified from Itoh,

that estimation of root densities by observation tubes and an endoscope offers a promise, at least in homogeneous systems, and thorough calibration is required for each set of experimental conditions.

Köpke[679] compared four root study methods (monolith, profile wall, auger, and glass tube) for studying root growth of oats on tilled and untilled loess soil. The root length density was highest with the monolith method and lower by a factor of 2.06 with the profile wall method. The auger and glass tube methods required more replications to obtain statistically significant differences between treatments.

Belford and Henderson[108] obtained a higher correlation between minirhizotron root counts and the root length separated from soil cores than between the length visible in minirhizotrons and the soil core measurements. The observing tube system was modified by including a black and white camera[1218] and was further modified by using a color video camera,[1219] for observing the roots intersecting the tubes on a video monitor and for evaluation of root system orientation.[1216] The use of a color camera allowed easier identification of the smaller diameter roots and provided an identification of relative root age.[1219] Recent developments in computer algorithms provide a new technology to extract more accurate information on roots from video-recorded images.[211,407,910,1090,1185,1340] The computer image analysis allows for identification of root diameter and patterns of branching and crossover.

The use of highly refined fiberoptics in minirhizotrons makes it possible to photograph the roots and use videotapes,[1038,1257] and a stereomicroscope[611] offers the possibility of detailed study of fine roots and root hairs. Inspection of aerial photographs in a three-dimensional image with a stereomicroscope can yield more information on direction of root growth and branching.[1234]

A root observation chamber (field installation) constructed by Arkin et al.[39] permits both rapid observations of root growth through the glass side wall and evapotranspiration measurements by weighing of the installation.

It was shown that with the use of glass- or acrylic-fronted containers, the rate of new root formation can be estimated by outlining all new roots at successive times.[109,565] Belgrand et al.[109] developed a semiautomated data acquisition and processing system that allows for the creation of data sets describing the complete structure of branching root systems. To characterize the root system orientation on the basis of the observations from the minirhizotrons, a mathematical model, based on probabilistic assumptions concerning root growth directions, was applied by Upchurch et al.[1216] Huck and Taylor[584] and Keng[662] suggest that rhizotrons can be more useful in studying specific questions than general patterns.

Relatively large coefficients of variations with the tube observation techniques could be reduced by increasing both the observation area and the number of replications.[662] Theoretical models to convert the number of roots that intersected a minirhizotron face to root length density (RLD) have been developed.[801,814,1218,1219] Bland and Dugas,[140] verifying the theoretical models in a field experiment, found that counts of cotton roots at the horizontally installed minirhizotrons can be converted to RLD with RLD = C × counts per area of the tube wall. In this case, the theoretical value C is equal to 2, provided the best estimate of RLD. However, counts of sorghum roots at minirhizotron walls and measured RLD were not simply related.

7.1.2 NUCLEAR MAGNETIC RESONANCE IMAGING (NMR)

Recently,[904,994,995] the use of NMR imaging to measure spatial distribution of roots was attempted. So far, results have suggested that NMR imaging is an effective method for measuring spatial distributions, both of roots and water in soil,[904] and as a nondestructive and noninvasive method for studying growth dynamics and physiological processes involving flow.[994,995] In addition, it was indicated that NMR computer tomography (CT) is more suitable for making three-dimensional images compared to X-ray CT.

7.1.3 RADIOGRAPHY

Couchat et al.[257] showed that neutron radiography is capable of supplying *in situ* root images, permitting qualitative and quantitative interpretation of roots and shoots. By comparing this method with γ-transmission measurements of soil moisture and bulk density, the influence of soil could be taken into account. In this method, the autoradiograms from soil sections are used for reconstruction of the spatial distribution and for measuring root quantity.[413] Soil sectionings can be obtained in some soils with the aid of a thin steel wire or by cutting a dried soil block and further imhibition in polyester resin. In the resin technique, the natural soil structure can be disturbed by the desiccation and infiltration of the soil with the resin. To avoid the above-mentioned inconveniences, Fusseder[413] proposed the autoradiographic method in which the rooted soil is deep-frozen with liquid nitrogen and cut into thin slices. The technique seems to be particularly effective for determination of the average distances between living roots *in situ*.

7.1.4 EXCAVATION TECHNIQUES

Recently, Steen[1115] proposed a mesh bag technique for field studies of root growth dynamics. Bags of nylon net are inserted into the soil and filled with root-free soil. The technique has been proven useful for studying the effects of soil compaction[1118] and soil treated with chemicals.[1116,1117]

Floris and van Noordwijk[404] and van Noordwijk et al.[1237] proposed technique and sampling schemes which make it possible to obtain pinboard samples from the field without digging a trench or complete soil column, permitting examination of the intact root systems.

An important factor in studying distribution of roots is their spatial variability. This necessitates the study of a large number of root populations to obtain statistical significance. Van Noordwijk et al.[1237] discussed the importance of sampling schemes including both the number of samples required and their position around plants.

7.1.5 ROOT COUNTING TECHNIQUES

To mechanize and accelerate calculation of the number of intersections, a photoelectric counting device with a moving platform,[910,985,1019] a computer-controlled digital scanning microdensitometer,[1257] a video camera,[543] image-processing equipment with a neutral red staining technique,[1058] and photography of roots and image analysis[1084] were used.

Scanning electron micrographs can be used not only for the description of roots, but also to determine dimensions of hairy roots. The method proved suitable for structurally complex proteoid roots of *Leucadendron laureolum* whose surface area was 10.6 multiplied by that of equivalent "normal" roots on a volume basis.[716] Because of technical difficulties thus far, the obtained results in the literature mostly refer to the epidermis without root hairs.

7.1.6 PREDICTION OF ROOT GROWTH BASED ON THE RELATIONSHIP WITHIN THE ROOT SYSTEM

Taking into consideration the existance of approximately constant relations between the total number, length, diameter, surface area, and volume of the root members within the root system,[381,382,500,501] methods for prediction of certain root parameters were proposed.[449,499,502] Hackett and Rose[502] proposed a model of the extension and branching of seminal roots of barley for studying and interpreting relations between root dimensions obtained from fragments of roots taken from the soil.

Recently, Habib[499] proposed and tested a method for estimating the total root length of a root system from subsamples which is based on measurement of the length and diameter of small pieces of roots and on the measurements of the bulk density of the roots. Gochev et al.[449] obtained and verified a formula permitting calculation of the surface area of some simple root systems without measuring the diameter of the root branches. Instead, an effective diameter is used as a function of the central root mean diameter and the lengths of the central and all side roots.

7.2 ROOT WATER POTENTIAL

Several techniques have been used to measure root water potential.[9,299,396,645,646,905] Measurements of the potential impose difficulties connected with hanging of roots in an atmosphere of humid air,[645] combined measurements of the average value of water potential of the soil plus roots,[299] and uncertainty as to which part of the root system was being measured.[396] Adeoye and Rawlins[9] proposed a technique in the divided root experiment. The measured water potential of an enclosed, nonabsorbing branch of the root system, to infer the water potential of the remaining roots growing in the soil, was used. The technique is nondestructive and *in situ* enables following of water potential changes continuously in a dynamic system by approximating the changes by a succession of steady states. Oosterhuis[905] proposed screen-caged thermocouple psychrometers to measure the water potential components of the roots. This technique enabled the demonstration of osmotic adjustment in roots and leaves.

7.3 ROOT CATION EXCHANGE CAPACITY (RCEC)

The methods of root cation exchange capacity may be divided into three groups: titration with a strong base after acidification; saturation with a cation followed by desorption with another cation; and isotopic exchange. Each method has its advantages, but they all have at least one common disadvantage: the variability of the results according to the details of the operating procedure.[337]

7.4 WATER USE BY ROOTS

Bar-Yosef and Lambert[102] adapted the split root system technique previously used by Brouwer[176] and Giskin and Kohnke[445] for root growth and water uptake in response to soil bulk density and water potential. The desired soil bulk densities were obtained by packing and soil water potential by application of suction. Hasegawa and Sato[525] used a root plane experiment to determine water uptake by roots. Water extraction rates were determined from the differences in weights of the soil blocks and the mean length of the roots between the individual periods. Recently, Sommer[1105] and Lipiec et al.[753] described and tested the ceramic tube negative water pressure technique for measurement of water use by roots at a given soil water potential and a continuous water supply. The technique can be satisfactorily implemented in precise greenhouse studies of water uptake as affected by various environmental factors. For evaluation of the effects of matric, osmotic, and total soil water potential on water uptake by roots and plant growth, a pot culture technique was proposed.[1050]

Tardieu[1161] and Tardieu and Manichon,[1166] studying in the field the vertical and horizontal distribution of maize roots by mapping, found that spatial arrangement of roots is of great importance for the availability of water reserves in the soil. They suggested that the study of water flow resistance between the soil and plant roots cannot be undertaken without taking the spatial arrangement of roots into account.

7.5 ROOT PROCESSES AND GROWTH

Some root processes can be measured in mist chamber cultures with a controlled atmosphere in the root zone.[149,1089,1325,1329] Smucker and Erickson[1089] designed a mist chamber which permits rapid gaseous exchange without altering root water relationships in a root zone and the measurement of root respiration, root exudation, and rhizosphere volatiles without disrupting the system. For measurements of the influence of bulk densities of soil on root growth and functions, several container experiments have been designed.[146] Asady et al.[50] proposed a rapid and nondestructive soil core seedling test to examine some primary physiological and morphological root responses to soil compaction. The system consists of an experimentally compacted soil having various air-filled porosities. The authors indicate that the test can be useful for breeding programs for determining the inheritance of root tolerance to compacted and other specific soil conditions.

In the system of Pfleger and Werner,9[43] the model soil cores of different density and moisture content were used to investigate the influence of holes of different diameter on the penetration of spring barley roots.

7.6 SEPARATION OF ROOTS FROM THE SOIL

One of the greatest limitations to the measurements of root responses to soil environmental conditions is a difficulty in separation of the roots from the soil without destroying them. Following this, Smucker et al.[1091] developed a rapid, inexpensive, and quantitative method which combines kinetic energy of pressurized spray jets and the low energy of air flotation (hydropneumatic elutriation system). Examples of root recovery rates of roots from soil materials of various textures are shown in Figure 2. The authors indicate that the system is very useful to separate roots from compacted soil. In the root samples separated by this technique, the morphological characteristics (i.e., length, root branching, nodulation, and porosity), content of bound nutrients, bacterial and fungal infections, ^{14}C partitioning of field-grown plants, genetic diversity of isoline root systems, and toxic metabolite and enzyme contents of roots, can be examined.

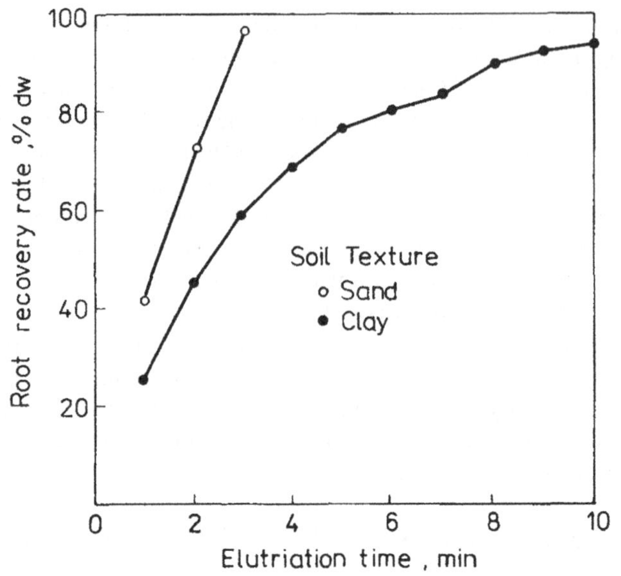

FIGURE 2. The time and recovery rates of dry bean roots separated by hydroelute system; 100% of the roots washed free of soil were those which traversed the primary sieve and were retained on a 75-μm sieve. Inlet pressure of water and air were 3.9 and 0.7 kg cm^{-2}, respectively. (Modified from Smucker, A. J. M., McBurney, S. L., and Srivastava, A. K., *Soil Sci. Soc. Am. J.*, 74, 500, 1982.)

Chapter 8

A LOOK AHEAD

Reviewing the literature on the various soil physical conditions and plant root relations, one can state that the effects of all of the main soil physical factors were studied. Among the great number of publications, the majority are related to three soil factors: water, aeration, and mechanical impedance. The publications also provide data on optimum and critical values of these characteristics for root growth of some plants and trees.

It was stressed that the effect of soil physical properties on root development includes many interactions between themselves and other factors. In spite of the importance of these interactive effects, there is a relatively small number of studies where all of these requirements were evaluated at the same time. Most of such studies were performed under laboratory conditions with atypical temperature and other factors. The responses of the roots to soil physical conditions in laboratory experiments are usually more sensitive than they are in natural conditions.

The effects of physical conditions on root growth and functions are closely related to climatic conditions, soil conditions, soil management, and the stage of plant development. For example, in the wetlands, the factor most correlated with root growth can be soil aeration, while in the northern parts of the world, it is soil temperature. Complex interactions result in no universal physical factor affecting root growth and functions. Many researchers indicate that in most conditions, soil water or soil moisture potential appears to be of particular importance since it directly affects both root growth and functions and indirectly influences other significant factors, such as aeration, mechanical impedance, and soil temperature. The importance of this property is enhanced by the fact that it can to some degree be manipulated and predicted by means of climatic factors. The interrelationships between soil water and other physical factors are expressed by the indicator called a "nonlimiting water range" at which soil aeration and mechanical impedance do not restrict root growth.

The pattern of root development is genetically controlled and can be greatly modified by physical and chemical factors in the soil environment and by plant breeding. However, there is no consistent opinion on the most favorable root size for maximum crop yield. The discrepancy is due to the fact that size is closely related to environmental conditions. An extensive root system causes plants to make more efficient use of fertilizers and water and to be less susceptible to chemical and physical stresses during vegetation season. Such a system is more important in soils of limited water storage capacity and nutritional status and in regions of uneven distribution of rains and in situations where it can release and absorb nutrients from natural resources in the soil and thus contribute to a reduction of fertilization. However, the plant with an extensive root system distributes more assimilates into root dry matter production than into the shoot system, which can result in a decreased yield. The smaller root system can be sufficient in soils of higher fertility or when such practices as irrigation and foliar fertilization are applied. One of the ways to avoid loss of assimilates for production of unneeded roots is the improvement of plants by breeding, taking into consideration not only greater availability of the soil moisture, but also increased yield per unit of water used.

With increasing vehicular traffic of heavy machinery on the fields, an important question is to define which situations and which range of soil compaction has a favorable or negative effect on root growth and yields and which factors or interactions between them are responsible for the reaction. Numerous experiments on crop response to vehicular traffic showed that a moderate compaction increased crop yield, but an intensive compaction decreased it again. This response needs to be studied more in regions with different physiographic properties which may change optimum density of soil.

The intensity of soil compaction still increases and its effects accumulate below the plow layer. Certainly, research is needed to look for new methods to alleviate this impact. Because loosening of the entire field is expensive, techniques with local loosening or other amelioration techniques on small areas should be further developed.

More research is needed to increase the use of natural phenomena alleviating physical stress. One of the ways is a selection of plant species which are more resistant to existing unfavorable conditions and which may alleviate stress conditions. As an example, certain plants which loosen the compact layer in the soil profile may be used. Another way is to breed genotypes with increased resistance for particular stress or with various size of root system. This approach requires consideration of what is the optimum size of a root system in a given condition. It is known that in an irrigated agriculture, a more shallow root system in fields with scarce precipitation develops. Selection of plants and breeding with the proper management of soil physical conditions may lead to higher water and nutrient use efficiency by the roots. Also earthworm activity may contribute considerably to alleviating compaction effects. These practices are still used to only a small degree.

For a long time, tillage was the most obvious way to improve soil structure for root and shoot growth, but within a few years it was shown that zero tillage or minimum tillage, provided weeds are controlled and germination is not restricted, can be succesfully used in numerous soil types without any decrease in crop yields. Recently, it was stressed in the literature that with the trend away from cultivation, root rot diseases may become a serious problem. Thus, to minimize the effects, we need to develop new management strategies where such tillage methods are being introduced.

Greater effort needs to be expanded toward the interrelationships of soil and root characteristics as a part of the entire plant in field studies under different conditions of soil, crop, and climate, including spatial variability of soil and root characteristics. Data so obtained will greatly increase the potential for modeling root growth, water and nutrient uptake, etc. Recent advances in computer and programming techniques would be very useful for that. To accomplish this, we require a team approach and closer cooperation in many scientific fields, including soil physics, root and plant physiology, mathematics, data processing, and others.

Root systems of certain plants are able to compensate for stress conditions in a part of the root system by increased activity of the roots in other nonstressed parts. Moreover, some plants can adapt to stress conditions. For example, waterlogged plants can form adventitious roots with aerenchyma through which oxygen from the atmosphere is transported to the roots. Roots of certain grasses such as Italian ryegrass and ryegrass (*Lolium perenne*) have an ability to decrease their diameter when penetrating rigid pores of smaller diameter than themselves, although the typical response is an increase of the root diameter.

Taking into consideration the compensatory abilities of the roots as a system and adaptations of plant to stress conditions, research is needed on which part of the roots may grow in the soil under unfavorable or physical stress conditions without any significant effect on crop yields and in which situations the root system controls crop productivity. Such knowledge can be useful in choosing cultural practices, alleviating stress conditions. Techniques for root studies in relation to soil physical conditions are still in a stage of development. New methods, especially nondestructive techniques, need to be improved. More research is needed on how morphological and anatomical changes of the root system due to unfavorable conditions affect root functions and plant development.

REFERENCES

1. **Abdalla, A. M., Hettiaratchi, D. R. P., and Reece, A. R.,** The mechanics of root growth in granular media, *J. Agric. Eng. Res.*, 14, 236, 1969.
2. **Abe, K. and Iwamato, M.,** An evaluation of tree-root effect on slope stability by tree-root strength, *J. Jpn. For. Soc.*, 68, 505, 1986.
3. **Abeles, F. B., Ed.,** *Ethylene in Plant Biology*, Academic Press, New York, 1973.
4. **Aceves-Novarro, E., Stolzy, L. H., and Mehuys, G. R.,** Response of three semidwarf Mexican wheats to different aeration in the rooting medium at a constant salinity level, *Soil Sci. Soc. Am. Proc.*, 39, 515, 1975.
5. **Acharya, C. L. and Bhagat, R. M.,** Infiltration behaviour, root development and yield of rainfed maize (*Zea mays* L.) under different soil management practices, *Proc. Indian Natl. Sci. Acad.*, 50, 441, 1984.
6. **Adams, D. O. and Yang, S. F.,** Ethylene biosynthesis: identification of 1-aminocyclopropane-1-carboxylic acid as an intermediate in the conversion of methionine to ethylene, *Proc. Natl. Acad. Sci. U. S. A.*, 76, 170, 1979.
7. **Adams, F.,** Alleviating chemical toxicities: liming acid soils, in *Modifying the Root Environment to Reduce Crop Stress*, Arkin, G. F. and Taylor, H. M., Eds., ASAE Monogr. No. 4, American Society of Agriculture Engineering, St. Joseph, MI, 1981.
8. **Adebayo, A. A. and Harris, R. F.,** Fungal growth responses to osmotic as compared to matric water potential, *Proc. Soil Sci. Soc. Am.*, 35, 465, 1971.
9. **Adeoye, K. B. and Rawlins, S. L.,** A split-root technique for measuring root water potential, *Plant Physiol.*, 68, 44, 1981.
10. **Agami, M. and Waisel, Y.,** The ecophysiology of roots of submerged vascular plants, *Physiol. Veg.*, 24, 607, 1986.
11. **Agamuthu, P. and Broughton, W. J.,** Factors affecting the development of the rooting system in young oil palms (*Elaesis guineensis* Jacq.), *Agric. Ecosyst. Environ.*, 17, 173, 1986.
12. **Agbani, M.,** The relation between root development and mineral nutrition of the sugar beet crop: its importance for fertilizer placement, in *Proc. 17th Colloq. Int. Potash Inst.* Bern, Switzerland, 1983, 249.
13. **Agnew, M. L. and Carrow, R. N.,** Soil compaction and moisture preconditioning in Kentucky bluegrass. I. Soil aeration, water use, and root responses, *Agron. J.*, 77, 872, 1985.
14. **Agraval, R. P. and Jhorar, B. S.,** Soil aggregates and growth of wheat, *J. Agron. Crop Sci.*, 158, 160, 1987.
15. **Agren, G. J., Axelsson, B. F., Flower-Ellis, J. G. K., Persson, H., Staaf, H., and Troeng, E.,** Annual carbon budget for a young Scots pine, in *Structure and Function of Northern Coniferous Forests — An Ecosystem Study*, Persson, T., Ed., Ecol. Bull., Stockholm, 1980, 32, 307.
16. **Ahlström, K., Persson, H., and Börjesson, I.,** Fertilization in a mature Scots pine (*Pinus sylvestris* L.) stand — effects on fine roots, *Plant Soil*, 106, 179, 1988.
17. **Aimi, R.,** Cell-physiological study on the function root. IV. Active oxygen supply into the root from leaves in rice plant, *Proc. Crop. Sci. Soc. Jpn.*, 29, 51, 1960.
18. **Aina, P. O. and Fapohunda, H. O.,** Root distribution and water uptake patterns of maize cultivars field-grown under differential irrigation, *Plant Soil*, 94, 257, 1986.
19. **Alaerts, M., Badji, M., and Feyen, J.,** Comparing the performance of root-water uptake models, *Soil Sci.*, 139, 289, 1985.
20. **Alejar, A. A.,** Growth of barley and wheat seedlings in compact soil with 2,4-D and ioxynil, *Kalikasan Philipp. J. Biol.*, 9, 19, 1980.
21. **Alejar, A. A. and Erasga, A. B.,** Growth of rice seedlings in compact and loose soil with 3,5-dichloro-4-hydroxybenzoic acid, *Kalikasan Philipp. J. Biol.*, 12, 295, 1983.
22. **Alejar, A. A. and Erasga, A. B.,** Varietal responses of rice seedlings in compact or loose soil with 2,4-dichlorophenoxy acetic acid, *Philipp. Agric.*, 67, 441, 1984.
23. **Allmaras, R. R., Gupta, S. C., Swan, J. B., and Nelson, W. W.,** Hydro-thermal effects on corn root development in a Mollisol, in *Trans. 13th Congr. Int. Soc. Soil Science*, Vol. 5, Hamburg, West Germany, 1986, 3.
24. **Allmaras, R. R. and Nelson, W. W.,** Corn (*Zea mays* L.) root configuration as influenced by some row-interrow variants of tillage and straw mulch management, *Soil Sci. Soc. Am. Proc.*, 35, 974, 1971.
25. **Allmaras, R. R. and Nelson, W. W.,** Corn root configuration response to soil temperature and matric suction, *Agron. J.*, 65, 725, 1973.
26. **Amoore, J. E.,** Dependence of mitosis and respiration in roots upon oxygen tension, *Proc. R. Soc. London*, 154, 109, 1961.
27. **Amory, D. E. and Dufey, J. E.,** Adsorption and exchange of Ca, Mg and K-ions on the root cell walls of clover and ryegrass, *Plant Soil*, 80, 181, 1984.
28. **Andersen, P. C., Lombard, P. B., and Westwood, M. N.,** Effect of root anaerobiosis on the water relations of several *Pyrus* species, *Physiol. Plant.*, 62, 245, 1984.

29. **Andersen, P. C., Montano, J. M., and Lombard, P. B.,** Root anaerobiosis, root respiration, and leaf conductance of peach, willow, quince, and several pear species, *HortScience,* 20, 248, 1985.
30. **Anderson, E. L.,** Corn root growth and distribution as influenced by tillage and nitrogen fertilization, *Agron. J.,* 79, 544, 1987.
31. **Anderson-Taylor, G. and Marshall, C.,** Root-tiller interrelationships in spring barley (*Hordeum distichum* (L.) Lam.), *Ann. Bot.,* 51, 47, 1983.
32. **Anderson, W. B. and Kemper, W. D.,** Corn growth as affected by aggregate stability, soil temperature, and soil moisture, *Agron. J.,* 5, 453, 1964.
33. **Anderson, W. P. and Collins, J. C.,** The exudation from excised maize roots bathed in sulphate media, *J. Exp. Bot.,* 20, 72, 1969.
34. **Ando, T., Yoshida, S., and Nishiyama, I.,** Nature of oxidizing power of rice roots, *Plant Soil,* 72, 57, 1983.
35. **André, M., Massimino, D., and Daguenet, A.,** Daily patterns under the life cycle of a maize crop. I. Photosynthesis, transpiration, respiration, *Physiol. Plant.,* 43, 397, 1978.
36. **Angus, J. F., Hasegawa, S., Hsiao, T. C., Liboon, S. P., and Zandstra, H. G.,** The water balance of post-monsoonal dryland crops, *J. Agric. Sci. Cambridge,* 101, 699, 1983.
37. **Arashi, K. and Nitta, H.,** Studies on the lysigenous intercellular space as the ventilating system in the culm of rice and some other graminaceous plants, *Proc. Crop Sci. Soc. Jpn.,* 24, 78, 1955.
38. **Arikado, H.,** Supplementary studies on the development of the ventilating system in various plants growing on lowland and on upland, *Bull. Fac. Agric. Mie Univ.,* 20, 1, 1959.
39. **Arkin, G. F., Blum, A., and Burnett, E.,** A root observation chamber field installation, *Tex. Agric. Exp. Stn. Misc. Publ.,* 1386, 10, 1978.
40. **Arkin, G. F. and Taylor, H. M.,** Systems, considerations and constraints, in *Modifying the Root Environment to Reduce Crop Stress,* Arkin, G. F. and Taylor, H. M., Eds., ASAE Monogr. No. 4, American Society of Agriculture Engineering St. Joseph, MI, 1981, 393.
41. **Armstrong, W.,** Oxygen diffusion from the roots of some British bog plants, *Nature (London),* 204, 801, 1964.
42. **Armstrong, W.,** The oxidizing activity of roots in waterlogged soils, *Physiol. Plant.,* 20, 920, 1967.
43. **Armstrong, W.,** Rhizosphere oxidizing in rice: an analysis of intervarietal differences in oxygen flux from the roots, *Physiol. Plant.,* 22, 296, 1969.
44. **Armstrong, W.,** Waterlogged soils, in *Environment and Plant Ecology,* Etherington, J. R., Ed., John Wiley & Sons, New York, 1975, 181.
45. **Armstrong, W.,** Root aeration in the wetland conditions, in *Plant/Life in Anaerobic Environments,* Hook, D. D. and Crawford, R. M. M., Eds., Ann Arbor Science, Ann Arbor, MI, 1978, 269.
46. **Armstrong, W.,** Aeration in higher plants, *Adv. Bot. Res.,* 7, 225, 1979.
47. **Armstrong, W. and Gaynard, T. J.,** The critical oxygen pressures for respiration in intact plants, *Physiol. Plant.,* 37, 200, 1976.
48. **Arndt, C. H.,** Temperature-growth relations of the root and hypocotyls of cotton seedlings, *Plant Physiol.,* 20, 200, 1945.
49. **Arya, L. M., Blake, G. R., and Farrell, D. A.,** Field study of soil water depletion patterns in presence of growing soybean roots. III. Rooting characteristics and root extraction of soil water, *Soil Sci. Soc. Am. Proc.,* 39, 437, 1975.
50. **Asady, G. H., Smucker, A. J. M., and Adams, M. W.,** Seedling test for the quantitative measurement of root tolerance to compacted soil, *Crop Sci.,* 25, 802, 1985.
51. **Asher, C. J. and Ozanne, P. G.,** The cation exchange capacity of plant roots, and its relationship to the uptake of insoluble nutrients, *Aust. J. Agric. Res.,* 12, 755, 1961.
52. **Aston, M. J. and Lawlor, D. W.,** The relationship between transpiration, root water uptake, and leaf water potential, *J. Exp. Bot.,* 30, 169, 1979.
53. **Atwell, B. J., Thomson, C. J., Greenway, H., Ward, G., and Waters, I.,** A study of the impaired growth of roots of *Zea mays* seedlings at low oxygen concentrations, *Plant Cell Environ.,* 8, 179, 1985.
54. **Aubertin, G. M. and Kardos, L. T.,** Root growth through porous media under controlled conditions. I. Effect of pore size and rigidity, *Soil Sci. Soc. Am. Proc.,* 29, 290, 1965a.
55. **Aubertin, G. M. and Kardos, L. T.,** Root growth through porous media under controlled conditions. II. Effect of aeration levels and rigidity, *Soil Sci. Soc. Am. Proc.,* 29, 363, 1965b.
56. **Aubertin, G. M., Rickman, R. W., and Letey, J.,** Plant ethanol content as an index of the soil-oxygen status, *Agron. J.,* 58, 305, 1966.
57. **Aujla, T. S. and Cheema, S. S.,** Water use, root density and yield of chickpea *Cicer arientinum* as affected by levels of soil-moisture storage with and without post-sowing irrigations, *Indian J. Ecol.,* 12, 91, 1985.
58. **Aura, E.,** Soil compaction by the tractor in spring and its effect on soil porosity, *J. Sci. Agric. Finl.,* 55, 91, 1983.
59. **Aycock, M. K. and McKee, C. G.,** Root size variability among several cultivars and breeding lines of Maryland tobacco, *Agron. J.,* 67, 604, 1975.
60. **Azcon, R. and Ocampo, J. A.,** Effect of root exudation on VA mycorrhizal infection at early stages of plant growth, *Plant Soil,* 82, 133, 1984.

61. **Babalola, O. and Lal, R.,** Subsoil gravel horizon and maize root growth.. I. Gravel concentration and bulk density effects, *Plant Soil,* 46, 337, 1977a.
62. **Babalola, O. and Lal, R.,** Subsoil gravel horizon and maize root growth. II. Effects of gravel size, inter-gravel texture and natural gravel horizon, *Plant Soil,* 46, 347, 1977b.
63. **Bacha, R. E. and Hossner, L. R.,** Characteristics of coatings formed on rice roots as affected by iron and manganese additions, *Soil Sci. Soc. Am. J.,* 41, 931, 1977.
64. **Baeumer, K.,** First experiments with direct drilling in Germany, *Neth. J. Agric. Sci.,* 18, 283, 1970.
65. **Baeumer, K.,** Tillage effects on root growth and crop yield, in *16th Colloq. Int. Potash Inst.,* 1981, 57.
66. **Bailey, A. C., Johnson, C. E., and Schafer, R. L.,** A model for agricultural soil compaction, *J. Agric. Eng. Res.,* 33, 257, 1986.
67. **Baker, J. M. and van Bavel, C. H. M.,** Resistance of plant roots to water loss, *Agron. J.,* 78, 641, 1986.
68. **Bakermans, W. A. P. and de Wit, C. T.,** Crop husbandry on naturally compacted soils, *Neth. J. Agric. Sci.,* 18, 225, 1970.
69. **Baldwin, J. P., Tinker, P. B., and Nye, P. H.,** Uptake of solutes by multiple root systems from soil. II. Theoretical effects of rooting density and pattern on uptake of nutrients from soil, *Plant Soil,* 36, 693, 1972.
70. **Baligar, V. C., Nash, V. E., Hare, M. L., and Price, J. A., Jr.,** Soybean root anatomy as influenced by soil bulk density, *Agron. J.,* 67, 842, 1975.
71. **Baligar, V. C., Nash, V. E., Whisler, F. D., and Myhre, D. L.,** Sorghum and soybean growth as influenced by synthetic pans, *Commun. Soil Sci. Plant Anal.,* 12, 97, 1981.
72. **Baligar, V. C., Whisler, F. D., and Nash, V. E.,** Soybean seedling root growth as influenced by soil texture, matric suction and bulk density, *Commun. Soil Sci. Plant Anal.,* 11, 903, 1980.
73. **Ball, B. C. and O'Sullivan, M. F.,** Soil strength and crop emergence in direct and ploughed cereal seedbeds in seven field experiments, *J. Soil Sci.,* 33, 609, 1982.
74. **Bandara, B. W. and Fritton, D. D.,** Directional response of corn roots to physical barriers, *Plant Soil,* 96, 359, 1986.
75. **Barber, D. A., Ebert, M., and Evans, N. T. S.,** The movement of $^{15}O_2$ through barley and rice plants, *J. Exp. Bot.,* 13, 397, 1962.
76. **Barber, S. A.,** A diffusion and mass-flow concept of soil nutrient availability, *Soil Sci.,* 93, 39, 1962.
77. **Barber, S. A.,** Effect of tillage practice on corn (*Zea mays* L.) root distribution and morphology, *Agron. J.,* 63, 724, 1971.
78. **Barber, S. A.,** Influence of the plant root on ion movement in soil, in *The Plant Root and Its Environment,* Carson, E. W., Ed., University Press of Virginia, Charlottesville, 1974, .
79. **Barber, S. A.,** *Soil Nutrient Bioavailability: A Mechanistic Approach,* John Wiley & Sons, New York, 1984.
80. **Barbetti, M. J. and MacNish, G. C.,** Root rot of subterranean clover in the irrigation areas of south-western Western Australia, *Aust. J. Exp. Agric. Anim. Husb.,* 18, 426, 1978.
81. **Barlett, R. J.,** Iron oxidation proximate to plant roots, *Soil Sci.,* 92, 372, 1961.
82. **Barley, K. P.,** Effects of root growth and decay on the permeability of synthetic sandy loam, *Soil Sci.,* 78, 205, 1954.
83. **Barley, K. P.,** The effects of mechanical stress on the growth of roots, *J. Exp. Bot.,* 13, 95, 1962.
84. **Barley, K. P.,** Mechanical resistance of the soil in relation to the growth of roots and emerging shoots, *Agrochimica,* XX, 173, 1976.
85. **Barley, K. P., Farrell, D. A., and Greacen, E. L.,** The influence of soil strength on the penetration of a loam by plant roots, *Aust. J. Soil Res.,* 3, 69, 1965.
86. **Barley, K. P. and Greacen, E. L.,** Mechanical resistance as a factor influencing the growth of roots and underground shoots, *Adv. Agron.,* 19, 1, 1967.
87. **Barlow, P. W.,** The root cap, in *The Development and Function of Roots,* Torrey, J. G. and Clarkson, D. T., Eds., Academic Press, London, 1975, 21.
88. **Barlow, P. W. and Hines, E. R.,** Regeneration of the root cap of *Zea mays* L. and *Pisum sativum* L.: a study with the scanning electron microscope, *Ann. Bot.,* 49, 521, 1982.
89. **Barlow, P. W. and Sargent, J. A.,** The ultrastructure of the regenerating root cap of *Zea mays* L., *Ann. Bot.,* 42, 791, 1978.
90. **Barnes, B. T. and Ellis, F. B.,** The effect of different methods of cultivation and direct drilling, and of contrasting methods of straw disposal, on population of earthworms, *J. Soil Sci.,* 30, 669, 1979.
91. **Barnes, K. K., Carleton, W. M., Taylor, H. M., Throckmorton, R. I., and van der Berg, G. E.,** in *Compaction of Agricultural Soils,* Barnes, K. K., Ed., American Society of Agriculture Engineering St. Joseph, MI, 1971, 471.
92. **Barney, C. W.,** Effects of soil temperature and light intensity on root growth of loblolly pine seedlings, *Plant Physiol.,* 26, 146, 1951.
93. **Barraclough, P. B.,** The growth and activity of winter wheat roots in the field: root growth of high-yielding crops in relation to shoot growth, *J. Agric. Sci. Cambridge,* 103, 439, 1984.
94. **Barraclough, P. B. and Leigh, R. A.,** The growth and activity of winter wheat roots in the field: the effect of sowing date and soil type on root growth of high-yielding crops, *J. Agric. Sci.,* 103, 59, 1984.

95. **Barraclough, P. B. and Weir, A. H.,** Effects of compacted subsoil layer on root and shoot growth, water and nutrient uptake of winter wheat, *J. Agric. Sci. Cambridge,* 110, 207, 1988.
96. **Barrs, H. D. and Klepper, B.,** Cyclic variations in plant properties under constant environmental conditions, *Physiol. Plant.,* 21, 711, 1968.
97. **Barry, D. A. J. and Miller, M. H.,** Corn response to restricted nodal root growth with relevance to zero tillage, *Can. J. Soil Sci.,* 66, 689, 1986.
98. **Barta, A. L.,** Regrowth and alcohol dehydrogenase activity on waterlogged alfalfa and birdsfoot trefoil, *Agron. J.,* 72, 1017, 1980.
99. **Barta, A. L.,** Ethanol synthesis and loss from flooded roots of *Medicago sativa* L. and *Lotus corniculatus* L., *Plant Cell Environ.,* 7, 187, 1984.
100. **Barta, A. L.,** Metabolic response of *Medicago sativa* L. and *Lotus corniculatus* L. roots to anoxia, *Plant Cell Environ.,* 9, 127, 1986.
101. **Barta, A. L.,** Supply and partitioning of assimilates to roots of *Medicago sativa* L. and *Lotus corniculatus* L. under anoxia, *Plant Cell Environ.,* 10, 151, 1987.
102. **Bar-Yosef, B. and Lambert, J. R.,** Corn and cotton root growth in response to soil impedance and water potential, *Soil Sci. Soc. Am. J.,* 45, 930, 1981.
103. **Bauder, W., Randall, G. W., and Schuler, R. T.,** Effects of tillage with controlled wheel traffic on soil properties and root growth of corn, *J. Soil Water Conserv.,* 40, 382, 1985.
104. **Baver, L. P., Gardner, W. H., and Gardner, W. R.,** *Soil Physics,* 4th ed., John Wiley & Sons, New York, 1972.
105. **Beckel, D. K. B,** Cortical disintegration in the roots of *Bouteloua gracilis* (H.B.K.) Lag., *New Phytol.,* 55, 183, 1956.
106. **Begg, J. E. and Turner, N. C.,** Water potential gradients in field tobacco, *Plant Physiol.,* 46, 343, 1970.
107. **Belford, R. K., Cannell, R. Q., Thomson, R. J., and Dennis, C. W.,** Effects of waterlogging and different stages of development on the growth and yield of peas (*Pisum sativum* L.), *J. Sci. Food Agric.,* 31, 857, 1980.
108. **Belford, R. K. and Henderson, F. K. G.,** Measurement of the growth of wheat roots using a TV camera system in the field, in *Wheat Growth and Modelling,* Day, W. and Arkin, R. K., Eds., Plenum Press, New York, 1985, 99.
109. **Belgrand, M., Dreyer, E., Joannes, H., Velter, C., and Scuiller, I.,** A semi-automated system for root growth analysis: application to a growing oak seedling, *Tree Physiol.,* 3, 393, 1987.
110. **Bell, T. S.,** Stunt Disorders in Arable Crops, Progress Report on Investigations in Eastern and East Midlands Regions, Aerial Photographic Unit, Agricultural Development and Advisory Services, Ministry of Agriculture, Fisheries and Food, London, 1979.
111. **Bengough, A. C. and Mullins, C. E.,** Use of a low-friction penetrometer to estimate soil resistance to root growth, in *Proc. 11th Conf. ISTRO,* Edinburgh, Scotland, 1988, 1.
112. **Bengston, C., Falk, C. O., and Larson, S.,** The aftereffect of water stress on transpiration rate and changes in abscisic acid content of young wheat plants, *Physiol. Plant.,* 41, 149, 1977.
113. **Bennet, R. J., Breen, C. M., and Fey, M. V.,** Aluminium uptake sites in the primary root of *Zea mays* L., *S. Afr. J. Plant Soil,* 2, 1, 1985a.
114. **Bennet, R. J., Breen, C. M., and Fey, M. V.,** The primary site of aluminium injury in the root of *Zea mays* L., *S. Afr. J. Plant Soil,* 2, 8, 1985b.
115. **Bennet, R. J., Breen, C. M., and Fey, M. V.,** Aluminium toxicity and induced nutrient disorders involving the uptake and transport of P, K, Ca and Mg in *Zea mays* L., *S. Afr. J. Plant Soil,* 3, 11, 1986c.
116. **Bennet, R. J., Breen, C. M., and Fey, M. V.,** Aluminium induced changes in the morphology of the quiescent centre, proximal meristem and growth region of the root of *Zea mays* L., *S. Afr. J. Bot.,* 51, 355, 1985d.
117. **Bennet, R. J., Breen, C. M., and Fey, M. V.,** Aluminium toxicity and regeneration of the root cap: preliminary evidence for a Golgi apparatus derived morphogen in the primary root of *Zea mays, S. Afr. J. Bot.,* 51, 363, 1985e.
118. **Bennett, O. L. and Doss, B. D.,** Effect of soil moisture level on root distribution cool-season forage species, *Agron. J.,* 52, 204, 1960.
119. **Bennett, R. A. and Lynch, J. M.,** Bacterial growth and development in the rhizosphere of gnobiotic cereal plants, *J. Gen. Microbiol.,* 125, 95, 1981.
120. **Bennie, A. T. P. and Botha, F. J. P.,** Effect of deep tillage and controlled traffic on root growth, water use efficiency and yield of irrigated maize and wheat, *Soil Till. Res.,* 7, 85, 1986.
121. **Bennie, A. T. P. and Burger, R. D. T.,** Root Characteristics of Different Crops as Affected by Mechanical Resistance in Fine Sandy Soils, Tech. Commun. No. 180, Department of Agriculture and Fisheries, Republic of South Africa, 1983, 29.
122. **Bennie, A. T. P. and van Antwerpen, R.,** An evaluation of empirical root growth models to describe root distribution in compacted sandy soils, in *Proc. 11th Conf. ISTRO,* Edinburgh, Scotland, 1988, 343.
123. **Berry, L. J.,** The influence of oxygen tension on the respiration rate in different segments of onion roots, *J. Cell. Comp. Physiol.,* 33, 41, 1949.

124. **Berry, L. J. and Norris, W. E.,** Studies of onion root respiration velocity of oxygen consumption in different segments of root of different temperatures as a function of partial pressure of oxygen, *Biochim. Biophys. Acta*, 3, 593, 1949.
125. **Bertani, A., Brambilla, I., and Menegas, F.,** Effect of anaerobiosis on rice seedlings growth, metabollic rate, and fate of fermentation products, *J. Exp. Bot.*, 31, 325, 1980.
126. **Bertrand, A. R. and Kohnke, H.,** Subsoil conditions and their effects on oxygen supply and growth of corn roots, *Soil Sci. Soc. Am. Proc.*, 21, 135, 1957.
127. **Beukes, D. J.,** Apple root distribution as affected by irrigation at different soil water levels on two soil types, *J. Am. Soc. Hortic. Sci.*, 109, 723, 1984.
128. **Bhat, K. K. S.,** Nutrient inflows into apple roots. II. Nitrate uptake rates measured on intact roots of mature trees under field conditions, *Plant Cell Environ.*, 5, 461, 1982.
129. **Bhat, K. K. S.,** Nutrient inflows into apple roots, *Plant Soil*, 71, 371, 1983.
130. **Bhat, K. K. S. and Nye, P. H.,** Diffusion of phosphate to plant roots in soil. I. Quantitative autoradiography of the depletion zone, *Plant Soil*, 38, 161, 1973.
131. **Bielorai, H.,** The effect of partial wetting of the root zone on yield and water use efficiency in a drip- and sprinkler-irrigated mature grapefruit grove, *Irrig. Sci.*, 3, 89, 1982.
132. **Bielorai, H.,** Moisture, salinity and root distribution of drip irrigated grapefruit, in *Proc. 3rd Int. Drip/Trickle Irrigation Congr., Vol. 2*, American Society of Agricultural Engineers, 1985, 562.
133. *Biologicheskaya Sovietskaya Enciklopedija* (Biological Encyclopedia), Moskva, Izd-vo 1986, 512 (in Russian).
134. **Black, C. A.,** *Soil-Plant Relationships*, 2nd ed., John Wiley & Sons, New York, 1968.
135. **Blacklow, W. M.,** Influence of temperature on germination and elongation of the radicle and shoot of corn (*Zea mays* L.), *Crop Sci.*, 12, 647, 1972.
136. **Blackwell, P. S., Ward, M. A., Lefevre, R. N., and Cowan, D. J.,** Compaction of a swelling clay soil by agricultural traffic; effects upon conditions for growth of winter cereals and evidence for some recovery of structure, *J. Soil Sci.*, 36, 633, 1985.
137. **Blackwell, P. S. and Wells, E. A.,** Limiting oxygen flux densities for oat root extension, *Plant Soil*, 73, 129, 1983.
138. **Blanchar, R. W., Edmonds, C. R., and Bradford, J. M.,** Root growth in cores formed from fragipan and B_t horizons of Hobson soil, *Soil Sci. Soc. Am.* J., 42, 437, 1978.
139. **Blanchet, R., Bosc, M., Maertens, C., and Puech, J.,** The influence of different hydrous systems on the absorption of water and mineral elements by crops. II. The influence of soil humidity and hydrous flows on the absorption of mineral elements by roots, *Ann. Agron.*, 25, 681, 1974 (French, English summary).
140. **Bland, W. L. and Dugas, W. A.,** Root length density from minirhizotron observations, *Agronomy*, 80, 271, 1988.
141. **Blevis, D. G., Lowe, R. H., and Staples, L.,** Nitrate reductase in barley root under sterile, low oxygen conditions, *Plant Physiol.*, 57, 458, 1976.
142. **Blizzard, W. E.,** Comparative resistance of the soil and the plant water transport, *Plant Soil*, 66, 809, 1980.
143. **Blum, A. and Arkin, S. F.,** Sorghum root growth and water-use as affected by water supply and growth duration, *Field Crop Res.*, 9, 131, 1984.
144. **Blum, A. and Ritchie, J. T.,** Effect of soil surface water content on sorghum root distribution in the soil, *Field Crops Res.*, 8, 169, 1984.
145. **Böhm, W., Ed.,** Minirhizotrons for root observations under field conditions, Z. Acker-, *Phlanzenbau*, 140, 282, 1974.
146. **Böhm, W., Ed.,** *Methods For Studying Root System, Ecological Studies*, Springer-Verlag, Berlin, 1979, 33.
147. **Bole, J. B.,** Influence of root hairs in supplying soil phosphorous to wheat, *Can. J. Soil Sci.*, 53, 169, 1973.
148. **Bolton, E. F., Dirks, V. A., and Findlay, W. I.,** Some relationships between soil porosity, leaf nutrient composition and yield for certain crop rotations at two fertility levels on Brookston clay, *Can. J. Soil Sci.*, 59, 1, 1979.
149. **Bolton, E. F. and Erickson, A. E.,** Ethanol concentration in tomato plants during soil flooding, *Agron. J.*, 62, 220, 1970.
150. **Boone, F. R.,** Notes on soil structure homogeneity and rootability, in *Trans. 7th Conf. ISTRO*, Sweden, 1976, 4.
151. **Boone, F. R.,** Towards soil compaction limits for crop growth, *Neth. J. Agric. Sci.*, 34, 349, 1986.
152. **Boone, F. R., Bouma, J., and de Smet, L. A. H.,** A case study on the effect of soil compaction on potato growth in a loamy sand soil. I. Physical measurements and rooting patterns, *Neth. J., Agric.*, 26, 405, 1978.
153. **Boone, F. R., de Smet, L. A. H., and van Loon, C. D.,** The effect of ploughpan in marine loam soils on potato growth. I. Physical properties and rooting patterns, *Potato Res.*, 28, 295, 1985.
154. **Boone, F. R., Kroesbergen, B., and Boers, A.,** Soil conditions and growth of spring barley on a tilled and untilled marine loam soil, in Experiments With Three Tillage Systems On a Marine Loam Soil, II, 1976-1979, Agric. Res. Rep., Pudoc,, Wageningen, The Netherlands 1984, 925, 124.

155. **Boone, F. R., van der Werf, H. M. G., Kroesbergen, B., ten Hag, B. A., and Boers, A.,** The effect of compaction on the arable layer in sandy soils on the growth of maize for silage. I. Critical matric potentials in relation to soil aeration and mechanical impedance, *Neth. J. Agric. Sci.,* 34, 155, 1986.
156. **Boone, F. R. and Veen, B. W.,** Influence of mechanical resistance and phosphate supply on morphology and function of maize roots, *Neth. J. Agric. Sci.,* 30, 179, 1982.
157. **Borisova, T.,** Effects of some exogenous factors on the water relations in maize roots, in *Abstr. Symp. Plant Roots and Their Environment,* Uppsala, Sweden 1988, 1.
158. **Borkert, C. M. and Barber, S. A.,** Soybean shoot and root growth and phosphorus concentration as affected by phosphorus placement, *Soil Sci. Soc. Am. J.,* 49, 152, 1985.
159. **Bowen, H. D.,** Alleviating mechanical impedance, in *Modifying the Root Environment to Reduce Crop Stress,* Arkin, G. F. and Taylor, H. M., Eds., ASAE Monogr. No. 4, American Society of Agriculture Engineering, St. Joseph, MI, 1981, 21.
160. **Boyer, J. S.,** Water transport in plants: mechanism of apparent changes in resistance during absorption, *Planta,* 117, 187, 1974.
161. **Bradford, J. M.,** The penetration resistance in a soil with well-defined structural units, *Soil Sci. Soc. Am. J.,* 44, 601, 1980.
162. **Bradford, J. M. and Blanchard, R. W.,** Profile modification of a Fragiudalf to increase crop production, *Soil Sci. Soc. Am. J.,* 41, 127, 1977.
163. **Bradford, K. J. and Dilley, D. R.,** Effects of root anaerobiosis on ethylene production, epinasty, and growth of tomato plants, *Plant Physiol.,* 61, 506, 1978.
164. **Bradford, K. J., Hsiao, T. C., and Yang, S. F.,** Inhibition of ethylene synthesis in tomato plants subjected to anaerobic root stress, *Plant Physiol.,* 70, 1503, 1982.
165. **Bradford, K. J. and Yang, S. F.,** Xylem transport of 1-aminocyclopropane-1-carboxylic acid, an ethylene precursor, in waterlogged tomato plants, *Plant Physiol.,* 65, 322, 1980a.
166. **Bradford, K. J. and Yang, S. F.,** Stress-induced ethylene production in the ethylene-requiring tomato mutant diageotropia, *Plant Physiol.,* 65, 327, 1980b.
167. **Bradford, K. J. and Yang, S. F.,** Physiological responses of plants to waterlogging, in Proc. of the Workshop on Adaptation to Water Stress in Plants, *HortScience* (Spec. Insert), 16(1), 25, 1981.
168. **Bragg, P. L.,** Genetic Differences in Root Morphology of Barley and Significance for Growth in Compact Soil, Thesis, Edinburgh University, Scotland, 1981, 190.
169. **Bragg, P. L., Govi, G., and Cannell, R. Q.,** Comparison of methods, including angle and vertical minirhizotrons for studying root growth and distribution in a spring oat crop, *Plant Soil,* 73, 435, 1983.
170. **Brams, E. A.,** The mucilagenous layer of citrus roots, its delineation in the rhizosphere and removal from roots, *Plant Soil,* 30, 105, 1969.
171. **Braun, H. J. and Marte, I.,** Concerning the water absorption of trees after the end of daily transpiration and thoughts on the transpiration — principle, *Allg. Forst Jagdztg.,* 158, 67, 1987.
172. **Braunack, M. V. and Freebairn, D. M.,** The effect of bulk density on root growth, in *Proc. 11th Conf. ISTRO,* Edinburgh, Scotland, 1988, 25.
173. **Breisch, H., Guckert, A., and Reisinger, O.,** Electron microscope study of the apical zone of maize roots, *Soc. Bot. Fr. Coll. Rhizosphere,* 122, 55, 1975 (in French, English summary).
174. **Breteler, H.,** A comparison between ammonium and nitrate nutrition of young sugarbeet plants grown in nutrient solution at constant acidity. I. Production of dry matter, ionic balance, chemical composition, *Neth. J. Agric. Sci.,* 21, 227, 1973.
175. **Breteler, H. and Smit, A. L.,** Effect of ammonium nutrition on uptake and metabolism of nitrate in wheat, *Neth. J. Agric. Sci.,* 22, 73, 1974.
176. **Brouwer, R.,** Nutritive influences on the distribution of dry matter in the plant, *Neth. J. Agric. Sci.,* 10, 399, 1962a.
177. **Brouwer, R.,** Influence of temperature of the root medium on the growth of seedlings of various crop plants, *Jaarb. Inst. Biol. Scheik. Onderz. Landb. Gewass.,* 11, 1962b.
178. **Brouwer, R. and van Vliet, G.,** The influence of root temperature on growth and uptake of peas, *Jaarb. Inst. Biol. Scheik. Onderz. Landb. Gewass.,* 23, 1960.
179. **Brown, E. M.,** Some effects of temperature on the growth and chemical composition of certain pasture grasses, *Mont. Agric. Exp. Stn. Res. Bull.,* 299, 1939.
180. **Buckman, H. C. and Brady, N. C., Eds.,** *The Nature and Properties of Soils,* 7th ed., Macmillan, London, 1969.
181. **Bullock, P., Newman, A. C. D., and Thomasson, A. J.,** Porosity aspects of the regeneration of soil structure after compaction, *Soil Till. Res.,* 5, 325, 1985.
182. **Bunce, J. A.,** Effects of shoot environment, apparent root resistance to water flow in soybean and cotton plants, *J. Exp. Bot.,* 29, 595, 1978.
183. **Burch, G. J.,** Soil and plant resistances to water absorption by plant root systems, *Aust. J. Agric. Res.,* 30, 279, 1979.

184. **Burke, D. W., Holmes, L. D., and Barker, A. W.,** Distribution of *Fusarium solani* f. sp. *phaseoli* and bean roots in relation to tillage and soil compaction, *Phytopathology,* 62, 500, 1972a.
185. **Burke, D. W., Miller, D. E., Holmes, L. D., and Barker, A. W.,** Counteracting bean root rot by loosening the soil, *Phytopathology,* 62, 306, 1972b.
186. **Busscher, W. J. and Fritton, D. D.,** Simulated flow through the root xylem, *Soil Sci.,* 125, 1, 1978.
187. **Busscher, W. J., Spivey, L. D., Jr., and Campbell, R. B.,** Estimation of soil strength properties for critical rooting conditions, *Soil Till. Res.,* 9, 377, 1987.
188. **Callot, G., Chamayon, H., Maertens, C., and Salsac, L.,** *Soil-Root Interaction. Aspects of Mineral Nutrition,* INRA, Paris, 1982 (in French).
189. **Cambus, J. P.,** Development of wheat roots under zinc deficiency, *Plant Soil,* 83, 313, 1985.
190. **Cameron, D. G.,** Lucerne in wet soils — the effect of stage of regrowth, cultivar, air temperature and root temperature, *Aust. J. Agric. Res.,* 24, 851, 1973.
191. **Camp, C. R. and Lund, Z. F.,** Effect of soil compaction on cotton roots, *Crops Soils,* 17, 13, 1964.
192. **Campbell, C. A. and Ferguson, W. S.,** Influence of air temperature, light intensity, soil moisture stress, and soil aeration on moisture use by wheat, *Can. J. Plant Sci.,* 49, 129, 1969.
193. **Campbell, G. S.,** *Soil Physics with Basic Development in Soil Science,* Elsevier, Amsterdam, 1985,14.
194. **Campbell, R. B., Reicosky, D. C., and Doty, C. W.,** Physical properties and tillage of Paleudults in the southeastern Coastal Plains, *J. Soil Water Conserv.,* 29, 220, 1974.
195. **Canarache, A. and Berindei, M.,** A synthetic index to describe soil physical status, *Stiinta Solului Noutati,* 1, 33, 1978 (in Romanian).
196. **Canarache, A., Colibas, I., Colibas, M., Horobeanu, I., Patru, V., Simota, H., and Trandafirescu, T.,** Effect of induced compaction by wheel traffic on soil physical properties and yield of maize in Romania, *Soil Till. Res.,* 4, 199, 1984.
197. **Cannell, R. Q.,** Reduced tillage in north-west Europe — a review, *Soil Till. Res.,* 5, 129, 1985.
198. **Cannell, R. Q., Belford, R. K., Blackwell, P. S., Govi, G., and Thomson, R. J.,** Effects of waterlogging on soil aeration and on root and shoot growth and yield of winter oats (*Avena sativa* L.), *Plant Soil,* 85, 361, 1985.
199. **Cannell, R. Q., Christian, D. G., and Henderson, F. K. G.,** A study of mole drainage with simplified cultivation for autumn-sown crops on a clay soil. IV A comparison of direct drilling and mouldboard ploughing on drained and undrained land on root and shoot growth, nutrient uptake and yield, *Soil Till. Res.,* 7, 251, 1986.
200. **Cannell, R. Q., Gales, K., Snaydon, R. W., and Suhail, B. A.,** Effects of short-term waterlogging on the growth and yield of peas (*Pisum sativum*), *Ann. Appl. Biol.,* 93, 327, 1979.
201. **Cannell, R. Q., Goss, M. J., Harris, G. L., Jarvis, M. J., Douglas, J. T., Howse, K. R., and Le Grice, S.,** A study of mole drainage with simplified cultivation for autumn-sown crops on a clay soil. I. Background experiment and site details drainage, system, measurement of drainflow and summary of results 1978-1980, *J. Agric. Sci. Cambridge,* 102, 539, 1984.
202. **Caradus, J. R.,** Selection for root hair length in white clover (*Trifolium repens* L.), *Euphytica,* 28, 849, 1979.
203. **Caradus, J. R.,** Effect of root hair length on white clover growth over a range of soil phosphorus levels, *N.Z. J. Agric. Res.,* 24, 353, 1981.
204. **Carey, R. W. and Berry, J. A.,** Effects of low temperature on respiration and uptake of rubidium ions by excised barley and corn roots, *Plant Physiol.,* 61, 858, 1978.
205. **Carlgren, K.,** Foliar Application of Plant Nutrients to Winter Wheat, Rep. 166, Swedish University of Agriculture, Uppsala, 1986 (in Swedish, English summary).
206. **Carlson, P. R., Jr.,** Oxygen Diffusion from the Roots of *Spartina alterniflora* and the Role of *Spartina* in the Sulfur Cycle of Salt Marsh Sediments, PhD. dissertation, The University of North Carolina, Chapel Hill, 1980.
207. **Carpenter, J. R. and Mitchell, C. A.,** Root respiration characteristics of flood-tolerant and intolerant tree species, *J. Am. Soc. Hortic. Sci.,* 105, 684, 1980a.
208. **Carpenter, J. R. and Mitchell, C. A.,** Flood-induced shift of electron flow between cyanide-sensitive and alternative respiratory pathways in roots of tolerant and intolerant tree species, *J. Am. Soc. Hortic. Sci.,* 105, 688, 1980b.
209. **Carr, M. K. V. and Dodds, S. M.,** Some effects of soil compaction on root growth and water use of lettuce, *Exp. Agric.,* 19, 117, 1983.
210. **Carrow, R. N.,** Influence of soil compaction on three turfgrass species, *Agron. J.,* 72, 1038, 1980.
211. **Casarin, M., Jacquey, S., and Fouere, A.,** Root picture processing software, in *Abstr. of Symp. on Plant Roots and Their Environment,* Uppsala, Sweden, 1988, 4.
212. **Cascio, B. L. and Venezia, G.,** Environmental quality of clay soil as affected by reduced tillage in a Mediterranean environment, *Riv. Agron.,* 20, 166, 1986 (in Italian, English summary).
213. **Case, V. W., Brady, N. C., and Lathwell, D. J.,** The influence of soil temperature and phosphorus fertilizers of different water-solubilities on the yield and phosphorus uptake by oats, *Soil Sci. Soc. Am. Proc.,* 28, 409, 1964.
214. **Cassel, D. K.,** Subsoiling, *Crop Soils,* 37, 7, 1979.

215. **Cassel, D. K.,** Tillage effects on soil bulk density and mechanical impedance, in *Predicting Tillage Effects on Soil Physical Properties and Processes,* Spec. Publ. 44, American Society Agronomy, Madison, WI, 1982, 45.
216. **Castillo, S. R., Dowdy, R. H., Bradford, J. M., and Larson, W. E.,** Effects of applied mechanical stress on plant growth and nutrient uptake, *Agron. J.,* 74, 526, 1982.
217. **Castle, W. S. and Krezdorn, A. H.,** Soil water use and apparent root efficiencies of citrus trees on four rootstocks, *J. Am. Soc. Hortic. Sci.,* 102, 403, 1977.
218. **Champion, R. A. and Barley, K. P.,** Penetration of clay by root hairs, *Soil Sci.,* 108, 402, 1969.
219. **Chamucah, G. S. and Dey, S. K.,** Cation-exchange of clonal tea roots and its implications on fertilizer responses, *J. Sci. Food Agric.,* 33, 309, 1982.
220. **Chancellor, W. J.,** *Compaction of Soil by Agricultural Equipment,* Bull. No. 1881, Division of Agricultural Sciences, University of California, 1977.
221. **Chang, L. A., Hammett, L. K., and Pharr, D. M.,** Carbon dioxide effects on ethanol production, pyruvate decarboxylase, and alcohol dehydrogenase activities in anaerobic sweet potato roots, *Plant Physiol.,* 71, 59, 1983.
222. **Charyulu, P. B. B. N. and Rao, V. R.,** Nitrogen fixation in some Indian rice soils, *Soil Sci.,* 128, 86, 1979.
223. **Chaudhary, M. R., Gajri, P. R., Prihar, S. S., and Khera, R.,** Effect of deep tillage on soil physical properties and maize yields on coarse textured soils, *Soil Till. Res.,* 6, 31, 1985.
224. **Chaudhary, M. R. and Prihar, S. S.,** Root development and growth response of corn following mulching, cultivation, or interrow compaction, *Agron. J.,* 66, 350, 1974.
225. **Chen, C. C., Dixon, J. B., and Turner, F. T.,** Iron coatings on rice roots: morphology and models of development, *Soil Sci. Soc. Am. J.,* 44, 1113, 1980.
226. **Chirkova, T. V.,** Some regulatory mechanisms of plant adaptation to temporal anaerobiosis, in *Plant Life in Anaerobic Environments,* Hook, D. D. and Crawford, R. M. M., Eds., Ann Arbor Science, Ann Arbor, MI, 1978, 137.
227. **Chirkova, T. V. and Gutman, T. S.,** Physiological role of branch lenticels in willow and poplar under conditions of root anaerobiosis, *Sov. Plant Physiol.,* 19, 289, 1972.
228. **Chirkova, T. V., Khoang, K. I., and Blyudzin, Y. A.,** Effect of anaerobic conditions on fatty acid composition of wheat and rice root phospholipids, *Fiziol. Rast.,* 28, 358, 1981.
229. **Chrominski, A., Khan, M. A., Weber, D. J., and Smith, B. N.,** Ethylene and ether productions in response to salinity stress, *Plant Cell Environ.,* 9, 687, 1986.
230. **Chu, D. B. and Toop, E. W.,** Effects of substrate potassium, substrate temperature and light intensity on growth and uptake of major cations by greenhouse tomato plants, *Can. J. Plant Sci.,* 55, 121, 1975.
231. **Ciamporova, M. and Luxova, M.,** The effect of polyethylene glycol induced water stress on plant growth and nutrient uptake, *Biol. Plant.,* 18, 173, 1976.
232. **Cieslinski, Z., Miatkowski, Z., and Soltysik, A.,** Influence of agromeliorative ploughing on some physical properties of soil and the distribution of the roots of wheat, *Zesz. Probl. Postepow Nauk Roln.,* 385, in press.
233. **Claassen, N. and Barber, S. A.,** Simulation model for nutrient uptake from soil by a growing plant root system, *Agron. J.,* 68, 961, 1976.
234. **Claassen, N. and Jungk, A.,** Potassium dynamics at the soil-root interface in relation to the uptake of potassium by maize plants, *Z. Pflanzenernaehr. Bodenkd.,* 145, 513, 1982 (in German, English summary).
235. **Claassen, N. and Jungk, A.,** Effects of K uptake rate, root growth and root hairs on potassium uptake efficiency of several plant species, *Z. Pflanzenernaehr. Bodenkd.,* 147, 276, 1984.
236. **Claassen, N., Syring, M., and Jungk, A.,** Verification of a mathematical model by simulating potassium uptake from soil, *Plant Soil,* 95, 209, 1986.
237. **Clarkson, D. T., Hopper, M. J., and Jones, L. H. P.,** The effect of root temperature on the uptake of nitrogen and the relative size of the root system in *Lolium perenne.* I. Solutions containing both NH_4^+ and NO_3^-, *Plant Cell Environ.,* 9, 535, 1986.
238. **Clarkson, D. T. and Warner, A.,** Relationship between root temperature and the transport of ammonium and nitrate ions by Italian and perennial ryegrass (*Lolium multiflorum* and *Lolium perenne*), *Plant Physiol.,* 64, 557, 1979.
239. **Clemens, J., Kirk, A. M., and Mills, P. D.,** The resistance to waterlogging of three *Eucalyptus* species. Effect of waterlogging and an ethylene-releasing growth substance on *E. robusta, E. grandis* and *E. saligna, Oecologia,* 34, 125, 1978.
240. **Clemens, J. and Pearson, C. J.,** The effect of waterlogging on the growth and ethylene content of *Eucalyptus robusta* Sm. (Swamp mahogany), *Oecologia,* 29, 249, 1977.
241. **Cline, R. A. and Erickson, A. E.,** The effect of oxygen diffusion rate and applied fertilizer on the growth, yield, and chemical composition of peas, *Soil Sci. Soc. Am. Proc.,* 23, 333, 1959.
242. **Clowes, F. A. L. and Juniper, B. E.,** The fine structure of the quiescent centre and neighbouring tissues in root meristems, *J. Exp. Bot.,* 15, 622, 1964.
243. **Coale, F. J. and Grove, J. H.,** Alterations in soybean root development due to cultural practices: a review, *Commun. Soil Sci. Plant Anal.,* 799, 1986.

244. **Coale, F. J., Meisinger, J. J., and Wiebold, W. J.,** Effects of plant breeding and selection on yields and nitrogen fixation in soybeans under two soil nitrogen regimes, *Plant Soil,* 86, 357, 1985.
245. **Cockroft, B., Barley, K. P., and Greacen, E. L.,** The penetration of clays by fine probes and root tips, *Aust. J. Soil Res.,* 7, 333, 1969.
246. **Cole, P. J. and Alston, A. M.,** Effect of transient dehydration on absorption of chloride by wheat roots, *Plant Soil,* 40, 243, 1974.
247. **Collins, J. C. and Kerrigan, A. P.,** The effect of kinetin and abscisic acid on water and ion transport in isolated maize roots, *New Phytol.,* 73, 309, 1974.
248. **Collis-George, N. and Yoganathan, P.,** The effect of soil strength on germination and emergence of wheat (*Triticum aestivum* L.). I. Low shear strength conditions, *Aust. J. Soil Res.,* 23, 577, 1985a.
249. **Collis-George, N. and Yoganathan, P.,** The effect of soil strength on germination and emergence of wheat (*Triticum aestivum* L.). II. High strength conditions, *Aust. J. Soil Res.,* 23, 589, 1985b.
250. **Cook, R. J. and Papendick, R. J.,** Influence of water potential of soils and plants on root diseases, *Annu. Rev. Phytopathol.,* 10, 349, 1972.
251. **Cooper, A. J.,** The effect of root pruning on the growth of tomato plants, *J. Hortic. Sci.,* 46, 111, 1971.
252. **Cooper, A. J.,** Root Temperature and Plant Growth, Res. Rev., No. 4, Commonwealth Bureau of Horticulture and Plantation Crops, East Malling, Maidstone, Kent, 1973, 73.
253. **Cornforth, I. S.,** The effect of size of soil aggregates on nutrient supply, *J. Agric. Sci. Cambridge,* 70, 83, 1968.
254. **Cornillon, P.,** Influence of substrate temperature on the composition of tomato roots and leaf blades. Consequences for water absorption, *Ann. Agron.,* 28, 277, 1977a (in French, English summary).
255. **Cornillon, P.,** Effects of root temperature on the absorption of mineral elements by tomatoes, *Ann. Agron.,* 28, 409, 1977b (in French, English summary).
256. **Cornillon, P. and Maisonneuve, B.,** Effect of low air or root temperatures on plant nutrition and pollen fertility in tomato, *Agronomie,* 5, 33, 1985 (in French, English summary).
257. **Couchat, Ph., Moutonnet, P., Honelle, M., and Picard, D.,** *In situ* study of corn seedling root and shoot growth by neutron radiography, *Agron. J.,* 72, 321, 1980.
258. **Coult, D. A. and Vallance, K. B.,** Observations on the gaseous exchanges which take place between *Manyanthes trifoliata* L. and its environment, *J. Exp. Bot.,* 9, 384, 1958.
259. **Coutts, M. P.,** Effects of waterlogging on water relations of actively growing and dormant Sitka spruce seedlings, *Ann. Bot.,* 47, 747, 1981.
260. **Coutts, M. P. and Armstrong, W.,** Role of oxygen transport in the tolerance of trees to waterlogging, in *Tree Physiology and Yield Improvement,* Cannell, M. G. R. and Last, F. T., Eds., Academic Press, New York, 1976, 361.
261. **Cowan, I. R.,** Transport of water in the soil-plant atmosphere system, *J. Appl. Ecol.,* 2, 221, 1965.
262. **Cowan, I. R. and Milthrope, F. L.,** Plant factors influencing the water status of plant tissues, in *Water Deficits and Plant Growth,* Academic Press, New York, 1968, 137.
263. **Cox, W. J. and Reisenauer, H. M.,** Growth and ion uptake by wheat supplied nitrogen as nitrate or ammonium or both, *Plant Soil,* 38, 363, 1973.
264. **Crawford, R. M. M.,** Physiological ecology: a comparison of adaptation to oxygen poor environment in plants and animals, *Flora,* 161, 209, 1972 (in German, English summary).
265. **Crawford, R. M. M.,** Tolerance of anoxia and the regulation of glycolysis in tree roots, in *Tree Physiology and Yield Improvement,* Cannell, M. G. R. and Last, E. T., Eds., Academic Press, London, 1976, .
266. **Crawford, R. M. M.,** Tolerance of anoxia and the metabolism of ethanol in tree roots, *New Phytol.,* 79, 519, 1977.
267. **Crawford, R. M. M.,** Metabolic indicators in the prediction of soil anaerobiosis, in *Nitrogen in the Environment,* Vol. 1, Nielsen, D. R. and McDonald, J. G., Eds., Academic Press, New York, 1978, 427.
268. **Crawford, R. M. M.,** Biochemical and ecological similarities in marsh plants and diving animals, *Izv. Akad. Nauk SSSR Ser. Biol.,* 1, 77, 1981 (in Russian, English summary).
269. **Crawford, R. M. M.,** The anaerobic retreat as a survival strategy for aerobic plants and animals, *Trans. Bot. Soc. Edinburgh,* 44, 57, 1982.
270. **Crawford, R. M. M. and Baines, M. A.,** Tolerance of anoxia and the metabolism of ethanol in intact roots and seeds, *New Phytol.,* 79, 519, 1977.
271. **Crawford, R. M. M. and Huxter, T. J.,** Root growth and carbohydrates metabolism at low temperatures, *J. Exp. Bot.,* 28, 917, 1977.
272. **Crawford, R. M. M. and Palin, M. A.,** Root respiration and temperature limits to the North-South distribution of four perennial maritime plants, *Flora,* 171, 338, 1981.
273. **Crooke, W. M. and Knight, A. H.,** Root cation-exchange capacity and organic acid content of tops as indices of varieted yields, *J. Sci. Food Agric.,* 228, 389, 1971.
274. **Crookston, R. K., O Toole, J., Lee, R., Ozbun, J. L., and Wallace, D. H.,** Photosynthetic depression in beans after exposure to cold for one night, *Crop Sci.,* 14, 457, 1974.
275. **Cruse, R. M., Cassel, D. K., and Averett, F. G.,** Effect of particle surface roughness on densification of coarse textured soil, *Soil Sci. Soc. Am. J.,* 44, 692, 1980.

276. **Crush, J. R. and Caradus, J. R.,** Effect of mycorrhizas on growth of some white clovers, *N.Z. J. Agric. Res.*, 23, 233, 1980.
277. **Curl, E. A. and Truelove, B.,** *The Rhizosphere*, Springer-Verlag, New York, 1986.
278. **Cushman, J. H.,** Numerical study of some age-dependent parameters in root nutrient uptake, *Plant Soil*, 79, 123, 1984.
279. **Cutforth, H. W., Shaykewich, C. F., and Cho, C. M.,** Effect of soil water and temperature on corn (*Zea mays* L.) root growth during emergence, *Can. J. Soil Sci.*, 66, 51, 1986.
280. **Dabholkar, A. R. and Saxena, M.,** Combining-ability analysis for cation- and anion-exchange capacity in sorghum, *Indian J. Agric. Sci.*, 57, 5, 1987.
281. **Dacey, J. W. H.,** Internal winds in water lilies: an adaptation for life in anaerobic sediments, *Science*, 210, 1017, 1980.
282. **Dacey, J. W. H.,** Pressurised ventilation in the yellow water lily, *Ecology*, 62, 1137, 1981.
283. **Dacey, J. W. H. and Klug, M. J.,** Methane efflux from lake sediment through water lilies, *Science*, 203, 1253, 1979.
284. **Dacey, J. M. H. and Klug, M. J.,** Tracer studies of gas circulation in Nuphar: oxygen-18 and carbon-14 dioxide transport, *Physiol. Plant.*, 56, 361, 1982.
285. **Danilova, M. F.,** The development of absorption and transport systems in the root: anatomy and ultrastructure, in *Abstr. Symp. Plant Roots and Their Environment*, Uppsala, Sweden, 1988, 1.
286. **Dart, P. J. and Day, J. M.,** Effects of incubation temperature and oxygen tension on nitrogenase activity of legume root nodules, *Plant Soil*, Spec. Vol., 167, 1971.
287. **Dart, P. J. and Mercer, F. M.,** The effect of growth temperature, level of ammonium nitrate and light intensity on the growth and nodulation of cowpea, *Aust. J. Agric. Res.*, 16, 321, 1965.
288. **Das, D. K. and Jat, R. L.,** Adaptability of maize to high soil water conditions, *Agron. J.*, 64, 849, 1972.
289. **Das, D. K. and Jat, R. L.,** Influence of three soil-water regimes on root porosity and growth of four rice varieties, *Agron. J.*, 69, 197, 1977.
290. **Davis, G. R.,** Effect of soil compaction on root growth of *Pinus radiata* D. Don., in *Proc. Symp. Site and Productivity of Fast Growing Plantations*, Vol. 2, South Africa, 1984, 871.
291. **Davis, G. R., Neilsen, W. A., and McDavitt, J. G.,** Root distribution of *Pinus radiata* related to soil characteristics in five Tasmanian soils, manuscript.
292. **Dawkins, T. C. K.,** Crop reaction to compaction, *Soil Water*, 11, 31, 1983.
293. **Dawkins, T. C. K., Roberts, J. A., and Brereton, J. C.,** Mechanical Impedance and Root Growth — The Role of Endogenous Ethylene, Monograph, British Plant Regulator Group, School of Agriculture, Sutton Bonnington, Loughborough, Leics., 10, 55, 1983.
294. **Debicki, R.,** Comparison of some soil structure indices and their relation to the physical properties of different soil types and plant yields, *Pol. J. Soil Sci.*, XII, 139, 1979.
295. **Dechnik, I., Lipiec, J., and Tarkiewicz, S.,** Dependence of penetration resistance on the water content of sandy soil compacted by tractor wheels, *Zesz. Probl. Postepow Nauk Roln.*, 304, 39, 1985.
296. **de Cockborne, A. M., Habib, R., and Lafolie, F.,** Transfer of water and nitrate in soil and absorption by root system of peaches: presentation of hypotheses of the model and results gathered for its evaluation, *Zesz. Probl. Postepow Nauk Roln.*, 385, in press (in French, English summary).
297. **de Haan, F. A. M. and van der Valk, G. G. M.,** Effects of soil compaction, restricted aeration and frost injury on flowerbulb crops. I. Effects of compaction on physical properties of soil and root growth of ornamental bulbs, in *Proc. 1st Int. Symp. on Flower Bulbs*, Noordwijk, The Netherlands, 1970, 326.
298. **Derera, N. F., Marshall, D. R., and Balaam, L. N.,** Genetic variability in root development in relation to drought tolerance in spring wheats, *Exp. Agric.*, 5, 327, 1969.
299. **de Roo, H. C.,** Water stress gradients in plants and soil-root systems, *Agron. J.*, 61, 511, 1969a.
300. **de Roo, H. C.,** Tillage and root growth, in *Proc. 15th School in Agricultural Science*, Wittington, W. J., Ed., University of Nottingham, England, 1969b, 339.
301. **de Willigen, P.,** Some theoretical aspects of the influence of soil-root contact on uptake and transport of nutrients and water, in *Proc.* ISSS *Symp. on Water and Solute Movement in Heavy Clay Soils*, Bouma, J. and Raats, P. A. C., Eds., Wageningen, The Netherlands, 1984a, 268.
302. **de Willigen, P. and van Noordwijk, M.,** Mathematical models on diffusion of oxygen to and within plant roots, with special emphasis on effects of soil-root contact. I. Derivation of the models, *Plant Soil*, 77, 215, 1984b.
303. **de Willigen, P. and van Noordwijk, M.,** Roots, Plant Production and Nutrient Use Efficiency, Ph.D. thesis, Agricultural University of Wageningen, The Netherlands, 1987.
304. **de Wit, M. C. J.,** Morphology and function of roots and shoot growth of crop plants under oxygen deficiency, in *Plant Life in Anaerobic Environment*, Hook, D. D. and Crawford, R. M. M., Eds., Ann Arbor Science, Ann Arbor, MI, 1978, 333.
305. **Dexter, A. R.,** Tunneling in soil by earthworms, *Soil Biol. Biochem.*, 10, 447, 1978a.
306. **Dexter, A. R.,** A stochastic model for the growth of roots in tilled soil, *J. Soil Sci.*, 29, 102, 1978b.

307. **Dexter, A. R.,** Model experiments on the behaviour of roots at the interface between a tilled seed-bed and a compacted sub-soil. I. Effect of seed-bed aggregate size and sub-soil strength on wheat roots, *Plant Soil*, 95, 123, 1986a.
308. **Dexter, A. R.,** Model experiments on the behaviour of roots at the interface between a tilled seed-bed and a compacted sub-soil. II. Entry of pea and wheat roots into sub-soil cracks, *Plant Soil*, 95, 235, 1986b.
309. **Dexter, A. R.,** Model experiments on the behaviour of roots at the interface between a tilled seed-bed and a compacted sub-soil. III. Entry of pea and wheat roots into cylidrical biopores, *Plant Soil*, 95, 149, 1986c.
310. **Dexter, A. R.,** Mechanics of root growth, *Plant Soil*, 98, 303, 1987a.
311. **Dexter, A. R.,** Comparison of soil around roots, *Plant Soil*, 97, 401, 1987b.
312. **Dexter, A. R. and Hewitt, J. S.,**The deflection of plant roots, *J. Agric. Eng. Res.*, 23, 17, 1978.
313. **Dickey, E. C., Peterson, T. R., Eisenhauer, D. E., and Jasa, P. J.,** Soil compaction. I. Where, how bad, a problem, *Crops Soils Mag.*, 12, 1985.
314. **Dijkshoorn, W., Lathwell, D. J., and de Wit, M. C. J.,** Temporal changes in carboxylate content of rye grass with stepwise change in nutrient, *Plant Soil*, 29, 369, 1968.
315. **Dobrzanski, B., Dechnik, I., Glinski, J., Pondel, H., and Stawinski, J.,** Surface area of arable soils of Poland, *Rocz. Nauk Roln. Ser. D*, 165, 1977 (in Polish, English summary).
316. **Dobrzanski, B., Witkowska, B., and Walczak.,** Soil-aggregation and water-stability index, *Pol. J. Soil Sci.*, VIII, 3, 1975.
317. **Donald, R. G., Kay, B. D., and Miller, M..,** The effect of soil aggregate size on early shoot and root growth of maize (*Zea mays* L.), *Plant Soil*, 103, 251, 1987.
318. **Doss, D. B., Ashley, D. A., and Bennett, O. L.,** Effect of soil moisture regime on root distribution of warm-season forage species, *Agron. J.*, 52, 569, 1960.
319. **Doty, C. W., Campbell, R. B., and Reicosky, D. C.,** Crop response to chiseling and irrigation in soils with compact A_t horizon, *Trans. ASAE*, 18, 668, 1975.
320. **Doty, C. W. and Reicosky, D. C.,** Chiseling to minimize the effects of drought, *Trans. ASAE*, 21, 495, 1978.
321. **Dowdell, R. J. and Cannell, R. Q.,** Effect of ploughing and direct drilling on soil nitrate content, *J. Soil Sci.*, 26, 53, 1975.
322. **Dowdell, R. J., Crees, R., Burford, J. R., and Cannell, R. Q.,** Oxygen concentrations in a clay soil after ploughing or drilling, *J. Soil Sci.*, 30, 239, 1979.
323. **Dowdy, R. H., Bidwell, A. M., Linden, D. R., and Allmaras, R. R.,** Corn root distributions as a function of tillage and residue management, in *Proc. 11th Conf. ISTRO*, Edinburgh, Scotland, 1988, 55.
324. **Drake, M., Vengris, J., and Colby, W. G.,** Cation exchange capacity of plant roots, *Soil Sci.*, 72, 139, 1951.
325. **Drakeford, D. R. and Reid, D. M.,** Changes in the ability of plant to alter the pH of the flooding medium as an early symptom of flooding stress in *Helianthus annuus, Can. J. Bot.*, 62, 2417, 1984.
326. **Drew, M. C.,** Plant responses to anaerobic conditions in soil and solution culture, *Curr. Adv. Plant Sci.*, 36, 1, 1979.
327. **Drew, M. C.,** Plant injury and adaptation to oxygen deficiency in the root environment: a review, *Plant Soil*, 75, 179, 1983.
328. **Drew, M. C., Channel, A., Garrec, J. P., and Fourcy, A.,** Critical air spaces (aerenchyma) in roots of corn subjected to oxygen stress: structure and influence on uptake and translocation of -Rb ions, *Plant Physiol.*, 65, 506, 1980.
329. **Drew, M. C. and Fourcy, A. R.,** Radial movement of cations across aerenchymatous roots of *Zea mays* measured by electron probe X-ray microanalysis, *J. Exp. Bot.*, 37, 823, 1986a.
330. **Drew, M. C., Jackson, M. B., and Giffard, S.,** Ethylene-promoted adventitious rooting and development of cortical air spaces (aerenchyma) in roots may be adaptive response to flooding in *Zea mays* L., *Planta*, 147, 83, 1979.
331. **Drew, M. C. and Nye, P. H.,** The supply of nutrient ions by diffusion to plant roots in soil. II. The effect of root hairs on the uptake of potassium by roots of ryegrass (*Lolium multiflorum*), *Plant Soil*, 31, 407, 1969.
332. **Drew, M. C. and Saker, L. R.,** Effects of direct drilling and ploughing on root distribution in spring barley, and on the concentrations of extractable phosphate and potassium in the upper horizons of a clay soil, *J. Sci. Food Agric.*, 29, 201, 1978b.
333. **Drew, M. C. and Saker, L. R.,** Ion transport to the xylem in aerenchymatous roots of *Zea mays* L., J. Exp. Bot., 37, 22, 1986b.
334. **Drew, M. C. and Sisworo, E. J.,** The development of waterlogging damage in young barley plants in relation to plant nutrient status and changes in soil properties, *New Phytol.*, 82, 301, 1979.
335. **Drew, M. C., Sisworo, E. J., and Saker, L. R.,** Alleviation of waterlogging damage to young barley plants by application of nitrate and a synthetic cytokinin, and comparison between the effects of waterlogging, nitrogen deficiency and root excision, *New Phytol.*, 82, 315, 1979.
336. **Dufey, J. E., Amory, D. E., and Braun, R.,** Electrical properties and ion exchange selectivity of roots, *Pedologie*, 35, 231, 1985 (in French, English summary).
337. **Dufey, J. E. and Braun, R.,** Cation exchange capacity of roots: titration, sum of exchange cations, copper adsorption, *J. Plant Nutr.*, 9, 1147, 1986.

338. **Dufey, J. E., Drimmer, D., Lambert, I., and Dupont, P.,** Aluminum-calcium adsorption on roots as related to soil acidity, in *Abstr. Symp. Plant Roots and Their Environment,* Uppsala, Sweden, 1988, 1.
339. **Duke, S. H., Schrader, L. E., Henson, C. A., Servaites, J. C., Vogelzang, R. D., and Pendleton, J. W.,** Low root temperature effects on soybean nitrogen metabolism and photosynthesis, *Plant Physiol.,* 63, 956, 1979.
340. **Dumas, W. T. and Doss, B. D.,** Correction of subsoil acidity in cotton production on Coastal Plain soils, *Highlights Agric. Res.,* 23(4), 1976.
341. **Dunham, R. J. and Nye, P. H.,** The influence of soil water content on the uptake of ions by roots. I. Soil water content gradients near a plane of onion roots, *J. Appl. Ecol.,* 10, 585, 1973.
342. **Duniway, J. M.,** Formation of sporangia by *Phytophtora drechsleri* in soil at high matric potentials, *Can. J. Bot.,* 53, 1270, 1975.
343. **Duniway, J. M.,** Movement of zoospores of *Phytophtora cryptogea* in soils of various texture and matric potentials, *Phytopathology,* 66, 877, 1976.
344. **Dworecki, K., Kargol, M., and Przestalski, S.,** Investigation of the instability diffusion layers in one-membrane osmotic systems, *Stud. Biophys.,* 122, 83, 1987.
345. **Dwyer, L. M., Stewart, D. W., and Balchin, D.,** Rooting characteristics of corn, soybeans and barley as a function of available water and soil physical characteristics, *Can. J. Soil Sci.,* 68, 121, 1988.
346. **Dyer, D. and Brown, D. A.,** Relationship of fluorescent intensity to ion uptake and elongation rates of soybean roots, *Plant Soil,* 72, 127, 1983.
347. **Eavis, B. W.,** Soil physical conditions affecting seedling root growth. I. Mechanical impedance, aeration and moisture availability as influenced by bulk density and moisture levels in a sandy loam soil, *Plant Soil,* 36, 613, 1972.
348. **Eavis, B. W. and Payne, D.,** Soil physical conditions and root growth, in *Proc. 15th Easter School in Agricultur Science,* Whittington, W.J., Ed., University of Nottingham, England, 1969, 315.
349. **Eavis, B. W., Ralliff, L. E., and Taylor, H. M.,** Use of a dead-load technique to determine axial root growth pressure, *Agron. J.,* 61, 640, 1969.
350. **Eavis, B. W. and Taylor, H. M.,** Transpiration of soybean as related to leaf area, root length, and soil water content, *Agron. J.,* 71, 441, 1979.
351. **Eck, H. V. and Unger, P. W.,** Soil profile modification for increasing crop production, *Adv. Soil Sci.,* 1, 66, 1985.
352. **Eckart, J. E. and Raguse, C. A.,** Effects of diurnal variation in light and temperature on acetylene reduction activity (nitrogen fixation) of subterranean clover, *Agron. J.,* 72, 519, 1980.
353. **Edling, P.,** *Soil Air Volume and Gas Exchange Mechanisms,* Report 151, Swedish University of Agricultur Science, Uppsala, Sweden, 1986.
354. **Edwards, W. M., Fehrenbacher, J. B., and Vavry, J. P.,** The effect of discrete ped density on corn root penetration in a Planosol, *Soil Sci. Soc. Am. Proc.,* 28, 560, 1964.
355. **Effer, W. R. and Ranson, S. L.,** Respiratory metabolism in buckwheat seedlings, *Plant Physiol.,* 42, 1042, 1967.
356. **Ehlers, W.,** Observations on earthworm channels and infiltration on tilled and untilled loess soil, *Soil Sci.,* 119, 242, 1975.
357. **Ehlers, W.,** Penetrometer soil strength and root growth in tilled and untilled loess soil, in *Proc. 9th Conf. ISTRO,* Yugoslavia, 1982, 458.
358. **Ehlers, W., Grimme, K., Stülpnagel, R., Köpke, U., and Böhm, W.,** Flow resistance in soils and plant during field growth of oats, *Geoderma,* 25, 1, 1981.
359. **Ehlers, W., Khosla, B. K., Köpke, U., Stülpnagel, R., Böhm, W., and Baeumer, K.,** Tillage effects on root development, water uptake and growth of oats, *Soil Till. Res.,* 1, 19, 1980/1981.
360. **Ehlers, W., Köpke, U., Hesse, F., and Böhm, W.,** Penetration resistance and root growth of oats in tilled and untilled loess soil, *Soil Till. Res.,* 3, 261, 1983.
361. **Ehlers, W. and van der Ploeg, R. R.,** Evaporation, drainage and unsaturated hydraulic conductivity of tilled and untilled fallow soil, *Z. Pflanzenernaehr. Bodenkd.,* 3, 373, 1976.
362. **Ekanayake, I. J., Garrity, D. P., Masajo, T. M., and O Toole, J. C.,** Root pulling resistance in rice: inheritance and association with drought tolerance, *Euphytica,* 34, 905, 1985.
363. **El-Beltagy, A. S. and Hall, M. A.,** Effect of water stress upon endogenous ethylene levels in *Vicia faba, New Phytol.,* 73, 47, 1974.
364. **Elkins, C. B.,** Plant roots as tillage tools, in *Traction and Transport as Related to Cropping Systems,* Int. Conf. on Soil Dynamic Proc., Vol. 3, Auburn, AL, 1985, 519.
365. **Elkins, C. B., Haaland, R. L., and Hoveland, C. S.,** Grass roots as a tool for penetrating soil hardpans and increasing crop yields, in *Proc. 34th Southern Pasture and Forage Crop Improvement Conf.,* Auburn, AL, 1977, 21.
366. **Elliot, L. F. and Lynch, J. M.,** Plant growth-inhibitory pseudomonads colonizing winter wheat (*Triticum aestivum* L.), *Plant Soil,* 84, 57, 1985.
367. **Elliot, L. F. and McCalla, T. M.,** The composition of the soil atmosphere beneath a beef cattle feedlot and a cropped field, *Soil Sci. Soc. Am. Proc.,* 36, 68, 1972.

368. **Ellis, F. B. and Barnes, B. T.,** Growth and development of root systems of winter cereals grown after different tillage methods including direct drilling, *Plant Soil*, 55, 283, 1980.

369. **Ellis, F. B., Christian, D. G., Bragg, P. L., Henderson, F. K. G., Prew, R. D., and Cannell, R. Q.,** A study of mole drainage with simplified cultivation for autumn sown crops on a clay soil. III. Agronomy, root and shoot growth of winter wheat 1978-80, *J. Agric. Sci. Cambridge*, 102, 583, 1984.

370. **Ellis, F. B., Elliot, J. G., Barnes, B. T., and Howse, K. R.,** Comparison of direct drilling reduced cultivation and ploughing on the growth of cereals. II. Spring barley on a sandy loam soil: soil physical conditions and root growth, *J. Agric. Sci.*, 89, 631, 1977.

371. **Emerson, W. W.,** Determination of the stability of soil crumbs, *J. Soil Sci.*, 5, 233, 1954.

372. **Engin, M. and Sprent, J. I.,** Effects of water stress on growth and nitrogen fixation activity of *Trifolium repens*, *New Phytol.*, 72, 117, 1973.

373. **Ennik, G. C. and Hofman, T. B.,** Variation in the root mass of ryegrass types and its ecological consequences, *Neth. J. Agric. Sci.*, 31, 325, 1983.

374. **Erickson, A. E.,** Short-term oxygen deficiencies and plant responses, in *Conf. Proc. American Society of Agricultur Engineer Drainage and Efficient Crop Production*, Chicago, 1965, 17.

375. **Ericsson, T.,** Root biomass distribution in willow stands grown on bog, in *Ecology and Management of Forest Biomass Production Systems*,. Perttu, K., Ed., Department of Ecology and Environmental Res. Rep., Swedish University of Agricultural Science, 1984, 15, 335.

376. **Eriksson, J.,** *Soil Compaction and Root Environment*, Rep. 126, Swedish University of Agricultural Science, Uppsala, 1982 (in Swedish, English summary),

377. **Eriksson, J., Håkansson, I., and Danfors, E.,** *The Effect of Soil Compaction on Soil Structure and Crop Yields*, Bull. No. 354, Swedish Institute of Agricultural Engineering, Uppsala, 1974, 15.

378. **Erlandsson, G. S., Pettersson, S., and Svensson, S.,** Rapid effects on ion uptake in sunflower roots, *Physiol. Plant.*, 43, 380, 1978.

379. **Esan, K.,** *Plant Anatomy*, 2nd ed., John Wiley & Sons, New York, 1965.

380. **Evans, N. T. S. and Ebert, M.,** Radioactive oxygen in the study of gas transport down the root of *Vicia faba*, *J. Exp. Bot.*, 11, 246, 1960.

381. **Evans, P. S.,** Root growth of *Lolium perenne* L. I. Effect of plant age, seed weight, and nutrient concentration on root weight, length, and number of apices, *N.Z. J. Bot.*, 8, 344, 1970.

382. **Evans, P. S.,** Root growth of *Lolium perenne* L. II. Effects of defoliation and shading, *N.Z. J. Agric. Res.*, 14, 552, 1971.

383. **Fageria, N. K., Baligar, V. C., and Wright, R. J.,** Aluminum toxicity in crop plants, *J. Plant Nutr.*, 11, 309, 1988.

384. **Faiz, S. M. A.,** Soil-Root Water Relations, Ph.D. thesis, University of Aberdeen, Scotland, 1973.

385. **Faiz, S. M. A. and Weatherley, P. E.,** Root contraction in transpiring plants, *New Phytol.*, 92, 333, 1982.

386. **Farrell, D. A. and Greacen, E. L.,** Resistance to penetration of fine probes in compressible soil, *Aust. J. Soil Res.*, 4, 1, 1966.

387. **Feddes, R. A.,** Water use models for assessing root zone modification, in *Modifying the Root Environment to Reduce Crop Stress*, ASAE Monogr. 4, Arkin, G. F. and Taylor, H. M., Eds., American Society of Agriculture Engineering, St Joseph, MI, 1981, 347.

388. **Feddes, R. A., Kowalik, P., and Zaradny, H.,** Simulation of Field Water Use and Crop Yield, Simulation Monograph, Pudoc, Wageningen, 1979.

389. **Fehrenbacher, J. B., Ray, B. W., and Edwards, W. M.,** Rooting volume of corn and alfalfa in shale-influenced soils in northwestern Illinois, *Soil Sci. Soc. Am. Proc.*, 29, 591, 1965.

390. **Feil, W., Kottke, I., and Oberwinkler, F.,** The effect of drought on mycorrhizal production and very fine root system development of Norway spruce under natural and experimental conditions, *Plant Soil*, 108, 221, 1988.

391. **Feldman, L. J.,** Regulation of root development, *Annu. Rev. Plant Physiol.*, 35, 223, 1984.

392. **Figliolia, A., Izza, C., and Pagliara, S.,** Effect of some nutritive elements on *Avena sativa* root respiration, Note I, Inst. Sperimentale Nutrizione Piante, Rome, 1977, 1.

393. **Findenegg, G. R, van Beusichem, M. L., and Keltjens, W. G.,** Proton balance of plants: physiological, agronomical and ecological implications, *Neth. J. Agric. Sci.*, 34, 371, 1986.

394. **Finney, J. R.,** Root growth in relation to tillage, *Chem. Ind.*, 13, 676, 1973.

395. **Finney, J. R. and Knight, B. A. G.,** The effect of soil physical conditions produced by various cultivation systems on the root development of winter wheat, *J. Agric. Sci. Cambridge*, 80, 435, 1973.

396. **Fiscus, E. L.,** *In situ* measurement of root-water potential, *Plant Physiol.*, 50, 191, 1972.

397. **Fiscus, E. L.,** Determination of hydraulic and osmotic properties of soybean root systems, *Plant Physiol.*, 59, 1013, 1977.

398. **Fiscus, E. L.,** Analysis of the components of area growth of bean root systems, *Crop Sci.*, 21, 909, 1981a.

399. **Fiscus, E. L.,** Effect of abscisic acid on the hydraulic conductance and the total ion transport through *Phaseolus* root systems, *Plant Physiol.*, 169, 1981b.

400. **Fiscus, E. L.,** Water transport and balance within the plant: resistance to water flow in roots, in *Limitations to Efficient Water Use in Crop Production,* Taylor, H. M., Jordan, W. R., and Sinclair, T. R., Eds., American Society of Agronomy, Crop Science Society of America, and Soil Science Society of America, Madison, WI, 1983, 183.
401. **Fiscus, E. L.,** Diurnal changes in volume and solute transport coefficients of *Phaseolus* roots, *Plant Physiol.,* 80, 752, 1986.
402. **Fiscus, E. L. and Markhart, A. H.,** Relationship between root system transport properties and plant size in *Phaseolus, Plant Physiol.,* 64, 770, 1979.
403. **Flocker, W. J. and Timm, H.,** Plant growth and root distribution in layered sand columns, *Agron. J.,* 61, 530. 1969.
404. **Floris, J. and van Noordwijk., M.,** Improved methods for the extraction of soil samples for root research, *Plant Soil,* 77, 369, 1984.
405. **Foil, R. R. and Ralson, C. W.,** The establishment and growth of Loblolly pine seedlings on compact soil, *Soil Sci. Soc. Am. Proc.,* 31, 365, 1967.
406. **Foster, R. C., Rovira, A. D., and Cockk, T, W.,** *Ultrastructure of the Root-Soil Interface,* American Phytopathological Society, St. Paul, MN, 1983.
407. **Fouere, A.,** A new field instrument: the rhizoscope, in Abstr. of *Symp. Plant Roots and Their Environment,* Uppsala, Sweden, 1988, 4.
408. **Frederickson, J. K. and Elliot, L. F.,** Colonization of winter wheat roots by inhibitory rhizobacteria, *Soil Sci. Soc. Am. J.,* 49, 1172, 1985.
409. **Frossard, J. S.,** Root temperature effect on root respiration and growth in maize seedlings of two grain types, *Agronomie,* 5, 719, 1985 (in French, English summary).
410. **Frota, J. N. E. and Tucker, T. C.,** Temperature influence on ammonium and nitrate absorption by lettuce, *Soil Sci. Soc. Am. Proc.,* 36, 97, 1972.
411. **Fulton, J. M. and Erickson, A. E.,** Relation between soil aeration and ethyl alcohol accumulation in xylem exudate of tomatoes, *Soil Sci. Soc. Am. Proc.,* 28, 610, 1964.
412. **Fulton, J. M., Martimore, C. G., and Hildebrand, A. A.,** Note on the relation of soil bulk density to the incidence of *Phytophtora* root and stalk root of soybean, *Can. J. Soil Sci.,* 41, 247, 1961.
413. **Fusseder, A.,** A method for measuring length, spatial distribution and distances of living roots *in situ, Plant Soil,* 73, 441, 1983.
414. **Fyson, A. and Sprent, J. J.,** The development of primary root nodules on *Vicia faba* L. grown at two temperatures, *Ann. Bot.,* 50, 681, 1982.
415. **Gaheen, S. A. and Njos, A.,** Effect of tractor traffic on timothy (*Phleum pratense* L.) root system in an experiment on a loam soil, *Meld. Nor. Landbrukshoegsk.,* 57, 1, 1978.
416. **Gajri, P. R. and Prihar, S. S.,** Rooting, water use and yield relations in wheat on loamy sand and sandy loam soils, *Field Crop Res.,* 12, 115, 1985.
417. **Gales, K., Ayling, S. M., and Cannell, R. Q.,** Effects of waterlogging and drought on winter barley grown on a clay and sandy loam soil. II. Soil-plant relationships, *Plant Soil,* 80, 67, 1984.
418. **Galligar, G. C.,** Temperature effects upon the growth of excised root tips, *Plant Physiol.,* 13, 835, 1938.
419. **Ganmore-Neumann, R. and Kafkafi, U.,** Root temperature and percentage NO_3/NH_4 effects on tomato plant development. II. Nutrient composition, *Agron. J.,* 72, 762, 1980.
420. **Ganmore-Neumann, R. and Kafkafi, U.,** The effect of root temperatrure and NO_3/NH_4 ratio on strawberry plants. I. Growth, flowering, and root development, *Agron. J.,* 75, 941, 1983.
421. **Ganmore-Neumann, R. and Kafkafi, U.,** Effects of root temperature and nitrate/ammonium ratio on strawberry plants. II. Nitrogen uptake, mineral ions, and carboxylate concentrations, *Agron. J.,* 77, 835, 1985.
422. **Ganmore-Neumann, R. and Kafkafi, U.,** Effects of root temperature, nitrate and ammonium on cation concentration and distribution in the strawberry plant (*Fragoria ananassa* Duch.), in *Proc. 9th Int. Plant Nutrient Colloq.,* Scaife, A., Ed., Commonwealth Agriculture Bureau, 1986, 181.
423. **Gantzer, C. J. and Blake, G. R.,** Physical characteristics of Le Sueur clay loam soil following no-till and conventional tillage, *Agron. J.,* 70, 853, 1978.
424. **Gardner, W. K., Parberry, D. G., and Barber, D. A.,** The acquisition of phosphorus by *Lupinus albus* L. I. Some characteristics of the soil/root interface, *Plant Soil,* 68, 19, 1982.
425. **Gardner, W. R.,** Dynamic apsects of water availability to plants, *Soil Sci.,* 84, 63, 1960.

425a. **Gardner, W. R.,** Relation of root distribution to water uptake and availability, *Agron. J.,* 56, 41, 1964.

426. **Gardner, W. R. and Ehlig, C. F.,** Impedance to water movement in soil and plant, *Science,* 138, 522, 1962.
427. **Garret, S. D.,** The soil-root interface in relation to disease, in *The Soil-Root Interface,* Harley, J. L. and Russell, R. S., Eds., Academic Press, London, 1979, 301.
428. **Geisler,G.,** Interactive effects of CO_2 and O_2 in soil on root and top growth of barley and peas, *Plant Physiol.,* 42, 305, 1967.
429. **Geisler, G.,** The influence of carbonic acid (CO_2, HCO_3) in the range of low oxygen concentration (O_2) in soil air on shoot and root growth of cultivated plants, *Z. Acker Pflanzenbau,* 130, 1, 1969a.

430. **Geisler, G.,** The influence of oxygen and carbon dioxide concentration on growth of roots and shoots of winter barley, maize and peas, *Bayer. Landwirtsch. Jahrb.*, 45, 259, 1969b (in German, English summary).
431. **Gerard, C. J. and Mehta, H. C.,** Influence of a root-crop on physical properties of a medium-textured soil, *Agron. J.*, 63, 889, 1971a.
432. **Gerard, C. J. and Mehta, H. C.,** Influence of a root crop on physical properties of a clay soil, *Agron, J.*, 63, 893, 1971b.
433. **Gerard, C. J., Mehta, H. C., and Hinojosa, E.,** Root growth in a clay soil, *Soil Sci.*, 114, 37, 1972.
434. **Gerard, C. J., Sexton, P., and Shaw, G.,** Physical factors influencing soil strength and root growth, *Agron. J.*, 74, 875, 1982.
435. **Gerik, T. J., Morrison, J. E., and Chichester, F. W.,** Effects of controlled-traffic on soil physical properties and crop rooting, *Agron. J.*, 79, 434, 1987.
436. **Gerwitz, A. and Page, E. R.,** An empirical mathematical model to describe plant root systems, *J. Appl. Ecol.*, 11, 773, 1974.
437. **Ghaderi, A., Smucker, A. J. M., and Adams, M. W.,** Expected correlated responses in selecting dry beans for tolerance to soil compaction, *Euphytica*, 33, 377, 1984.
438. **Ghildyal, B. P.,** Nature, physical properties and management of submerged rice soils, in *Vertisols and Rice Soils of the Tropics*, Trans. 12th Int. Congr. Soil Science, New Delhi, India, Symp. Papers II, Shri S. N. Mehta, New Delhi, India, 1982, 121.
439. **Gibson, A. H.,** Physical environment and symbiotic nitrogen fixation. I. The effect of root temperature on recently nodulated *Trifolium subterraneum* L. plants, *Aust. J. Biol. Sci.*, 16, 28, 1963.
440. **Gibson, A. H.,** Physical environment and symbiotic nitrogen fixation. II. Root temperature effects on the relative nitrogen assimilation rate, *Aust. J. Biol. Sci.*, 18, 295, 1965.
441. **Gill, C. J.,** The flooding tolerance of woody species — a review, *For. Abstr.*, 31, 671, 1970.
442. **Gill, W. R. and Bolt, G. H.,** Pfeffer's studies of the root growth pressures exerted by plants, *Agron. J.*, 47, 166, 1955.
443. **Gill, W. R. and Miller, R. D.,** A method for study of the influence of mechanical impedance and aeration on the growth of seedling roots, *Soil Sci. Soc. Am. Proc.*, 20, 154, 1956.
444. **Ginrich, J. R. and Russell, M. B.,** Effect of soil moisture tension and oxygen concentration on the growth of corn roots, *Agron. J.*, 48, 517, 1956.
445. **Giskin, M. L. and Kohnke, H.,** Measuring root responses to soil properties with a single plant, *Agron. J.*, 57, 96, 1965.
446. **Gliński, J. and Stepniewska, Z.,** An evaluation of soil resistance to reduction processes, *Pol. J. Soil Sci.*, XIX, 15, 1986.
447. **Gliński, J. and Stepniewski, W.,** *Soil Aeration and Its Role for Plants*, CRC Press, Boca Raton, FL, 1985.
448. **Gliński, J. and Stepniewski, W.,** Influence of soil oxygen on root growth and functioning (review of the literature), *Zesz. Probl. Postepow Nauk Roln.*, 385, in press.
449. **Gochev, A. D., Babalakova, N., and Radichkova, D.,** Optimal surface-mass relation of some root systems, *Int. Agrophys.*, 1, 219, 1985.
450. **Goff, E. S.,** Influence of depth of tillage upon the depth of roots, *N. Y. Agric. Exp. Stn. Annu. Rep.*, 7, 171, 1988.
451. **Good, B. J., Faulkner, S. P., and Patrick, W. H., Jr.,** Evaluation of green ash root responses as a soil wetness indicator, *Soil Sci. Soc. Am. J.*, 50, 1570, 1986.
452. **Good, B. J. and Patrick, E. H., Jr.,** Gas composition and respiration of water oak (*Quercus nigra* L.) and green ash (*Fracinus pennsylvanica* Marsh.) roots after prolonged flooding, *Plant Soil*, 97, 419, 1987.
453. **Gooderhaam, P. T.,** Soil Physical Conditions and Plant Growth, Ph.D. thesis, Reading University, England, 1973.
454. **Gooderhaam, P. T.,** Some aspects of soil compaction, root growth and crop yield, *Agric. Prog.*, 52, 33, 1977.
455. **Goss, M. J.,** Effect of mechanical impedance on root growth in barley (*Hordeum vulgare* L.). I. Effects on the elongation and branching of seminal root axes, *J. Exp. Bot.*, 28, 96, 1977.
456. **Goss, M. J., Dexter, A. R., and Evans, M.,** Mechanics of root elongation and the effects of 3,5-diiodo-4-hydroxybenzoic acid (DIHB), *Plant Soil*, 99, 211, 1987.
457. **Goss, M. J., Ehlers, W., Boone, F. R., White, I., and Howse, K. R.,** Effects of soil management practice on soil physical conditions affecting root growth, *J. Agric. Eng. Res.*, 30, 131, 1984.
458. **Goss, M. J., Howse, K. R., and Harris, W.,** Effects of cultivation on soil water retention and water use by cereals in clay soils, *Soil Sci.*, 29, 475, 1978.
459. **Goss, M. J. and Russell, R. S.,** Effects of mechanical impedance on root growth in barley (*Hordeum vulgare* L.). III. Observation on the mechanism of response, *J. Exp. Bot.*, 31, 577, 1980.
460. **Goss, M. J. and Walter, C. H. S.,** Effect of Mechanical Resistance on the Growth of Plant Roots, Annu. Rep., Letcombe Laboratory, Agriculture Research Council (G.B.), 1968, 27.
461. **Gosselin, A. and Trudel, M. J.,** Interactions between air and root temperatures on greenhouse tomato. II. Mineral composition of plants, *J. Am. Soc. Hortic. Sci.*, 108, 905, 1983.

462. **Gosselin, A. and Trudel, M. J.,** Interactions between root-zone temperaure and light levels on growth, development and photosynthesis of *Lycopersicon esculentum* Mill. cultivar "Vender", *Sci. Hortic.*, 23, 313, 1984.
463. **Gosselin, A. and Trudel, M.,** Influence of root-zone temperature on growth, development and yield of cucumber plants cv. Toska, *Plant Soil,* 85, 327, 1985.
464. **Gosselin, A. and Trudel, M. J.,** Root-zone temperature effects on pepper, *J. Am. Soc. Hortic. Sci.*, 111, 220, 1986.
465. **Gotoh, K., Chang, T. T., O Toole, J. C., Riley, R., and Murty, B. R.,** Crop adaptation, in *Plant Breeding in Perspective,* Sneep, J. and Hendriksen, A. J. T., Eds., Pudoc, Wageningen, The Netherlands, 1979, 234.
466. **Gradwell, M. W.,** Soil physical conditions of winter and the growth of ryegrass plants. I. Effects of compaction and puddling, *N Z. J. Agric. Res.*, 8, 238, 1965.
467. **Graham, J. M., Leonard, R. T., and Menge, J. A.,** Interaction of light intensity and soil temperature with phosphorus inhibition of vesicular-arbuscular mycorrhizal formation, *New Phytol.*, 91, 683, 1982.
468. **Graham, J. P., Blackwell, P. S., Armstrong, J. V., Christian, D. G., Howse, K. R., Bawson, C. J., and Butler, A. R.,** Compaction of a silt loam by wheeled agricultural vehicles. II. Effects on growth and yield of direct-drilled winter wheat, *Soil Till. Res.*, 7, 189, 1986.
469. **Gram, W. J. and Pitman, M. G.,** The action of abscisic acid on ion uptake and water flow in plant roots, *Aust. J. Biol. Sci.*, 25, 1125, 1972.
470. **Grant, E. A., MacLean, A. A., and Gupta, U. C.,** Effects of rate and placement of phosphorus, pH, and temperature on the early growth, phosphorus, boron, and zinc content of corn, *Can. J. Plant Sci.*, 52, 35, 1972.
471. **Gray, L. E. and Pope, R. A.,** Influence of soil compaction on soybean stand, yield, and *Phytophtora* root rot incidence, *Agron. J.*, 78, 189, 1986.
472. **Greacen, E. L.,** Mechanisms and models of water transfer, in *Soil Factors in Crop Product in a Semi-arid Environment,* Russell, J. S. and Greacen, E. L., Eds., Queensland University Press, Brisbane, Australia, 1977.
473. **Greacen, E. L.,** Root response to soil mechanical properties, in *Trans. 13th Congr. Int. Soc. Soil Science,* Vol. 5, Hamburg, West Germany, 1986, 20.
474. **Greacen, E. L., Barley, K. P., and Farrell, D. A.,** The mechanics of root growth in soils with particular reference to the implications for root distribution, in *Root Growth,* Whittington, W. J., Ed., Butterworths, London, 1968a, .
475. **Greacen, E. L., Farrell, D. A., and Cockroft, B.,** Soil resistance to metal probe and plant roots, in *Trans. 9th Int. Congr. Soil Science,* Vol. 1, Adelaide, Australia, 1968b, 769.
476. **Greacen, E. L. and Oh, J. S.,** Physics of root growth, *Nature London New Biol.*, 235, 24, 1972.
477. **Greacen, E. L. and Sands, R.,** Compaction of forest soils. A review, *Aust. J. Soil Res.*, 18, 163, 1980.
478. **Greaves, M. P. and Darbyshire, J. F.,** The ultrastructure of the mucilaginous layer on plant roots, *Soil Biol. Biochem.*, 4, 443, 1972.
479. **Green, M. S. and Etherington, J. R.,** Oxidation of ferrous iron by rice (*Oryza sativa* L.) roots: a mechanism for waterlogging tolerance, *Exp. Bot.*, 28, 678, 1977.
480. **Greenwood, D. J.,** The effect of oxygen concentration on the decomposition of organic materials in soils, *Plant Soil,* 14, 360, 1961.
481. **Greenwood, D. J.,** Studies on the transport of oxygen through the stems and roots of vegetable seedlings, *New Phytol.*, 66, 337, 1967.
482. **Greenwood, D. J.,** Root growth and oxygen distribution in soil, in *Trans. 9th Int. Congr. Soil Science,* Vol. 1, Adelaide, Australia, 1968, 823.
483. **Greenwood, D. J. and Goodman, D.,** Studies on the supply of oxygen to the roots of mustard seedlings (*Sinapis alba* L.), *New Phytol.*, 70, 85, 1971.
484. **Gregory, P. J.,** A periscope method for observing root growth and distribution in field soil, *J. Exp. Bot.*, 30, 205, 1979.
485. **Gregory, P. J.,** Response to temperature in a stand of pearl millet (*Penisetum typhoides* S. and H.). III. Root development, *J. Exp. Bot.*, 34, 744, 1983.
486. **Gregory, P. J., Biscoe, P. V., and Hunter, B.,** Water relations of winter wheat. I. Growth of the root system, *J. Agric. Sci. Cambridge,* 91, 91, 1978.
487. **Gregory, P. J., McGowan, M., Biscoe, P. V., and Hunter, B.,** Water relations of winter wheat. II. Soil water relations, *J. Agric. Sci. Cambridge,* 91, 103, 1978.
488. **Grier, C. C., Vogt, K. A., Keyes, M. R., and Edmonds, R. L.,** Biomass distribution and above- and below-ground production in young and mature *Abies amabilis* zone ecosystems of the Washington Cascades, *Can. J. For. Res.*, 11, 155, 1980.
489. **Grimes, D. W., Miller, R. J., and Wiley, P. L.,** Cotton and corn root development in two field soils of different strength characteristics, *Agron. J.*, 67, 519, 1975.
490. **Grineva, G. M.,** Alcohol formation and excretion by plant roots under anaerobic conditions, *Fiziol. Rast.*, 10, 361, 1963 (in Russian, English summary).

491. **Grinsted, M. J., Hedley, M. J., White, R. E., and Nye, P. H.,** Plant-induced changes in the rhizosphere of rape (*Brassica napus* var. Emerald) seedlings. I. pH changes and increase in P concentration in the soil solution, *New Phytol.*, 91, 19, 1982.
492. **Groenevelt, P. H.,** Impact of external forces on soil structure, in impact of water and external forces on soil structure, *Catena Suppl.* 11, 163, 1988.
493. **Groenevelt, P. H., Kay, B. D., and Grant, C. D.,** Physical assessment of a soil with respect to rooting potential, *Geoderma*, 34, 101, 1984.
494. **Grossmann, K., Kwiatkowski, J., Siebecker, H., and Jung, J.,** Regulation of plant morphology by growth retardants, *Plant Physiol.*, 84, 1018, 1987.
495. **Guckert, A., Breisch, H., and Reisinger, O.,** Electron microscope study of the mucigel-clay-microorganisms relations, *Soil Biol. Biochem.*, 7, 241, 1975 (in French, English summary).
496. **Guidi, G., Poggio, G., and Petruzelli, G.,** The porosity of soil aggregates from bulk soil and from soil adhering to roots, *Plant Soil.*, 87, 311, 1985.
497. **Guinn, G.,** Nutritional stress and ethylene evolution by young cotton balls, *Crop Sci.*, 16, 89, 1976.
498. **Gupta, S. C. and Larson, W. E.,** Predicting soil mechanical behavior during tillage, in *Predicting Tillage Effects on Soil Physical Properties and Processes*, Spec. Publ. 44, Unger, P, et al. Eds., American Society of Agronomy, 1982, 151.
499. **Habib, R.,** Total root length as estimated from small sub-samples, *Plant Soil*, 108, 267, 1988.
500. **Hackett, C.,** A study of the root system of barley. II. Relationship between root dimensions and nutrient uptake, *New Phytol.*, 68, 1023, 1969.
501. **Hackett, C.,** Relations between the dimensions of the barley root system: effects of mutilating the root axes, *Aust. J. Biol. Sci.*, 24, 1057, 1971.
502. **Hackett, C. and Rose, D. A.,** A model of the extension and branching of a seminal root of barley, and its use in studying relations between root dimensions, *Aust. J. Biol. Sci.*, 25, 669, 1972.
503. **Hagemeyer, J., Kahle, H., Breckle, S. W., and Waisel, Y.,** Cadmium in *Fagus sylvatica* L. trees and seedlings: leaching, uptake and interconnection with transpiration, *Water Air Soil Pollut.*, 29, 347, 1986.
504. **Håkansson, I.,** Swedish experiments on subsoil compaction by vehicles with high axle load, *Soil Use Manage.*, 1 (4), 113, 1985.
505. **Håkansson, I.,** A method for characterizing the state of compactness of an arable soil, in impact of water and external forces on soil structure, *Catena Suppl.* 11, 1988, 101.
506. **Håkansson, I., Voorhees, W. B., and Riley, H.,** Vehicle and wheel factors influencing soil compaction and crop response in different traffic regimes, *Soil Till. Res.*, 11, 239, 1988.
507. **Hale, M. G. and Moore, L. D.,** Factors affecting root exudation. II. 1970-1978, *Adv. Agron.*, 31, 93, 1979.
508. **Hallmark, W. B. and Barber, S. A.,** Root growth and morphology, nutrient uptake, and nutrient status of soybeans as affected by soil K and bulk density, *Agron. J.*, 73, 779, 1981.
509. **Hamblin, A. P.,** The influence of soil structure on water movement, crop root growth, and water uptake, *Adv. Agron.*, 38, 95, 1985.
510. **Hamblin, A. P. and Tennant, D.,** Interactions between soil type and tillage level in a dryland situation, *Aust. J. Soil Res.*, 17, 171, 1979.
511. **Hammond, C., Reid, J. T., and Seigler, W. E.,** Seedbed preparation behind row subsoilers, *Trans. ASAE*, 24, 897, 1981.
512. **Hansen, G. K.,** A dynamic continuous simulation model of water state and transportation in the soil-plant-atmosphere system. 1. The model and its sensitivity, *Acta Agric. Scand.*, 25, 129, 1975.
513. **Hansen, G. K.,** Utilization of photosynthates for growth, respiration, and storage in tops and roots of *Lollium multiflorum*, *Physiol. Plant.*, 42, 5, 1978.
514. **Hansen, G. K. and Jensen, C. R.,** Growth and maintenance respiration in whole plants, tops and roots of *Lollium multiflorum*, *Physiol. Plant.*, 39, 275, 1977.
515. **Harding, S. C. and Sheehy, J. E.,** Influence of shoot and root temperature on leaf growth, photosynthesis and nitrogen fixation of lucerne, *Ann. Bot.*, 45, 229, 1980.
516. **Hargrove, W. L., Radcliffe, D. E., and Box, J. E., Jr.,** Influence of long-term no-tillage on rooting in an Ultisol, in *Proc. 11th Conf. ISTRO*, Edinburgh, Scotland, 1988, 61.
517. **Harkes, P. A. A.,** Structure and dynamics of the root cap of *Avena sativa*, *Acta Bot. Neerl.*, 22, 321, 1973.
518. **Harris, D. G. and van Bavel, C. H. M.,** Root respiration of tobacco, corn, and cotton plants, *Agron. J.*, 49, 182, 1957.
519. **Harris, G. L., Goss, M. J., Dowdell, R. J., Howse, K. R., and Morgan, P.,** A study of mole drainage with simplified cultivation for autumn-sown crops on a clay soil. II. Soil water regimes, water balances and nutrient losses in drain water, 1978-80, *J. Agric. Sci. Cambridge*, 102, 561, 1984.
520. **Harris, W. F., Kinerson, R. S., and Edwards, N. T.,** Comparison of belowground biomass of natural deciduous forest and loblolly pine plantations, *Pedobiologia*, 7, 369, 1978.
521. **Harrison, M. A. and Walton, D. C.,** Abscisic acid metabolism in water stressed bean leaves, *Plant Physiol.*, 56, 250, 1975.

522. **Hasegawa, S.,** A simplified model of water uptake by dryland rice root system, *Soil Physical Conditions Plant Growth,* 44, 14, 1981 (in Japanese, English summary).
523. **Hasegawa, S.,** Changes in soil water contents in the vicinity of soybean roots, in *Trans. 13th Congr. Int. Soc. Soil Science,* Vol. 2, Hamburg, West Germany, 1986, 73.
524. **Hasegawa, S., Parao, F. T., and Yoshida, S.,** Root development and water uptake under field conditions, Special Seminar, March 21, 1979, Plant Physiological Department, Tsukuba.
525. **Hasegawa, S. and Sato, T.,** Soil water movement in the vicinity of soybean roots determined by root plane experiment, *Trans. Jpn. Soc. Irrig. Drain. Reclam. Eng. (Nogyo Doboku Gakkai Ronbunshu),* 17, 1985a.
526. **Hasegawa, S. and Sato, T.,** Water uptake by roots in cracks and water movenent in clayey subsoil, *Soil Sci.,* 143, 381, 1987.
527. **Hasegawa, S,. Thangaraj, M., and O Toole, J. C.,** Root behavior: field and laboratory studies for rice and nonrice crops, in *Soil Physics and Rice,* International Rice Research Institute, 1985b, 83.
528. **Hasegawa, S. and Yoshida, S.,** Water uptake by dryland rice root system during soil drying cycle, *Soil Sci. Plant Nutr.,* 28, 191, 1982.
529. **Haynes, R. J.,** Effects of soil management practices on soil physical properties, earthworm population and tree root distribution in a commercial apple orchard, *Soil Till. Res.,* 1, 269, 1981.
530. **Haynes, R. J. and Ludecke, T. E.,** Yield, root morphology and chemical composition of two pasture legumes as affected by lime and phosphorus applications to an acid soil, *Plant Soil,* 62, 241, 1981.
531. **Head, G. C.,** Shedding of roots, in *Shedding of Plant Parts,* Kozlowski, T.T., Ed., Academic Press, New York, 1973, 237.
532. **Hecht-Buchholz, C. and Foy, C. D.,** Effect of aluminium toxicity on root morphology of barley, *Plant Soil,* 63, 93, 1981.
533. **Heen, A.,** Root development and water use in some Norwegian barley, wheat and oat varieties, Meld. Nor. Landbrukshoegsk., 59, 1, 1980.
534. **Heinonen, R.,** The notion of "normal bulk density" in arable soils, in *Proc. 8th Conf. ISTRO,* Stuttgart-Hohenheim, West Germany, 1979, I, 87.
535. **Heinonen, R.,** *Soil Management and Crop Water Supply,* 4th ed., Swedish University of Agricultural Science, Uppsala, 1985, 105.
536. **Heinonen, R.,** Alleviation of soil compaction by natural forces and cultural practices, in *Land Clearing and Development in the Tropics,* Lal, R., Sanchez, P. A., and Cummings, R. W., Jr., Eds., Balkema, A. A., Rotterdam, 1986, 285.
537. **Heinrichs, D. H.,** Root-zone temperature effects on flooding tolerance of legumes, *Can. J. Plant Sci.,* 52, 985, 1972.
538. **Heinrichs, D. H. and Nielsen, K. F.,** Growth responses of alfalfa varieties of diverse genetic origin to different root zone temperatures, *Can. J. Plant Sci.,* 46, 291, 1966.
539. **Heintze, S.,** Studies on cation-exchange capacities of roots, *Plant Soil,* 13, 365, 1961.
540. **Hemsath, D. L. and Mazurak, A. P.,** Seedling growth of sorghum in clay — sand mixtures at various compactions and water contents, *Soil Sci. Soc. Am. Proc.,* 38, 387, 1974.
541. **Hendrickson, O. Q. and Robinson, J. B.,** Effects of roots and litter on mineralization processes in forest soil, *Plant Soil,* 80, 391, 1984.
542. **Hendriks, I., Claassen, N., and Jungk, A.,** Phosphate depletion at the soil-root interface and the phosphate uptake of maize and rape, *Z. Pflanzenernaehr. Bodenkd.,* 144, 486, 1981.
543. **Heringa, J. W., Groenwold, J., and Schoonderbeek, D.,** An improved method for the isolation and the quantitative measurement of crop roots, *Neth. J. Agric. Sci.,* 28, 127, 1980.
544. **Herkelrath, W. N., Miller, E. E., and Gardner, W. R.,** Water uptake by plants. I. Divided root experiments, *Soil Sci. Soc. Am. J.,* 41, 1033, 1977a.
545. **Herkelrath, D. N., Miller, E. E., and Gardner, W. R.,** Water uptake by plants. II. The root contact model, *Soil Sci. Soc. Am. J.,* 41, 1039, 1977b.
546. **Hernandez, L. F. and Orioli, G. A.,** Relationships between root permeability to water, leaf conductance and transpiration rate in sunflower (*Helianthus annuus* L.) cultivars, *Plant Soil,* 85, 229, 1982.
547. **Hetherington, A. M., Hunter, M. I. S., and Crawford, R. M. M.,** Contrasting effects of anoxia on rhizome lipids in *Iris* species, *Phytochemistry,* 21, 1275, 1982.
548. **Hettiaratchi, D. P. R. and Ferguson, C. A.,** Stress-deformation behaviour of soil in root growth mechanics, *J. Agric. Eng. Res.,* 18, 309, 1973.
549. **Hewitt, J. S. and Dexter, A. R.,** An improved model of root growth in structured soil, *Plant Soil,* 52, 325, 1979.
550. **Hewitt, J. S. and Dexter, A. R.,** The behaviour of roots encountering cracks in soil. II. Development of a predictive model, *Plant Soil,* 79, 11, 1984a.
551. **Hewitt, J. S. and Dexter, A. R.,** Statistical distribution of root maximum growth pressures, root buckling stresses, and soil penetration strength, *Plant Soil,* 77, 39, 1984b.
552. **Hillel, D.,** *Application of Soil Physics,* Academic Press, New York, 1980, 147.

553. **Hillel, D. and Tolpaz, H.,** Simulation of root growth and its effect on the pattern of soil water uptake by a non-uniform root system, *Soil Sci.*, 121, 307, 1976.
554. **Hinsinger, Ph., Dufey, J. E., and Jaillard, B.,** Biological weathering of micas in the rhizosphere as related to potassium absorption by plant roots, in *Abstr. Symp. Plant Roots and Their Environment,* Uppsala, Sweden, 1988, 1.
555. **Hladynowski, J., Kargol, M., Dulewicz, Z., and Przestalski, S.,** The problem of diffusion in the field of gravity, *Stud. Biophys.*, 113, 39, 1986.
556. **Hoffman, J. G.,** Alleviating salinity stress, in *Modifying the Root Environment to Reduce Crop Stress,* ASAE Monogr. No. 4, Arkin, G. F. and Taylor, H. M., Eds., American Society of Agriculture Engineering, St. Joseph, MI, 1981, 269.
557. **Hoffman, W. F. and Barber, S. A.,** Phosphorous uptake by wheat (*Triticum aestivum*) as influenced by ion accumulation in the rhizocylinder, *Soil Sci.*, 112, 256, 1971.
558. **Hoffmann, F.,** Simulation of yield formation of sugar beet, *Syst. Anal. Model. Simul.*, 2, 55, 1985.
559. **Holder, C. B. and Brown, K. W.,** The relationship between oxygen and water uptake by roots of intact bean plants, *Soil Sci. Soc. Am. J.*, 44, 21, 1980.
560. **Holthausen, R. S. and Caldwell, M. M.,** Seasonal dynamics of root system respiration in *Atriplex confertifolia, Plant Soil,* 55, 307, 1980.
561. **Holzhei, D. E. and Burkhardt, T. H.,** *Modeling the Seed Zone in the Soil,* ASAE Pap. No. 85-2020, American Society of Agriculture Engineering, St. Joseph, MI, 1985.
562. **Hoogenboom, G. and Huck, M. G.,** ROOT-SIMU V 4.0, *A Dynamic Simulation Root Growth, Water Uptake, and Biomass Partitioning in a Soil-Plant-Atmosphere, Continuum Update and Documentation,* Seri. No. 109, Agronomy and Soils Department, Auburn University, Auburn, AL, 1986a, 83.
563. **Hoogenboom, G., Huck, M. G., and Hillel, D.,** Modification and testing of a model simulating root and shoot growth as related to soil water dynamics, *Adv. Irrig.*, 4, 331, 1987a.
564. **Hoogenboom, G., Huck, M. G., and Peterson, C. M.,** Measured and simulated drought stress on daily shoot and root growth rates of soybean, *Neth. J. Agric. Sci.*, 34, 497, 1986b.
565. **Hoogenboom, G., Huck, M. G., and Peterson, C. M.,** Root growth rate of soybean as affected by drought stress, *Agron. J.*, 79, 607, 1987b.
566. **Hoogland, J. C., Feddes, R. A., and Belmans, C.,** Root water uptake model depending on soil water pressure head and maximum extraction rate, *Acta Hortic.*, 119, 123, 1981.
567. **Hook, D. D. and Brown, C. L.,** Root adaptations and relative flood tolerance of five hardwood species, *For. Sci.*, 19, 225, 1973.
568. **Hook, D. D., Brown, C. L., and Kormanik, P. P.,** Inductive flood tolerance in swamp tupelo (*Nyssa sylvatica* var. *biflora* (Walt.) Sagr.), *J. Exp. Bot.*, 22, 78, 1971.
569. **Hook, D. D., Brown, C. L., and Wetmore, R. H.,** Aeration in trees, *Bot. Gaz.*, 133, 443, 1972.
570. **Hook, D. D. and Scholtens, J. R.,** Adaptations and flood tolerance of tree species, in *Plant Life in Anaerobic Environments,* Hook, D. D. and Crawford, R. M. M., Eds., Ann Arbor Science, Ann Arbor, MI, 1978, 299.
571. **Hopkins, R. M. and Patrick, W. H.,** Combined effects of oxygen concentration and soil compaction on root penetration, *Soil Sci.*, 108, 408, 1969.
572. **Horlacher, R. and Poskuta, J.,** The influence of oxygen deficiency in roots on contents of ethanol, lactate and glucose and activities of ADH and LDH of wheat seedlings, *Biol. Plant.*, 28, 130, 1986 (in German, English summary).
573. **Horst, W. J., Wagner, A., and Marschner, H.,** Mucillage protects meristems from aluminium injury, *Z. Pflanzenphysiol.*, 105, 435, 1982.
574. **Horst, W. J., Wagner, A., and Marschner, H.,** Effect of aluminum on root growth, cell-division rate and mineral element contents in roots of *Vigna unguiculata* genotypes, *Z. Pflanzenphysiol.*, 109, 95, 1983.
575. **Hosner, J. F. and Boyce, S. G.,** Tolerance to water saturated soil of various bottomland hardwoods, *For. Sci.*, 8, 180, 1962.
576. **Hozumi, K.,** Water tolerance of vegetable crops, *Nogyo Gijyutsu,* 26, 352, 1966 (in Japanese).
577. **Hsiao, T. C. and Acevedo, E.,** Plant responses to water deficits, water use efficiency and drought resistance, *Agric. Meteorol.*, 14, 59, 1974.
578. **Hsiao, T. C. and Bradford, K. J.,** Physiological consequences of cellular water deficits, in *Limitations to Efficient Water Use, in Crop Production,* Taylor, H. M., Jordan, W. R., and Sinclair, T. R., Eds., American Society of Agronomy, Madison, WI, 1983, 227.
579. **Huck, M. G.,** Variation in taproot elongation rate as influenced by composition of the soil air, *Agron. J.*, 62, 815, 1970a.
580. **Huck, M. G., Hageman, R. H., and Hanson, J. B.,** Diurnal variation in root respiration, *Plant Physiol.*, 37, 371, 1962.
581. **Huck, M. G. and Hillel, D.,** A model of root growth and water uptake accounting for photosynthesis, respiration, transpiration, and soil hydraulics, *Adv. Irrig.*, 2, 273, 1983.
582. **Huck, M. G., Klepper, B., and Taylor, H. M.,** Diurnal variations in root diameter, *Plant Physiol.*, 45, 529, 1970b.

583. **Huck, M. G., Peterson, C. M., Hoogenboom, G., and Busch, C. D.,** Distribution of dry matter between shoots and roots of irrigated and nonirrigated determinate soybeans, *Agron. J.*, 78, 807, 1986.
584. **Huck, M. G. and Taylor, H. M.,** The rhizotron as a tool for root research, *Adv. Agron.*, 35, 1, 1982.
585. **Hulugalle, N. R. and Willatt, S. T.,** The role of soil resistance in determining water uptake by plant root systems, *Aust. J. Soil Res.*, 21, 571, 1983.
586. **Humphries, E. C.,** Effect of removal of a part of the root system on the subsequent growth of the root and shoot, *Ann. Bot.*, 22, 251, 1958.
587. **Hunt, E. R., Jr. and Nobel, P. S.,** A two-dimensional model for water uptake by desert succulents: implications of root distribution, *Ann. Bot.*, 59, 559, 1987a.
588. **Hunt, E. R., Jr., Zakir, N. J. D., and Nobel, P. S.,** Water costs and water revenues for establishing rain-induced roots of *Agave deserti*, *Functional Ecol.*, 1, 125, 1987b.
589. **Hurd, E. A.,** Growth of roots of spring wheat at high and low moisture levels, *Agron. J.*, 60, 201, 1968.
590. **Hurd, E. A.,** Phenotype and drought tolerance in wheat, *Agric. Meteorol.*, 14, 39, 1974.
591. **Hurd, R. G. and Graves, C. J.,** Some effects of air and root temperatures on the yield and quality of glasshouse tomatoes, *J. Hortic. Sci.*, 60, 359, 1985.
592. **Ibanga, I. J., Bidwell, O. W., Powers, W. L., Feyerhern, A. M., and Williams, W. W.,** Soil consistence: effect of particle size, *Soil Sci. Soc. Am. J.*, 44, 1124, 1980.
593. **Ide, G., Hofman, G., Ossemerct, C., and van Ruymbeke, M.,** Influence of subsoiling on the growth of cereals, *Pedologie*, 22, 193, 1982.
594. **Ide, G., Hofman, G., Ossemerct, C., and van Ruymbeke, M.,** Root-growth response of winter barley to subsoiling, *Soil Till. Res.*, 4, 419, 1984.
595. **Idso, S. B., Jackson, R. D., Pinter, P. J., Jr., Reginato, R. J., and Hatfield, J. L.,** Normalizing the stress-degree-day parameter for environmental variability, *Agric. Meteorol.*, 24, 45, 1981.
596. **Ishii, T., Fusao, K., Mizutani, K., and Kadoya, K.,** Ethylene evolution in the rhizosphere of grapevines, and growth disorders as affected by ethylene, *Mem. Coll. Agric. Ehime Univ.*, Japan, 2, 97, 1982.
597. **Itoh, S.,** *In situ* measurement of rooting density in microrhizotron, *Soil Sci. Plant Nutr.*, 31, 653, 1985.
598. **Itoh, S. and Barber, S. A.,** A numerical solution of whole plant nutrient uptake for soil-root systems with root hairs, *Plant Soil*, 70, 403, 1983a.
599. **Itoh, S. and Barber, S. A.,** Phosphorus uptake by six plant species as related to root hairs, *Agron. J.*, 75, 457, 1983b.
600. **Jackson, M. B.,** Ethylene as a growth promoting hormone under flooded conditions, in *Plant Growth Substances*, Wareing, P. F., Ed., Academic Press, London, 1982, 291.
601. **Jackson, M. B. and Campbell, D. J.,** Waterlogging and petiole epinasty in tomato: the role of ethylene and low oxygen, *New Phytol.*, 76, 21, 1976.
602. **Jackson, M. B. and Campbell, D. J.,** Effects of benzyladenine and gibberelic acid on the response of tomato plants to anaerobic root environments and to ethylene, *New Phytol.*, 82, 331, 1979.
603. **Jackson, M. B., Drew, M. C., and Giffard, S. C.,** Effects of applying ethylene to the root system of *Zea mays* on growth and nutrient concentration in relation to flooding tolerance, *Physiol. Plant.*, 52, 23, 1981.
604. **Jackson, M. B., Gales, K., and Campbell, D. J.,** Effect of waterlogged soil conditions on the production of ethylene and on water relationships in tomato plants, *J. Exp. Bot.*, 29, 183, 1978.
605. **Jackson, M. B. and Goss, M. J.,** Can Stress Experienced by Root System Be Alleviated by Endogenous Growth Regulators? Monogr. No. 21, British Crop Protection Council, 1978, 95.
606. **Jackson, M. B., Herman, B., and Goodenough, A.,** An examination of the importance of ethanol in causing injury to flooded plants, *Plant Cell Environ.*, 5, 163, 1982.
607. **Jackson, W. T.,** The role of adventitious roots in recovery of shoots following flooding of the original root systems, *Am. J. Bot.*, 42, 816, 1955.
608. **Jacobsen, L., Cooper, B. R., and Volz, M. G.,** The interaction of pH and aeration in Cl uptake by barley roots, *Physiol. Plant.*, 25, 432, 1971.
609. **Jaillard, B.,** Water dynamics, root activity and soil structure in carbonate environment, *Zesz. Probl. Postepow Nauk Roln.*, 312, 229, 1986 (in French).
610. **Jaillard, B.,** Rhizomorphic Carbonate Structures: Model of Rearrangement of Soil Minerals by Roots, Ph.D. thesis, Montpellier, 1987, 221 (in French).
611. **James, B. R., Bartlett, R. J., and Amadon, J. F.,** A root observation and sampling chamber (rhizotron) for pot studies, *Plant Soil*, 85, 291, 1985.
612. **Jamison, C. V. and Domby, C. W.,** The effect of a dense soil layer and varying air-water relations on the growth, root development, and nutrient uptake of cotton in commerce silt loam, *Soil Sci. Soc. Am. Proc.*, 447, 1956.
613. **Jarvis, R. J.,** The effect of cultivation depth and timing on rhizoctonias infection of wheat, in *Research on Root and Crown Roots of Wheat in Australia*, Proc. of the Wheat Root and Crown Workshop, Waite Agricultural Research Institute, Adelaide, Australia, 1984, 22.

614. **Jarvis, R. J. and Brennan, R. F.**, Timing and intensity of surface cultivation and depth of cultivation affect *Rhizoctonia* patch and wheat yield, *Aust. J. Exp. Agric.*, 26, 703, 1986.

615. **Jat, R. L., Dravid, M. S., Das, D. K., and Goswami, N. N.**, Effect of flooding and high soil water conditions on root porosity and growth of maize, *J. Indian Soc. Soil Sci.*, 23, 291, 1975.

616. **Jennison, J. R., Shank, D. B., and Penny, L. H.**, Root characteristics of 44 maize inbreds evaluated in four environments, *Crop Sci.*, 21, 233, 1981.

617. **Jenny, H. and Grossenbacher, K.**, Root-soil boundary zones as seen in the electron microscope, *Soil Sci. Soc. Am. Proc.*, 27, 273, 1963.

618. **Jensen, C. R., Letey, J., and Stolzy, L. H.**, Labelled oxygen transport through growing corn roots, *Science*, 144, 530, 1964.

619. **Jensen, C. R., Stolzy, L. H., and Letey, J.**, Tracer studies of oxygen diffusion through roots of barley, corn and rice, *Soil Sci.*, 103, 23, 1967.

620. **John, C. D. and Greenway, H.**, Alcoholic fermentation and activity of some enzymes in rice roots under anaerobiosis, *Aust. J. Plant Physiol.*, 3, 325, 1976.

621. **Johnson, C. R., Graham, J. M., Leonard, R. T., and Menge, A.**, Effect of flower and development in chrisanthemum on vesicular-arbuscular mycorrhiza formation, *New Phytol.*, 90, 671, 1982.

622. **Johnson, J. and Hartman, R. E.**, Influence of soil environment on the root-rot of tobacco, *J. Agric. Res.*, 17, 41, 1919.

623. **Jones, H., Tomos, A. D., Leight, R. A., and Wyn Jones, R. G.**, Water-relation parameters of epidermal and cortical cells in the primary root of *Triticum aestivum* L., *Planta*, 158, 230, 1983.

624. **Jordan, P. W. and Nobel, P. S.**, Thermal and water relations of roots of desert succulents, *Ann. Bot.*, 54, 705, 1984.

625. **Jordan, W. R., McCrary, M., and Miller, F. R.**, Compensatory growth in the crown root system of sorghum, *Agron. J.*, 71, 803, 1979.

625a. **Jordan, W. R. and Miller, F. R.**, Genetic variability in sorghum root systems: implications for drought tolerance, in *Adaptation of Plants to Water and High Temperature Stress*, Turner, N. C. and Kramer, P. J., Eds., John Wiley & Sons, New York, 1980, 383.

626. **Jugsujinda, A. and Patrick, W. H.**, Growth and nutrient uptake by rice in a flooded soil under controlled aerobic-anaerobic and pH conditions, *Agron. J.*, 69 ,705, 1977.

627. **Jung, J., Luib, M., Sauter, H., Zeeh, B., and Rademacher, W.**, Growth regulation in crop plants with new types of triazole compounds, *J. Agron. Crop. Sci.*, 158, 324, 1987.

628. **Jung, Y. S.**, Water Uptake and Transport of Soybeans as a Function of Rooting Pattern, Ph.D. thesis, Iowa State University, Ames, 1980.

629. **Jungk, A. and Claassen, N.**, Availability of phosphate and potassium as the result of interactions between root and soil in the rhizosphere, *Z. Pflanzeneraehr. Bodenkd.*, 149, 411, 1986.

630. **Jung La, W., Callebaut, F., Gabriels, D., and de Boodt, M.**, Dependence of the cone index on the cone type and the soil structure, *Meded. Fac. Landbouwwet. Rijksuniv. Gent*, 50, 55, 1985.

631. **Juniper, B.**, Mechanisms of perception and patterns of organisation in root caps, in *The Dynamics of Meristem Cell Population*, Vol. 18, Advances in Experimental Medicine and Biology, Miller, M. W. and Kuehnert, C. C., Eds., Plenum Press, New York, 1972, .

632. **Juo, A. S. R. and Lal, R.**, The effect of fallow and continuous cultivation on the chemical and physical properties of an Alfisol in Western Nigeria, *Plant Soil*, 47, 567, 1977.

633. **Jupp, A. P. and Newman, E. I.**, Morphological and anatomical effects of severe drought on the roots of *Lolium perenne* L., *New Phytol.*, 105, 393, 1987.

634. **Kadej, F.**, Regeneration of apex meristem of root, *Acta Soc. Bot. Pol.*, 39, 373, 1970 (in Polish).

635. **Kahnt, G., Hijazi, L. A., and Rao, M.**, Effect of homogeneous soil compaction on shoot and root growth of field bean and soybean, *J. Agron. Crop Sci.*, 157, 105, 1986.

636. **Kalir, A. and Poljakoff-Mayber, A.**, Effect of salinity on respiratory pathways in root tips of *Tamarix tetragyna*, *Plant Physiol.*, 57, 167, 1976.

637. **Kamazawa, K.**, Physiological specificity of rice root in relation to oxidizing power and nutrient uptake, in *Biology of Rice*, Tsunoda, S. and Takahashi, N., Eds., Japanese Scientific Society Press, Tokyo, and Elsevier, Amsterdam, 1984, 117.

638. **Kamprath, J., Cassel, D. K., Gross, H. D., and Dill, D. W.**, Tillage effects on biomass production and moisture utilization by soybeans on coastal plain soils, *Agron. J.*, 71, 1001, 1979.

639. **Kar, S., Varade, S. B., Subramanyam, T. K., and Ghildyal, P. B.**, Soil physical conditions affecting rice root growth: bulk density and submerged soil temperature regime effects, *Agron. J.*, 68, 23, 1976.

640. **Karczewski, K.**, The influence of passing speed of the agricultural machinery wheels on soil compaction, *Zesz. Probl. Postepow Nauk Roln.*, 201, 69, 1978 (in Polish, English summary).

641. **Kargol, M., Dworecki, K., and Przestalski, S.**, Interferometric investigations of diffusion boundary layers in a graviosmotic system, *Stud. Biophys.*, 113, 31, 1986.

642. **Kargol, M., Przestalski, S., and Suchanek, G.**, Practical description of passive transport through membranes separating multi-component solution, *Stud. Biophys.*, 121, 143, 1987.

643. **Karmoker, J. L. and van Steveninck, R. F. M.,** Simulation of volume flow and ion flux by abscisic acid in excised root systems of *Phaseolus vulgaris* L. c.v. Redland Pioneer, *Planta,* 141, 37, 1978.
644. **Katayama, T.,** Studies on the intercellular spaces in rice. I. *Proc. Crop Sci. Soc. Jpn.,* 29, 229, 1961.
645. **Kaufmann, M. R.,** Water relations of pine seedlings in relation to root and shoot growth, *Plant Physiol.,* 43, 281, 1968a.
646. **Kaufmann, M. R.,** Evaluation of the pressure chamber technique for estimating plant water potential of forest tree species, *For. Sci.,* 14, 369, 1968b.
647. **Kaufmann, M. R. and Fiscus, E. L.,** Water transport through plants — internal integration of edaphic and atmospheric effects, *Acta Hortic.,* 171, 83, 1985.
648. **Kawase, M.,** Effect of flooding on the ethylene concentration in horticultural plants, *J. Am. Soc. Hortic. Sci.,* 97, 584, 1972.
649. **Kawase, M.,** Role of ethylene in induction of flooding damage in sunflower, *Physiol. Plant.,* 31, 29, 1974.
650. **Kawase, M.,** Ethylene accumulation in flooded plants, *Physiol. Plant.,* 36, 236, 1976.
651. **Kawase, M.,** Anaerobic elevation of ethylene concentrations in waterlogged plants, *Am. J. Bot.,* 65, 736, 1978.
652. **Kawase, M.,** Role of cellulase in aerenchyma development in sunflower, *Am. J. Bot.,* 66, 183, 1979.
653. **Kawase, M.,** Anatomical and morphological adaptation of plants to waterlogging, *HortScience,* 16, 31, 1981.
654. **Kawase, M. and Whitmoyer, R. E.,** Aerenchyma development in waterlogged plants, *Am. J. Bot.,* 67, 18, 1980.
655. **Kawata, S., Soejima, M., and Yamazaki, K.,** The superficial root formation and yield of hulled rice, *Jpn. J. Crop Sci.,* 47, 617, 1978 (in Japanese, English summary).
656. **Kayambo, B. and Lal, R.,** Effects of soil compaction by rolling on soil structure and development of maize in no-till and disc ploughing systems on a tropical Alfisols, *Soil Till. Res.,* 7, 117, 1986.
657. **Kays, S. J., Nicklow, C. W., and Simons, D. H.,** Ethylene in relation to the response of roots to physical impedance, *Plant Soil,* 40, 565, 1974.
658. **Keisling, T. C. and Smittle, D. A.,** Soil bulk density corrections for providing a better relationship with root growth in gravelly soil, *Commun. Soil Sci. Plant Anal.,* 12, 91, 1981.
659. **Kellock, A. W.,** A Study of Pathological and Ecological Factors Causing Root-Rot of Subterranean Clover (*T. subterraneum* L.), Ph.D. thesis, University of Melbourne, Australia, 1975.
660. **Kemper, W. D., Stewart, B. A., and Porter, L. K.,** Effects of compaction on soil nutrient status, in *Compaction of Agricultural Soils,* American Society of Agriculture Engineering, St. Joseph, MI, 1971, 178.
661. **Kenefick, D. G.,** Formation and elimination of ethanol in sugar beet roots, *Plant Physiol.,* 37, 434, 1962.
662. **Keng, J. C. W.,** Monitoring corn root development with a mini-rhizotron in confined growth boxes, *Agron. J.,* 80, 287, 1988.
663. **Kesik, T.,** Influence of loess soil compaction on the crop and some morphological and physical features of the carrot roots, *Zesz. Probl. Postepow Nauk Roln.,* 385, in press.
664. **Khanna, P. K., Virman, S. M., and Safaya, N. M.,** Phosphorus uptake and root activity in maize as affected by bulk density of soil, *Indian J. Agric. Sci.,* 44, 60, 1974.
665. **Kibrab, T. and Danielsen, R. E.,** Measurement of radish root enlargement under mechanical stress, *Agron. J.,* 69, 857, 1977.
666. **Kirkham, M. B.,** Physical model of water in a split-root system, *Plant Soil,* 75, 153, 1983.
667. **Kirkham, M. B.,** Simulation of oxygen movement to plant roots as affected by tillage width and depth, *Soil Till. Res.,* 7, 221, 1986.
668. **Kiyosawa, W.,** Studies on the effects of alcohols on membrane water permeability of *Nitella, Protoplasma,* 86, 243, 1975.
669. **Kleinendorst, A. and Veen, B. W.,** Responses of young cucumber plants to root and shoot temperatures, *Neth. J. Agric. Sci.,* 31, 47, 1983.
670. **Klepper, B.,** Managing root systems for efficient water use: axial resistance to flow in root systems — anatomical considerations, in *Limitations to Efficient Water Use in Crop Production,* Taylor, H. M., Jordan, W. R., and Sinclair, T. R., Eds., American Society of Agronomy, Crop Science Society of America, and Soil Science Society of America, Madison, WI, 1983, 115.
670a. **Klepper, B., Belford, R. K., and Rickman, R. W.,** Root and shoot development in winter wheat, *Agron. J.,* 76, 117, 1984.
671. **Kloepper, J. W. and Schroth, M. N.,** Relationship of in vitro antibiosis of plant growth-promoting rhizobacteria to plant growth and the displacement of root microflora, *Phytopathology,* 71, 1078, 1981.
672. **Kobayashi, H.,** Studies on soil holding function with pasturage grass-roots, *Soil Physical Conditions Plant Growth,* 34, 2, 1976 (in Japanese, English summary).
673. **Kobayashi, H. and Otake, Y.,** Studies on physical interaction of forage roots on grassland soil. III. Effect of growth of grass root-system on soil structure at plow layer, *J. Jpn. Grass. Sci.,* 23, 235, 1977.
674. **Kobayashi, H. and Otake, Y.,** Studies on physical interaction of forage roots on grassland soil. IV. Growth of root-system in the fixed pore model, *J. Jpn. Grass., Sci.,* 23, 241, 1977 (in Japanese, English summary).

675. **Konings, H. and Jackson, M. B.,** A relationship between rates of ethylene production by roots and the promoting or inhibiting effects of exogenous ethylene and water on root elongation, *Z. Pflanzenphysiol.*, 92, 385, 1979.

676. **Konstankiewicz, K.,** The influence of pore distribution on water potential in soils subjected to consolidation, *Zesz. Probl. Postepow Nauk Roln.*, 197, 79, 1977.

677. **Koolen, A. J. and Kuipers, H.,** *Agricultural Soil Mechanics*, Springer-Verlag, Berlin, 1983.

678. **Koorevaar, P., Menelik, G., and Dirkens, C.,** *Elements of Soil Physics, Development in Soil Science*, Elsevier, Amsterdam, 1983, 13.

679. **Köpke, U.,** A comparison of methods for measuring root growth of field crops, *Z. Acker Pflanzenbau*, 150, 39, 1981.

680. **Köpke, U., Böhm, W., and Jachman, Th.,** Rooting patterns of three winter wheat cultivars in a field and greenhouse experiment, *Z. Acker Pflanzenbau*, 151, 42, 1982.

681. **Kowalik, P. J.,** *Influence of Land Improvement on Soil Oxidation*, Rep. 42, Swedish University of Agricultural Science, Uppsala, 1985.

682. **Kramer, P. J.,** Causes of injury to plants resulting from flooding of the soil, *Plant Physiol.*, 26, 722, 1951.

683. **Kramer, P. J.,** *Plant and Soil Water Relationships: A Modern Synthesis*, McGraw-Hill, New York, 1969.

684. **Kramer, P. J., Ed.,** *Water Relations of Plants*, Academic Press, Orlando, FL, 1983.

685. **Kramer, P. J. and Bullock, H. C.,** Seasonal variations in the proportions of suberized and unsuberized roots of trees in relation to the absorption of water, *Am. J. Bot.*, 53, 200, 1966.

686. **Kramer, P. J. and Jackson, W. T.,** Causes of injury to flooded tobacco plants, *Plant Physiol.*, 29, 241, 1954.

687. **Krizek, D. T.,** Plant response to atmospheric stress caused by waterlogging, in *Breeding Plants for Less Favourable Environments*, John Wiley & Sons, 1982, 293.

688. **Krizek, D. T., Carmi, A., Mirecki, R. M., Snyder, F. W., and Bunce, J. A.,** Comparative effects of soil moisture stress and restricted root zone volume on morphogenic and physiological responses of soybean [*Glycine max* (L.) Merr.], *J. Exp. Bot.*, 36, 25, 1985.

689. **Kuan, T. L. and Erwin, D. C.,** Predisposition effect of water saturation of soil on *Phytophtora* root rot of alfalfa, *Phytopathology*, 70, 981, 1980.

690. **Kubota, T., Branco, J. L. C., and Ike, M.,** Soybean root penetration as affected by the compact layer in Cerrado Oxisol, *Jpn. J. Soil Sci. Plant Nutr.*, 54, 389, 1983.

691. **Kubota, T. and Williams, R. J. B.,** The effects of changes in soil compaction and porosity on germination, establishment and yield of barley and globe beet, *J. Agric. Sci.* Cambridge, 68, 227, 1967.

692. **Kuchenbuch, R. and Jung, J.,** Changes in root-shoot ratio and ion uptake of maize (*Zea mays* L.) from soil as influenced by a plant growth regulator, *Plant Soil*, 109, 151, 1988.

693. **Kuhlman, H. and Barraclough, P. B.,** Comparison between the seminal and nodal root systems of winter wheat in their activity for N and K uptake, *Z. Pflanzenernaehr. Bodenkd.*, 150, 24, 1987.

694. **Kuiper, F. and Kuiper, P. J. C.,** Permeability and self-induction as factors in water transport through bean roots, *Physiol. Plant.*, 31, 159, 1974.

695. **Kuipers, H.,** Soil compaction in arable farming, in *Trans. 13th Congr. Int. Soc. Soil Science*, Vol. 5, Hamburg, West Germany, 1986, 310.

696. **Kulkarni, B. K. and Savant, N. K.,** Effect of soil compaction on root-cation exchange capacity of crop plants, *Plant Soil*, 48, 269, 1977.

697. **Kunelins, H. T. and Clark, K. W.,** Influence of root temperature on the early growth and symbiotic nitrogen fixation of nodulated *Lotus corniculatus* plants, *Can. J. Plant Sci.*, 50, 569, 1970.

698. **Labanauskas, C. K., Letey, J., Stolzy, L. H., and Valoras, N.,** Effects of soil-oxygen and irrigation on the accumulation of macro- and micronutrients in citrus seedlings (*Citrus sinensis* var. Osbeck), *Soil Sci.*, 101, 378, 1966.

699. **Labanauskas, C. K., Stolzy, L. H., and Handy, M. F.,** Protein and nonprotein amino acids in citrus leaves as affected by *Phytophtora* sp. root infestation and soil oxygen content, *J. Am. Soc. Hortic. Sci.*, 97, 433, 1972a.

700. **Labanauskas, C. K., Stolzy, L. H., and Handy, M. F.,** Concentrations and total amounts of nutrients in citrus seedlings (*Citrus sinensis* var. Osbeck) and in soil as influenced by differential soil oxygen treatments, *Soil Sci. Soc. Am. Proc.*, 36, 454, 1972b.

701. **Labanauskas, C. K., Stolzy, L. H., Klotz, L. J., and de Wolfe, T. A.,** Effects of soil temperature and oxygen on the amounts of macronutrients in citrus seedlings (*Citrus sinensis* var. *Bessie*), *Soil Sci. Soc. Am. Proc.*, 29, 60, 1965.

702. **Labanauskas, C. K., Stolzy, L. H., Klotz, L. J., and de Wolfe, T. A.,** Soil oxygen diffusion rates and mineral accumulation in citrus seedlings (*Citrus sinensis* var. *Osbeck*), *Soil Sci.*, 11, 386, 1971.

703. **Labanauskas, C. K., Stolzy, L. H., Zentmyer, G. A., and Szuszkiewicz, T. E.,** Influence of soil oxygen and soil water on the accumulation of nutrients in avocado seedlings (*Persea ameraicana* Mill.), *Plant Soil*, 29, 391, 1968.

704. **Labuda, S.,** Plant mineral content as affected by the soil oxygen stress, *Probl. Agrofizyki*, 52, 1988 (in Polish, English summary).

705. **Lachno, D. R., Harrison-Murray, R. S., and Andus, L. J.,** The effects of mechanical impedance to growth on the levels of ABA and LAA in root tips of *Zea mays* L., *J. Exp. Bot.*, 33, 943, 1982.
706. **Lafolie, F. and Bruckler, L.,** Modelling in plane of water absorption by roots when non-uniform root system, *Zesz. Probl. Postepow Nauk Roln.*, 385, in press (in French, English summary).
707. **Lal, R. D.,** No-tillage effects on soil properties under different crops in Western Nigeria, *Soil Sci. Soc. Am. Proc.*, 40, 762, 1976.
708. **Lal, R., Wilson, G. F., and Okigbo, B. N.,** No-til farming after various grasses and leguminous cover crops in tropical Alfisol. I. Crop performance, *Field Crop Res.*, 1, 71, 1978.
709. **Lal, R., Wilson, G. F., and Okigbo, B. N.,** Changes in properties of an Alfisol produced by various crop covers, *Soil Sci.*, 27, 377, 1979.
710. **Lambers, H.,** Respiration and NADH-oxidation of the roots of flood-intolerant *Senecio* species as affected by anaerobiosis, *Physiol. Plant.*, 37, 117, 1976.
711. **Lambers, H. and Smakman, G.,** Respiration of the roots of flood-tolerant and flood-intolerant *Senecio* species: affinity for oxygen and resistance to cyanide, *Physiol. Plant.*, 42, 163, 1978.
712. **Lambers, H. and Steingröver, E.,** Efficiency of root respiration of a flood-tolerant and a flood-intolerant *Senecio* species as affected by low oxygen tension, *Physiol. Plant.*, 42, 179, 1978a.
713. **Lambers, H. and Steingröver, E.,** Growth respiration of a flood-intolerant *Senecio* species: correlation between calculated and experimental values, *Physiol. Plant.*, 43, 219, 1978b.
714. **Lambers, H., Szaniawski, R. K., and de Visser, R.,** Respiration for growth, maintenance and ion uptake. An evaluation of concepts, methods, values and their significance, *Physiol. Plant.*, 58, 556, 1983.
715. **Lamont, B.,** Mechanism for enhancing nutrient uptake with particular reference to Mediterranean North Africa and Western Australia, *Bot. Rev.*, 48, 597, 1982.
716. **Lamont, B.,** Root hair dimensions and surface (volume) weight ratios of roots with the aid of scanning electron microscopy, *Plant Soil*, 74, 149, 1983.
717. **Langridge, J.,** Biochemical aspects of temperature response, *Annu. Rev. Plant Physiol.*, 13, 411, 1963.
718. **Larney, F. J., Fortune, R. A., and Collins, J. F.,** Soil physical factors influencing sugar emergence, root shape, and root yield on a range of Irish soils, in *Proc. 9th Conf. ISTRO*, Edinburgh, Scotland, 1988, 79.
719. **Larque-Saavedra, A., Wilkins, H., and Wain, R. L.,** Promotion of cross root elongation in white light by 3,5-diiodo-4-hydroxybenzoic acid, *Planta*, 126, 269, 1975.
720. **Larson, M. M.,** Effects of soil moisture on early growth of oak seedlings, *For. Res. Rev. Ohio Agric. Res. Dev. Center*, 74, 10, 1974.
721. **Larson, W. E. and Allmaras, R. R.,** *Management Factors and Natural Forces as Related to Compaction in Compacting of* Agricultural Soils, American Society of Agriculture Engineering, St. Joseph, MI, 1971, 367.
722. **Larson, W. E., Gupta, S, C., and Useche, R. A.,** Compression of agricultural soils from eight soil orders, *Soil Sci. Soc. Am. J.*, 44, 450, 1980.
723. **Larson, W. E., Gupta, S. C., and Voorhees, W. B.,** A model for predicting soil compaction, in *Trans. 13th Congr. Int. Soc. Soil Science*, Vol. 5, Hamburg, West Germany, 1986, 290.
724. **Lawlor, D. W.,** Growth and water absorption of wheat with parts of the roots at different water potentials, *New Phytol.*, 72, 297, 1973.
725. **Layzell, D. B., Rainbird, R., Atkins, C. A., and Pate, J. S.,** Economy photosynthate use in N-fixing legume nodules: observations on two contrasting symbioses, *Plant Physiol.*, 64, 888, 1979.
726. **Layzell, D. B., Rochman, P., and Canvin, D. T.,** Low root temperatures and nitrogenase activity in soybean, *Can. J. Bot.*, 62, 965, 1984.
727. **Lee, K. K., Holst, R. W., Watanabe, I., and App, A.,** Gas transport through rice, *Soil Sci. Plant Nutr.*, 27, 151, 1981.
728. **Lee, R. B.,** The effect of nitrite on root growth of barley and maize, *New Phytol.*, 83, 615, 1979.
729. **Leiser, A. T.,** A mucilaginous root sheath in Ericaceae, *Am. J. Bot.*, 55, 391, 1968.
730. **Lemon, E. R.,** Soil aeration and plant relations. I. Theory, *Agron. J.*, 54, 167, 1962.
731. **Lemon, E. R. and Erickson, A. E.,** The measurement of oxygen diffusion in the soil with a platinum microelectrode, *Soil Sci. Soc. Am. Proc.*, 16, 160, 1952.
732. **Lemon, E. R. and Erickson, A. E.,** Principles of the platinum microelectrode as a method of characterizing soil aeration, *Soil Sci.*, 79, 383, 1955.
733. **Lemon, E. R. and Wiegand, C. L.,** Soil aeration and plant relation. II. Root respiration, *Agron. J.*, 54, 171, 1962.
734. **Leppard, G. G. and Ramamoorthy, S.,** The aggregation of wheat rhizosphere fibrils and the accumulation of soil bound cations, *Can. J. Bot.*, 53, 1729, 1975.
735. **Leshem, B.,** The annual activity of intermediary roots of the Aleppo pine, *For. Sci.*, 11, 291, 1965.
736. **Letey, J.,** Relationship between soil physical properties and crop production, *Adv. Soil Sci.*, I, 1985, 277.
737. **Letey, J., Morgan, W. C., Richards, S. J., and Valoras, N.,** Physical soil amendments, soil compaction, irrigation, and wetting agents in turfgrass management. III. Effects on oxygen diffusion rate and root growth, *Agron. J.*, 58, 531,. 1966.

738. **Letey, J., Stolzy, L. H., Blank, G. B., and Lunt, O. R.,** Effect of temperature on oxygen diffusion rates and subsequent shoot growth, root growth and mineral content of two plant species, *Soil Sci.*, 92, 314, 1961.
739. **Letey, J., Stolzy, L. H., Lunt, O. R., and Younger, V. B.,** Growth and nutrient uptake of Newport bluegrass as affected by soil oxygen, *Plant Soil*, 20, 143, 1964.
740. **Letey, J., Stolzy, L. H., Valoras, N., and Szuszkiewicz, T. E.,** Influence of oxygen diffusion rate on sunflower growth at various soil and air temperatures, *Agron. J.*, 54, 316, 1962.
741. **Levan, M. A. and Riha, S. J.,** The precipitation of black coatings on flooded conifer roots of low internal porosity, *Plant Soil*, 95, 33, 1986a.
742. **Levan, M. A. and Riha, S. J.,** Response of root systems of northern conifer transplants to flooding, *Can. J. For. Res.*, 16, 42, 1986b.
743. **Levin, I., Bravdo, B., and Assaf, R.,** Relation between apple root distribution and soil water extraction in different irrigation regimes, in *Physical Aspects of Soil Water and Salts in Ecosystems*, Hadas, A., Swartzendruber, D., Rijtema, P.E., Fuchs, M., and Yaron, B., Eds., Springer-Verlag, Berlin, 1973.
744. **Levy, G.,** Importance of soil properties for the rooting of *Picea excela* and *Pinus silvestris*, *Ann. Sci. For.*, 25, 157, 1968 (in French, English summary).
745. **Leyshon, A. J. and Sheard, R. W.,** Influence of short-term flooding on the growth and plant nutrient composition of barley, *Can. J. Soil Sci.*, 54, 463, 1974.
746. **Liberman, M., Kunishi, A., Mapson, L. W., and Wardale, D. A.,** Stimulation of ethylene production in apple tissue slices by methionine, *Plant Physiol.*, 41, 376, 1966.
747. **Lindberg, S. and Pettersson, S.,** Effects of mechanical stress on uptake and distribution of nutrients in barley, *Plant Soil*, 83, 295, 1985.
748. **Lindemann, W. C. and Ham, G. E.,** Soybean plant growth, nodulation, and nitrogen fixation as affected by root temperature, *Soil Sci. Soc. Am. J.*, 43, 1134, 1979.
749. **Lindemann, W. C., Ham, G. E., and Randall, G. W.,** Soil compaction effects on soybean nodulation, $N_2(C_2H_2)$ fixation and seed yield, *Agron. J.*, 74, 307, 1982.
750. **Lipiec, J.,** Possibilities of hydraulic conductivity's estimation of soils on the base of their basic properties, *Probl. Agrofizyki*, 40, 1983 (in Polish).
751. **Lipiec, J.,** The effect of soil compaction on root growth of spring barley, in *Abstr. Symp. Plant Roots and Their Environment*, Uppsala, Sweden, 1988, 3.
752. **Lipiec, J., Kania, W., and Tarkiewicz, S.,** Effect of wheeling on the physical characteristics of soils and rooting of some cereals, *Zesz. Probl. Postepow Nauk Roln.*, 385, in press.
753. **Lipiec, J., Kubota, T., Iwama, H., and Hirose, J.,** Measurement of plant water use under controlled soil moisture conditions by negative water circulation technique, *Soil Sci. Plant Nutr.*, 34, 417, 1988.
754. **Lipiec, J. and Tarkiewicz, S.,** Influence of compaction of loess soil on root system of spring barley and some physical properties of soil, in *Proc.* INRA, in press (in French).
755. **Logsdon, S. D., Parker, J. C., and Reneau, R. B., Jr.,** Root growth as influenced by aggregate size, *Plant Soil*, 99, 267, 1987a.
756. **Logsdon, S. D., Reneau, R. B., Jr., and Parker, J. C.,** Corn seedling root growth as influenced by soil physical properties, *Agron. J.*, 79, 221, 1987b.
757. **Lopushinsky, W.,** Effect of water movement on ion movement into the xylem of tomato roots, *Plant Physiol.*, 39, 494, 1964.
758. **Lundegårdh, H. G.,** Carbon dioxide evolution of soil and crop growth, *Soil Sci.*, 23, 417, 1927.
759. **Lungley, D. R.,** The growth of root systems — a numerical computer simulation model, *Plant Soil*, 38, 145, 1973.
760. **Lunt, O. R., Letey, J., and Clark, S. B.,** Oxygen requirements for root growth in three species of desert shrubs, *Ecology*, 54, 1356, 1973.
761. **Luxmoore, R. J., Sojka, R. E., and Stolzy, L. H.,** Root porosity and growth responses of wheat to aeration and light intensity, *Soil Sci.*, 113, 354, 1972.
762. **Luxmoore, R. J. and Stolzy, L. H.,** Root porosity and growth responses of rice and maize to oxygen supply, *Agron. J.*, 61, 202, 1969.
763. **Luxmoore, R. J. and Stolzy, L. H.,** Oxygen diffusion in the soil-plant system. V. Oxygen concentration and temperature effects on oxygen relations predicted for maize roots, *Agron. J.*, 64, 720, 1972a.
764. **Luxmoore, R. J. and Stolzy, L. H.,** Oxygen diffusion in the soil-plant system. VI. A synopsis with commentary, *Agron. J.*, 64, 725, 1972b.
765. **Luxmoore, R. J., Stolzy, L. H., and Letey, J.,** Oxygen diffusion in the soil-plant system. I. A model, *Agron. J.*, 62, 317, 1970a.
766. **Luxmoore, R. J., Stolzy, L. H., and Letey, J.,** Oxygen diffusion in the soil-plant system. II. Respiration rate, permeability, and porosity of consecutive excised segments of maize and rice roots, *Agron. J.*, 62, 322, 1970b.
767. **Luxmoore, R. J., Stolzy, L. H., and Letey, J.,** Oxygen diffusion in the soil-plant system. III. Oxygen concentration profiles, respiration rates, and the significance of plant aeration predicted for maize roots, *Agron. J.*, 62, 325, 1970c.

768. **Luxmoore, R. J., Stolzy, L. H., and Letey, J.,** Oxygen diffusion in the soil-plant system. IV. Oxygen concentration profiles, respiration rates and radial oxygen losses predicted for rice roots, *Agron. J.*, 62, 329, 1970d.
769. **Luxova, M.,** Growth regions of the primary roots of maize (*Zea mays* L.), in *Structure and Function of Plant Roots*, Brouwer, et al. Eds., 2nd Int. Symp., Chechoslovakia, Martinus Nijhoff, The Hague, 1980, 9.
770. **Lyklema, J. C.,** Absorption of ammonium and nitrate by perennial ryegrass, *Acta Bot. Neerl.*, 12, 361, 1963.
771. **Lyda, S. D.,** Alleviating pathogen stress, in *Modifying the Root Environment to Reduce Crop Stress*, ASAE Monogr. No. 4, Arkin, G. F. and Taylor, H. M., Eds., American Society of Agriculture Engineering St. Joseph, MI, 1981, 195.
772. **Lynch, J. M.,** The rhizosphere, in *Experimental Microbial Ecology*, Burns, R.G. and Slater, J.H., Eds., Blackwell Scientific, Oxford, 1982, 395.
773. **Lyr, G. and Hoffman, G.,** Growth rates and growth periodicity of tree roots, in *International Review of Forestry Research*, Vol. 2, Rombereger, J. A. and Mikola, P., Eds., Academic Press, New York, 1967, 181.
774. **MacDonald, J. D. and Duniway, J. M.,** Influence of soil texture and temperature on the mobility of *Phytophtora cryptogea* and *P. megasperma* zoospores, *Phytopathology*, 68, 1627, 1978.
775. **MacDowell, F. D. H.,** Reversible effect of chemical treatments on the rhythmic exudation of sap by tobacco roots, *Can. J. Bot.*, 42, 115, 1964.
776. **Macduff, J. H. and Hopper, M. J.,** Effects of root temperature on nutrient uptake of nitrate and ammonium ions by barley (*Hordeum vulgare* L.) grown in flowing solution culture, *Plant Soil*, 91, 303, 1985.
777. **Macduff, J. H., Wild, A., Hopper, M. J., and Dhanoa, M. S.,** Effects of temperature on parameters of root growth relevant to nutrient uptake: measurements on oilseed rape and barley grown in flowing nutrient solution, *Plant Soil*, 94, 321, 1986.
778. **Machlis, L.,** The respiratory gradient in barley roots, *Am. J. Bot.*, 31, 281, 1944.
779. **Mackay, A. D. and Barber, S. A.,** Soil temperature effects on root growth and phosphorus uptake by corn, *Soil Sci. Soc. Am. J.*, 48, 818, 1984.
780. **Mackay, A. D. and Barber, S. A.,** Effect of soil moisture and phosphate level on root hair growth of corn roots, *Plant Soil*, 86, 321, 1985.
781. **MacNish, G. C.,** *Studies of Rhizoctonia Patch Disease Cereals in Western Australia*, Abstr. No. 651, 4th Int. Congr. of Plant Pathology, Melbourne, Australia, 1983a.
782. **MacNish, G. C.,** *Rhizoctonia* patch in Western Australian grain belt, *Aust. Plant Pathol.*, 12, 49, 1983b.
783. **MacNish, G. C.,** Methods of reducing *Rhizoctonia* patch of cereals in Western Australia, *Plant Pathol.*, 34, 175, 1985.
784. **MacNish, G. C., Barbetti, M. J., Gillespie, D., and Hawley, K.,** Root Rot of Subterranean Clover, Bull. No. 3977, West Australia Department of Agriculture, ,1976.
785. **Madsen, H. B.,** Distribution of spring barley roots in Danish soils of different texture and under different climatic conditions, *Plant Soil*, 88, 31, 1985.
786. **Madsen, H. B.,** The construction of root zone capacity maps based on computerized soil maps and pedological data, *Pedologie*, XXXVI(2), 133, 1986.
787. **Maertens, C.,** Relationships between the uptake of nutrients and water and the root system, in *Proc. 16th Colloq. of Int. Potash Inst.*, Bern, 1981, 77.
788. **Maertens, C.,** Soybean rooting system and exploitation, in Le Soja, Physiologie de Plante et d'Adaptation aux Conditions Françaises, CETIOM, INRA Supplement à Information Techniques (1er trimestre 1986).
789. **Maertens, C., Blanchet, R., and Puech, J.,** The influence of different hydrous systems on the absorption of water and mineral elements by crops. I. Hydrous systems, root systems and modalities of water supply, *Ann. Agron.*, 25, 575, 1974 (in French., English summary).
790. **Maher, M. J.,** Effect of root zone warming, flow rate and propagation system on tomato crop performance in NFT, *Acta Hortic.*, 98, 45, 1980.
791. **Malzer, G. L. and Barber, S. A.,** Precipitation of calcium and strontium sulfates around plant roots and its evaluation, *Proc. Soil Sci. Soc. Am.*, 39, 492, 1975.
792. **Mambani, B. and Lal, R.,** Response of upland rice varieties to drought stress. I. Relation between root system development and leaf water potential, *Plant Soil*, 73, 59, 1983a.
793. **Mambani, B. and Lal, R.,** Response of upland rice varieties to drought stress. II. Screening rice varieties by means of variable moisture regimes along a toposequence, *Plant Soil*, 73, 73, 1983b.
794. **Mambani, B. and Lal, R.,** Response of upland rice varieties to drought stress. III. Estimating root system configuration from soil moisture data, *Plant Soil*, 73, 95, 1983c.
795. **Mapson, L. W.,** Biogenesis of ethylene, *Biol. Rev.*, 44, 155, 1969.
796. **Markhart, A. H., Fiscus, E. L., Naylor, A. W., and Kramer, P. J.,** Effect of abscisic acid on root hydraulic conductivity, *Plant Physiol.*, 64, 611, 1979.
797. **Maronek, D. M. and Wott, J. A.,** Morphological and physiological responses of *Acer rubrum* L., and *Acer saccharum* Marsh. to inundation, *HortScience*, 10, 316, 1975.
798. **Marschner, H. and Römheld, V.,** *In vivo* measurement of root-induced pH changes at the soil-root interface: effect of plant species and nitrogen source, *Z. Pflanzenphysiol.*, 111, 241, 1983.

799. **Marschner, H., Römheld, V., Horst, W. J., and Martin, P.,** Root-induced changes in the rhizosphere: importance for the mineral nutrition of plants, *Z. Pflanzenernaehr. Bodenkd.*, 149, 441, 1986.
800. **Marschner, H., Römheld, V., and Ossenberg–Neuhaus, H.,** Rapid method for measuring changes in pH and reducing processes along roots of intact plants, *Z. Pflanzenphysiol.*, 105, 407, 1982.
801. **Marsh, B. B.,** Measurement of length in random arrangements of lines (short communication), *J. Appl. Ecol.*, 8, 265, 1971.
802. **Marshall, J. D.,** Drought and shale interact to cause fine-root mortality in Douglas-fir seedlings, *Plant Soil*, 91, 51, 1986.
803. **Marshall, T. J. and Holmes, J. W.,** *Soil Physics*, Cambridge University Press, Cambridge, 1979.
804. **Martin, G. C. and Wilcox, G. E.,** Critical soil temperature for tomato plant growth, *Proc. Soil Sci. Soc. Am.*, 27, 565, 1963.
805. **Mason, W. K., Rowse, H. R., Bennie, A. T. P., Kaspar, T. C., and Taylor, H. M.,** Responses of soybeans to two row spacings and two soil water levels. II. Water use, root growth, and plant water status, *Field Crop Res.*, 5, 15, 1982.
805a. **Masse, J., Tardieu, F., and Colnenne, C.,** Kinetics of rooting depth and roots distribution on winter wheat in field conditions: effects of soil compaction, in *Abstr. of Symp. Plant Roots and Their Environment*, Uppsala, Sweden, 1988, 3.
806. **Matthews, S. and Whitebread, R.,** Factors influencing preemergence mortality in peas, *Plant Physiol.*, 17, 11, 1968.
807. **Mbagwu, J. S.,** Effects of bulk density and irrigation frequency on the root growth and dry matter yields of corn and cowpea for three Nigerian topsoil and subsoil profiles, *Beitr. Trop. Landwirtsch. Veterinaermed.*, 23, 277, 1985.
808. **McCoy, E. L. and Boersma, L.,** A mathematical model of axial root growth and soil penetration, in *Trans. 13th Congr. Int. Soc. Soil Science*, Vol. 2, Hamburg, West Germany, 1986, 116.
809. **McKee, K. L. and Mendelssohn, I. A.,** Root metabolism in the black mangrove (*Avicennia germinous* (L.) L): response to hypoxia, *Environ. Exp. Bot.*, 27, 147, 1987.
810. **McLean, E. O., Adams, D., and Franklin, R. E.,** Cation exchange capacities of plant roots as related to their nitrogen contents, *Soil Sci. Soc. Am. Proc.*, 20, 345, 1956.
811. **McManmon, M. and Crawford, R. M. M.,** A metabolic theory of flooding tolerance: the significance of enzyme distribution and behavior, *New Phytol.*, 70, 299, 1971.
812. **McPherson, D. C.,** Cortical air spaces in the roots of *Zea mays* L., *New Phytol.*, 38, 190, 1939.
813. **McSweney, K. and Jansen, I. J.,** Soil structure and associated rooting behaviour in minesoils, *Soil Sci. Soc. Am. J.*, 48, 607, 1984.
814. **Melhuish, F. M. and Lang, A. R. G.,** Quantitative studies of roots in soil. I. Length and diameters of cotton roots in a clay-loam soil by analysis of surface-ground blocks of resin-impregnated soil, *Soil Sci.*, 106, 16, 1968.
815. **Mench, M., Morel, L., and Guckert, A.,** Cadmium binding with molecular weight soluble root exudates from maize (*Zea mays* L.), C. R. Acad. Sci. Paris, t. 301, serie III, No. 7, 379, 1985.
816. **Mench, M., Morel, J. L., and Guckert, A.,** Metal binding properties of high molecular weight soluble exudates from maize (*Zea mays* L.) roots, *Biol. Fertil. Soils*, 3, 165, 1987.
817. **Mendelssohn, I. A. and Burdick, D. M.,** The relationship of soil parameters and root metabolism to primary production in periodically inundated soils, in *The Ecology of Wetlands*, Vol. 1, Hook, D. D. and al. Eds., Croom Helm, Ltd., London, 1988, 398.
818. **Mendelssohn, I. A. and McKee, K. L.,** Root metabolic response of *Spartina alterniflora* to hypoxia, in *Plant Life in Aquatic and Amphibious Habitats*, Spec. Publ. No. 5, British Ecology Society Special Symp., Crawford, R. M. M., Ed., Blackwell Scientific, Oxford, 1987, 239.
819. **Mendelssohn, A. and Postek, M. T.,** Elemental analysis of deposits on the roots of *Spartina alterniflora* Loisel, *Am. J. Bot.*, 69, 904, 1982.
820. **Mengel, K. and Steffens, D.,** Potassium uptake of rye-grass (*Lolium perenne*) and red clover (*Triflolium pratense*) as related to root parameters, *Biol. Fertil. Soils*, 1, 53, 1985.
821. **Meredith, H. L. and Patrick, W. H., Jr.,** Effects of soil compaction on subsoil root penetration and physical properties of three soils in Louisiana, *Agron. J.*, 53, 163, 1961.
822. **Merget, A.,** Sur la reproduction artificielle des phénoménes de thermodiffusion gareuse des feuilles, par les corps poreaux et pulvérulents humides, *C. R. Acad. Sci.*, 78, 884, 1874.
823. **Merrill, S. D. and Rawlins, S. L.,** Distribution of growth of sorghum roots in response to irrigation frequency, *Agron. J.*, 71, 738, 1979.
824. **Mettauer, H., Tual, Y., Haeflinger, R., Conesa, A. P., and Trendel, R.,** Influence of the physical and mechanical properties of the soil on the sugar beet morphology and root yield, *Ann. Agron.*, 29, 147, 1978 (in French, English summary).
825. **Meyer, W. S. and Alston, A. M.,** Wheat responses to seminal geometry and subsoil water, *Agron. J.*, 70, 981, 1978.

826. **Meyer, W. S. and Barrs, H. D.**, Non-destructive measurement of wheat roots in large undisturbed and repacked clay soil cores, *Plant Soil*, 85, 232, 1985.
827. **Meyer, W. S., Barrs, H. D., Smith, R. C. G., White, N. S., Heritage, A. D., and Short, D. L.**, Effect of irrigation on soil oxygen status and root and shoot growth of wheat in a clay soil, *Aust. J. Agric. Res.*, 36, 171, 1985.
828. **Meyer, W. S. and Ritchie, J. T.**, Water status of cotton as related to tap root length, *Agron. J.*, 72, 577, 1980.
829. **Mihalić, V., Butorac, A., and Folivarski, I.**, Some results of the basic soil tillage for sugar beet in semi-humid climate, in *Proc. 8th Conf. ISTRO*, Stuttgart, West Germany, 1979, 229.
830. **Miller, D. E.**, Water retention and flow in layered soil profiles, *Field Soil Water Regime*, SSSA Spec. Publ., 5, 107, 1973.
831. **Miller, D. M.**, Pressure-flow characteristics of the roots of *Zea mays*, *Plant Soil*, 63, 15, 1981.
832. **Mingeau, M.**, Root porosity and tolerance to flooding, *Ann. Agron.*, 28, 171, 1977 (in French, English summary).
833. **Taylor, H. M., Ed.**, *Minirhizitron Observation Tubes; Methods and Applications for Measuring Rhizosphere Dynamics*, ASA Spec. Publ. No. 50, American Society of Agronomy, Crop Science Society of America, Soil Science Society of America, Madison, WI, 1987.
834. **Mirreh, H. F. and Ketcheson, J. W.**, Influence of soil bulk density and matric pressure on soil resistance to penetration, *Can. J. Soil Sci.*, 52, 477, 1972.
835. **Mirreh, H. F. and Ketcheson, J. W.**, Influence of soil water matric potential and resistance to penetration on corn root elongation, *Can. J. Soil Sci.*, 53, 383, 1973.
836. **Misra, R. K.**, Relation of root development to drought resistance of plants, *Indian J. Agron.*, 1, 41, 1956.
837. **Misra, R. K., Alston, A. M., and Dexter, A. R.**, Root growth and phosphorus uptake in relation to the size and strength of soil aggregates. I. Experimental studies, *Soil Till. Res.*, 11, 103, 1988a.
838. **Misra, R. K., Alston, A. M., and Dexter, A. R.**, Role of root hairs in phosphorus depletion from a macrostructured soil, *Plant Soil*, 107, 11, 1988.
839. **Misra, R. K., Dexter, A. R., and Alston, A. M.**, Penetration of soil aggregates of finite size. II. Plant roots, *Plant Soil*, 94, 59, 1986a.
840. **Misra, R. K., Dexter, A. R., and Alston, A. M.**, Maximum axial and radial growth pressure of plant roots, *Plant Soil*, 95, 315, 1986b.
841. **Misra, R. K., Alston, A. M., and Dexter, A. R.**, Root growth and phosphorus uptake in relation to the size and strength of soil aggregates. II. Prediction by a stochastic model, *Soil Till. Res.*, 11, 117, 1988b.
842. **Arkin, G. F. and Taylor, H. M., Eds.**, Modifying the Root Environment to Reduce Crop Stress, American Society of Agriculture Engineering, St. Joseph, MI, 1981.
843. **Mollenhauer, H. H.**, Transition forms of Golgi apparatus secretion vesicles, *J. Ultrastruct. Res.*, 12, 439, 1965.
844. **Molz, F. J.**, Potential distributions in the soil-root system, *Agron. J.*, 67, 726, 1975.
845. **Molz, F. J.**, Models of water transport in the soil-plant system: a review, *Water Res.*, 17, 1245, 1981.
846. **Molz, F. J. and Peterson, C. M.**, Water transport from root to soil, *Agron. J.*, 68, 904, 1976.
847. **Molz, F. J. and Remson, I.**, Application of an extraction term model to the study of moisture flow to plant roots, *Agron. J.*, 63, 72, 1971.
848. **Monroe, C. D. and Kladivko, E. J.**, Aggregate stability of a silt loam soil as affected by roots of corn, soybeans and wheat, *Commun. Soil Sci. Plant Anal.*, 18, 1077, 1987.
849. **Moorby, H. and Nye, P. H.**, The effect of temperature variation over the root system on root extension and phosphate uptake by rape, *Plant Soil*, 78, 283, 1984.
850. **Morel, J. L., Mench, M., and Guecket, A.**, Measurement of Pb- and Cd- binding with mucilage exudates from maize (*Zea mays* L.) roots, *Biol. Fertil. Soils*, 2, 29, 1986.
851. **Morishima, H., Hirata, K., and Oka, H. J.**, Floating ability and drought resistance in wild and cultivated species of rice, *Indian J. Genet. Plant Breed.*, 22, 1, 1962.
852. **Morita, S. and Aoki, A.**, Studies on cation exchange capacity and cation adsorption of fruit tree root, *Soil Sci. Plant Nutr.*, 7, 8, 1961.
853. **Moss, D. N., Woolley, J. T., and Stone, J. F.**, Plant modification for more efficient water use: the challenge, *Agric. Meteorol.*, 14, 311, 1974.
854. **Moss, G. J.**, Root-zone warming of greenhouse tomatoes in nutrient film as a means of reducing heating requirements, *J. Hortic. Sci.*, 58, 103, 1983.
855. **Mouat, M. C. H.**, Competitive adaptation by plants to nutrient shortage through modification of root growth and surface charge, *N.Z. J. Agric. Res.*, 26, 327, 1983.
856. **Mozhaeva, L. V. and Pilshchikova, H. B.**, Relationship between the value of root pressure components and the rate of water pumping by the root, *Dokl. Akad. Nauk SSSR*, 239, 1005, 1978.
857. **Mugwira, L. M. and Elgawhary, S. M.**, Aluminum accumulation and tolerance of triticale and wheat in relation to root cation exchange capacity, *Soil Sci. Soc. Am. J.*, 43, 736, 1979.
858. **Mulligan, M. F., Smucker, A. J. M., and Safir, G. F.**, Tillage modifications of dry edible bean root colonization by VAM fungi, *Agron. J.*, 77, 140, 1985.

859. **Mulqueen, J., Stafford, J. V., and Tanner, D. W.,** Evaluation of penetrometers for measuring soil strength, *J. Terramech.*, 14, 137, 1977.

860. **Muneer, M., Ahmed, I. U., Faiz, S. M. A., and Bhuiya, M. R.,** Effect of soil bulk density and moisture level interactions on the growth of soybean roots, in *Trans. 12th Congr. Int. Soc. Soil Science,* New Delhi, India, 1982, 630.

861. **Munevar, F. and Wollum, A. G.,** Response of soybean plants to high root tremperature as affected by plant cultivar and *Rhizobium* strain, *Agron. J.*, 74, 138, 1982.

862. **Murray, D. I. L.,** *Rhizoctonia solani* causing barley stunt disorder, in *Trans. Br. Mycol. Soc.*, 76, 383, 1981.

863. **Musgrave, A., Jackson, M. B., and Ling, E.,** Callitriche stem elongation is controlled by ethylene and gibberellin, *Nature (London) New Biol.*, 238, 93, 1972.

864. **Nagarajah, S.,** Effect of soil texture on the rooting patterns of Thompson seedless vines on own roots and on Ramsey rootstock in irrigated vineyards, *Am. J. Enol. Viticult.*, 38, 54, 1987.

865. **Nambiar, E. K. S., Bowen, G. D., and Sands, R.,** Root regeneration and plant water status of *Pinus radiata* D. Don seedlings transplanted to different soil temperatures, *J. Exp. Bot.*, 30, 1119, 1979.

866. **Neumann, H. H., Thurtell, G. W., and Stevenson, K. R.,** *In situ* measurements of leaf water potential and resistance to water flow in corn, soybean, and sunflower at several transpiration rates, *Can. J. Plant Sci.*, 54, 175, 1973.

867. **Newman, E. I.,** Relationship between root growth of flax (*Linum usitatissimum*) and soil water potential, *New Phytol.*, 65, 273, 1966a.

868. **Newman, E. I.,** A method for estimating the total length of root in a sample, *J. Appl. Ecol.*, 3, 139, 1966b.

869. **Newman, E. I.,** Resistance to water flow in soil and plant. I. Soil resistance in relation to amounts of root; theoretical estimates, *J. Appl. Ecol.*, 6, 1, 1969a.

870. **Newman, E. I.,** Resistance to water flow in soil and plant. II. A review of experimental evidence on the rhizosphere resistance, *J. Appl. Ecol.*, 6, 261, 1969b.

871. **Newman, E. I.,** Permeability to water of the roots of five herbaceous species, *New Phytol.*, 72, 547, 1973.

872. **Newman, E. I.,** Root and soil water relations, in *The Plant Root and Its Environment,* Carson, E.W., Ed., University Press of Virginia, Charlottesville, 1974, .

873. **Newman, E. I.,** Water movement through root systems, *Philos. Trans. R. Soc. London Ser. B,* 273, 463, 1976.

874. **Newman, E. I. and Bowen, H. J.,** Patterns of distribution of bacteria on root surfaces, *Soil Biol. Biochem.*, 6, 205, 1974.

875. **Newman, E. I. and Kramer, P. J.,** Effects of decenylsuccinic acid on the permeability and growth of bean roots, *Plant Physiol.*, 41, 606, 2966.

876. **Newman, E. I. and Watson, A.,** Microbial abundance in the rhizosphere in computer model, *Plant Soil,* 48, 17, 1977.

877. **Nielsen, K. F.,** Roots and root temperatures, in *The Plant Root and Its Environment,* Carson, E. W., Ed., University Press of Virginia, Charlottesville, 1974, .

878. **Nielsen, K. F., Halstead, R. L., MacLean, A. J., Holmes, R. M., and Bourgest, S. J.,** The influence of soil temperature on the growth and mineral composition of oats, *Can. J. Soil Sci.*, 40, 255, 1960.

879. **Nielsen, K. F. and Humphries, E. C.,** Effects of root temperature on plant growth, *Soils Fert.*, 29, 1, 1966.

880. **Nielsen, N. E.,** A transport kinetic concept for ion uptake by plants. III. Test of the concept by results from water culture and pot experiments, *Plant Soil,* 45, 659, 1976.

881. **Nimah, M. N. and Hanks, R. J.,** Model for estimating soil water, plant and atmospheric interrelations. I. Description and sensitivity, *Soil Sci. Soc. Am. Proc.*, 37, 522, 1973a.

882. **Nimah, M. N. and Hanks, R. J.,** Model for estimating soil water, plant and atmospheric interrelations. II. Field test of model, *Soil Sci. Soc. Am. Proc.*, 37, 528, 1973b.

883. **Norris, W. E., Wiegand, C. L., and Johanson, J.,** Effect of carbon dioxide on respiration of excised onion root tips in high oxygen atmosphere, *Soil Sci.*, 88, 144, 1959.

884. **Nour, M. and Weibel, D. E.,** Evaluation of root characteristics in grain sorghum, *Agron. J.*, 70, 217, 1978.

885. **Novak, V.,** Estimation of soil-water extraction patterns by roots, *Agric. Water Manage.*, 12, 271, 1987.

886. **Nutman, P. S.,** Perspectives in biological nitrogen fixation, *Sci. Prog. Oxford,* 59, 55, 1971.

887. **Nuttall, M. and May, M. J.,** Sugar beet. Cultivations for sugar beet on light compactable soil 1983-1985, in 78th Annu. Rep. 1985-1986, Norfolk Agricultural Station, Wymontham, England, 1986, 33.

888. **Nye, P. H.,** The effect of the nutrient intensity and buffering power of a soil, and the absorbing power size and root hairs of a root on nutrient absorption by diffusion, *Plant Soil,* 25, 81, 1966.

889. **Nye, P. H.,** Changes of pH across the rhizosphere induced by roots, *Plant Soil,* 61, 7, 1981.

890. **Nye, P. H., Brewster, J. L., and Bhat, K. K. S.,** The possibility of predicting solute uptake and plant growth response from independently measured soil and plant characteristics. I. Theoretical bases of the experiments, *Plant Soil,* 42, 161, 1975.

891. **Nye, P. H. and Marriott, F. H. C.,** A theoretical study of the distribution of substances around roots resulting from simultaneous diffusion and mass flow, *Plant Soil,* 30, 459, 1969.

892. **Nye, P. H. and Tinker, P. B.,** *Solute Movement in the Soil-Root System, Studies in Ecology,* Vol. 4, Blackwell Scientific, Oxford, 1977.

893. **Oades, J. M.,** Mucilages at the root surface, *J. Soil Sci.*, 29, 1, 1978.
894. **O'Brien, L.,** Effect of root media on growth of wheat seminal roots, *Crop Sci.*, 18, 685, 1978.
895. **Obroucheva, N.,** Physiology of growing root cells, in *The Development and Function of Roots*, Academic Press, New York, 1975, 279.
896. **Ogata, S., Saneoka, H., Fujita, K., and Matsumoto, K.,** Nutritio-physiological evaluation of the drought resistance of warm season forage species. III. CO_2 assimilation and distribution of ^{14}C photosynthate, *J. Jpn. Grass. Sci.*, 31, 159, 1985a (in Japanese, English summary).
897. **Ogata, S., Saneoka, H., Fujita, K., and Matsumoto, K.,** Nutritio-physiological evaluation of the drought resistance of warm season forage species. IV. Comparative studies on root development, water and nutrient levels, *J. Jpn. Grass. Sci.*, 31, 263, 1985b (in Japanese, English summary),
898. **Ogata, S., Saneoka, H., and Matsumoto, K.,** Nutritio-physiological evaluation of the drought resistance of warm season forage species. I. Species differences of drought resistance under the field conditions, *J. Jpn. Grass. Sci.*, 31, 34, 1985c (in Japanese, English summary).
899. **Ogata, S., Saneoka, H., and Matsumoto, K.,** Nutritio-physiological evaluation of the drought resistance of warm season forage species. II. The comparative growth and water-nutrient absorption of warm season forage species grown under various soil water levels, *J. Jpn. Grass. Sci.*, 31, 43, 1985d (in Japanese, English summary).
900. **Ohmura, T. and Howell, R. W.,** Inhibitory effect of water on oxygen consumption by plant materials, *Plant Physiol*, 35, 184, 1960.
901. **Old, K. M. and Nicolson, T. H.,** Electron microscopical studies of the microflora of roots of sand dune grasses, *New Phytol.*, 74, 51, 1975.
902. **Olsthoorm, A. F. M., van Baren, B., and Hopman, M.,** Ammonium influence on root growth and rhizosphere pH of Douglas-fir seedlings, in *Abstr. Symp. Plant Roots and Their Environment*, Uppsala, Sweden, 1988, 2.
903. **Olszta, W. and Zawadzki, S.,** Influence of the ground water table on the growth and harvesting of grass, *Zesz. Probl. Postepow Nauk Roln.*, 385, in press.
904. **Omesa, K., Onoe, M., and Yamada, H.,** NMR imaging for measuring root system and soil water content, *Environ. Control Biol.*, 23, 99, 1985.
905. **Oosterhuis, D. M.,** A technique to measure the components of root water potential using screen-caged thermocouple psychrometers, *Plant Soil*, 103, 285, 1987.
906. **Oosterhuis, D. M. and Wiebe, H. H.,** Water stress preconditioning and cotton root pressure-flux relationships, *Plant Soil*, 95, 69, 1986.
907. **Opara-Nadi, O. A. and Lal, R.,** Effects of land clearing and tillage methods on soil properties and maize root growth, *Field Crop Res.*, 15, 193, 1987.
908. **Orchard, B.,** Solution heating for the tomato crop, *Acta Hortic.*, 98, 19, 1980.
909. **O'Toole, J. C. and Soemaratano,** Evaluation of a simple technique for characterizing rice root systems in relation to drought resistance, *Euphytica*, 30, 283, 1981.
910. **Ottman, M. J. and Timm, H.,** Measurement of viable plant roots with the image analyzing computer, *Agron. J.*, 76, 1018, 1984.
911. **Ottow, J. C. G.,** Mechanisms of bacterial iron-reducing in flooded soils, in *Proc. Symp. Paddy Soils China*, Nanjing, China, 1980, Institute Soil Sci., Academia Sinica, Science Press Beijing, Springer-Verlag, Berlin, 1981, 330.
912. **Pages, L.,** Modelling the root system architecture, Zesz. Probl. Postepow Nauk Roln., 385, in press (in French, English summary).
913. **Paleg, L. G. and Aspinall, D.,** *The Physiology and Biochemistry of Drought Resistance in Plants*, Academic Press, New York, 1981.
914. **Pankhurst, C. E. and Gibson, A..,** *Rhizobium* strain influence on disruption of clover nodule development at high root temperature, *J. Gen. Biol.*, 74, 219, 1973.
915. **Pankhurst, C. E. and Layzell, D. B.,** The effect of bacterial strain and temperature changes on the nitrogenase activity of *Lotus pedunculatus* root nodules, *Physiol. Plant.*, 62, 404, 1984.
916. **Pankhurst, C. E. and Sprent, J. I.,** Effect of temperature and oxygen tension on the nitrogenase and respiratory activities of turgid and water-stressed soybean and French bean root nodules, *J. Exp. Bot.*, 27, 96, 1976.
917. **Parish, D. H.,** Soil conditions as they affect plant establishment, root development and yield. Effects of compaction on nutrient supply to plants, in *Compaction of Agricultural Soils*, Monogr. American Society of Agriculture Engineering, St. Joseph, MI, 1971, 277.
918. **Parsons, L. R. and Kramer, P. J.,** Diurnal cycling in root resistance to water movement, *Physiol. Plant.*, 30, 19, 1974.
919. **Passioura, J. B.,** The effect of root geometry on the yield of wheat growing on stored water, *Aust. J. Agric. Res.*, 23, 745, 1972.
920. **Passioura, J. B.,** Water collection by roots, in *The Physiology and Biochemistry of Drought Resistance in Plants*, Paleg, G. and Aspinall, D., Eds., Academic Press, New York, 1981, 39.

921. **Passioura, J. B.,** Roots and drought resistance, *Agric. Water Manage.*, 7, 265, 1983.
922. **Pate, J. S.,** Temperature characteristics of bacterial variation in legume symbiosis, *Nature, London* 192, 637, 1961.
923. **Pate, J. S.,** Nodulation studies in legumes. V. The effects of temperature on symbiotic performances of bacterial associations of *Medicago tribuloides* Desr. and *Vicia atropurpurea* Desf., *Phyton,* 18, 65, 1962.
924. **Patel, M. S. and Singh, N. T.,** The effects of soil compaction on growth and water use efficiency of rice, *Indian Agron.*, 24, 429, 1979.
925. **Paterson, D. R., Earhart, D. R., and Fuqua, M. C.,** Effects of flooding level on storage root formation, ethylene production and growth of sweet potato, *HortScience,* 14, 739, 1979.
926. **Patrick, W. H., Delaune, E., and Engler, R. M.,** *Soil Oxygen Content and Root Development of Cotton in Mississippi River Alluvial Soils,* Bull. 673, Lousiana Agriculture Experiment Station, 1973.
927. **Patrick, W. H., Turner, F. T., and Delaune, R. D.,** Soil Oxygen Content and Root Development of Sugar Cane, Bull. 641, Louisiana Agriculture Experiment Station, 1969.
928. **Patterson, R.., Grunes, D. L., and Lathwell, D. J.,** Influence of root-zone temperature and P supply on total and inorganic P, free sugars, aconitane, and soluble amino N in corn, *Crop Sci.*, 12, 227, 1972.
929. **Pawlowski, F. and Malicki, L.,** The relationship between soil humidity, bulk density and root mass of winter wheat and grass mixture with red clover cultivated in undulated loess soils, *Ann. Univ. Mariae Curie Sklodowska Sect. E,* 19, 263, 1964 (in Polish, English summary).
930. **Peacock, C. H. and Dudeck, A. E.,** Effect of irrigation interval on St. Augustinegrass rooting, *Agron. J.*, 77, 813, 1985.
931. **Pearson, R. W., Childs, J., and Lund, Z. F.,** Uniformity of limestone mixing in acid subsoil as a factor in cotton root penetration, *Soil Sci. Soc. Am. Proc.*, 37, 727, 1973.
932. **Pearson, R. W., Ratliff, L. F., and Taylor, H. M.,** Effect of soil temperature, strength, and pH on cotton seedlings root elongation, *Agron. J.*, 62, 43, 1970.
933. **Pellerin, S. and Tardieu, F.,** Distribution of maize root system. Some results obtained for elaboration of the concept of the three dimensional model, *Zesz. Probl. Postepow Nauk Roln.*, 385, in press (in French, English summary).
934. **Penny, L. H.,** Vertical pull resistance of maize inbreds and their test crosses, *Crop Sci.*, 21, 237, 1981.
935. **Pera, A., Vallini, G., Ines Sireno, Lorella Bianchin, M., and de Bertoldi, M.,** Effect of organic matter on rhizosphere microorganisms and root development of sorghum plants in two different soils, *Plant Soil,* 74, 3, 1983.
936. **Pereira, J. S. and Kozlowski T. T.,** Variations among woody angiosperms in response to flooding, *Physiol. Plant.*, 41, 184, 1977.
937. **Perrochet, P.,** Water uptake by plant roots — a simulation model. I. Conceptual model, *J. Hydrol. Neth.*, 95, 55, 1987.
938. **Persson, H.,** Fine root production, mortality, and decomposition in forest ecosystems, *Vegetatio,* 41, 101, 1979.
939. **Petelkau, H., Gätke, R., Dannowski, M., and Seidel, M.,** Relationship Between Bulk Density of Soil and Crop Production, Report of Soil Science Research Center, Müncheberg, East Germany, 1985, 24 (in German).
940. **Peterson, W. K. and Barber, S. A.,** Soybean root morphology and K uptake, *Agron. J.*, 73, 316, 1981.
941. **Petersson, H., Messing, I., and Steen, E.,** Influence of root mass on saturated hydraulic conductivity in arid soils of central Tunisia, *Arid Soil Res. Rehabil.*, 1, 149, 1987.
942. **Pfender, W. F., Hine, R. B., and Stanghellini, M. E.,** Production of sporongia and release of zoospores by *Phytophtora megasperma* in soil, *Phytopathology,* 67, 657, 1977.
943. **Pfleger, I. and Werner, D.,** Effects of tube-shaped hollow spaces in soil pastes of different density and moisture content on the penetrating power of spring barley roots, *Arch. Acker Pflanzenbau Bodenkd.*, 30, 259, 1986 (in German).
944. **Pfleger, I., Werner, D., and Unger, H.,** Root growth in biopores formed by preceding plants — results from model studies, *Arch. Acker Pflanzenbau Bodenkd.*, 32, 311, 1988 (in German).
945. **Phene, C. J. and Beale, O. W.,** High-frequency irrigation for water nutrient management in humid regions, *Soil Sci. Soc. Am. J.*, 40, 430, 1976.
946. **Phillips, H. L. and Torrey, J. G.,** The ultrastructure of the root cap in cultured roots of *Convulvulus arvensis* L., *Am. J. Bot.*, 61, 879, 1974.
947. **Philips, R. E., Belcher, C. R., Blevins, R. L., Miller, H. F., Thomas, G. W., NaNagara, T., Quisenberry, V., and McMahon, ,** No-Tillage Farming, Agron. Res. Misc. Publ. No. 394, Kentucky Agriculture Experiment Station, 1971, 36.
948. **Philips, R. E. and Brown, D. A.,** Ion diffusion. III. The effect of soil compaction on self diffusion of rubidium-86 and strontium-89, *Soil Sci. Soc. Am.* Proc., 29, 657, 1965.
949. **Philipson, J. J. and Coutts, M. P.,** The tolerance of tree roots to waterlogging. III. Oxygen transport in lodgepole pine and Sitka spruce roots of primary structure, *New Phytol.*, 80, 341, 1978.
950. **Pidgeon, J. D. and Soane, B. D.,** Effects of tillage and direct drilling on soil properties during the growing season in a long-term barley monoculture system, *J. Agric. Sci.* Cambridge, 88, 431, 1977.

951. **Pirone, P. P.,** *Tree Maintenance,* 4th ed., Oxford University Press, New York, 1972, 574.
952. **Pitman, M. G.,** Adaptation of barley roots to low oxygen supply and its relation to potassium and sodium uptake, *Plant Physiol.,* 44, 1233, 1969.
953. **Pitman, M. G. and Welfare, D.,** Inhibition of ion transport in excised barley roots by abscisic acid: relation to water permeability of the roots, *J. Exp. Bot.,* 29, 1125, 1978.
954. **Ponnamperuma, F. N.,** The chemistry of submerged soils, *Adv. Agron.,* 24, 29, 1972.
955. **Portas, C. A. M. and Taylor, H. M.,** Growth and survival of young plant roots in dry soil, *Soil Sci.,* 121, 170, 1976.
956. **Porter, J. R., Bragg, P. L., Rayner, J. H., Weir, A. H., and Landsberg, J. J.,** The ARC winter wheat simulation model — principles and progress, in *Opportunities for Manipulation of Cereal Productivity,* Monogr. 7, British Plant Growth Regulator Group, 1981, 97.
957. **Prihar, S. S.,** Management of physical environment of soil for increasing crop production, in *Proc. 12th Congr. Int. Soc. Soil Science,* New Delhi, India, 1982, 3.
958. **Prihar, S. S. and Gajri, P. R.,** Magnitude and mode of response of dryland crops to limited irrigation, in *Trans. Symp. Rainwater and Dryland Agriculture,* Indian National Science Academy, New Delhi, 1982, 115.
959. **Prummel, J.,** Effect of soil structure on phosphate nutrition of crop plants, *Neth. J. Agric. Sci.,* 23, 62, 1975.
960. **Przestalski, S. and Kargol, M.,** Graviosmotic volume flow through membrane systems, *Stud. Biophys.,* 34, 7, 1972.
961. **Przestalski, S. and Kargol, M.,** Graviosmosis, *Comments Mol. Cell. Biophys.,* 4, 249, 1987.
962. **Pumphrey, F. V., Wilkins, D. E., Hane, D. C., and Smiley, R. W.,** Influence of tillage and nitrogen fertilizer on *Rhizoctonia* root rot (bare patch) of winter wheat, *Plant Dis.,* 71, 125, 1987.
963. **Raffineau-Delile, A.,** Evidence du mode respiratoire des feuilles de *Nelambium, Ann. Sci. Nat.* Ser. II, 16, 328, 1841.
964. **Raghavan, G. S. V., McKyes, E., and Beaulieu, B.,** Prediction of clay soil compaction, *J. Terramech.,* 14, 31, 1977a.
965. **Raghavan, G. S. V., McKyes, E., Chassé, M., and Mérineau, F.,** Development of compaction patterns due to machinery operation in an orchard soil, *Can. J. Plant Sci.,* 56, 505, 1976.
966. **Raghavan, G. S. V., McKyes, E., Stemshorn, E., Gray, A., and Beaulieu, B.,** Vehicle compaction patterns in clay soil, *Trans. ASAE,* 20, 218, 1977b.
967. **Rainbird, R. M., Atkins, C. A., and Pate, J. S.,** Diurnal variations in the functioning of cowpea nodules, *Plant Physiol.,* 73, 392, 1983.
968. **Rajkai Vegh, K.,** Effect of soil water and plant nutrient supply on root characteristics and nutrient uptake, in *Abstr. Symp. Plant Roots and Their Environment,* Uppsala, Sweden, 1988, 1.
969. **Ramos, C. and Kaufmann, M. R.,** Hydraulic resistance of rough lemon roots, *Physiol. Plant.,* 45, 311, 1979.
970. **Rao, V. R. and Rao, J. L. N.,** Nitrogen fixation (C_2H_2 reduction) in soil samples from rhizosphere of rice grown under alternate flooded and nonflooded conditions, *Plant Soil,* 81, 111, 1984.
971. **Raskin, I. and Kende, H.,** How does deep water rice solve its aeration problem, *Plant Physiol.,* 72, 447, 1983.
972. **Rasmunssen, K. J.,** Plant growth and soil physical properties by reduced tillage, in *Proc. 9th Conf. ISTRO,* Osijek, Yugoslavia, 1982, 526.
973. **Raymond, P., Bruceau, E., and Pradet, A.,** Study of oxygen transport of above ground parts of roots by the metabolic parameter. The energetic charge, *C. R. Acad. Sci.,* 286, 1061, 1978 (in French).
974. **Raymond, P. and Pradet, A.,** Stabilization of adenine nucleotide ratios at various values by an oxygen limitation of respiration in germination lettuce (*Lactuca sativa*) seeds, *Biochem. J.,* 190, , .
975. **Reicosky, D. C. and Ritchie, J. T.,** Relative importance of soil resistance and plant resistance in root water absorption, *Soil Sci. Soc. Am. J.,* 40, 293, 1976.
976. **Reid, D. M. and Grozier, A.,** The effects of flooding on the export of gibberellins from the root to the shoot, *Planta,* 89, 376, 1969.
977. **Reid, J. B.,** Observations on root hair production by lucerne, maize and perennial ryegrass grown in a sandy loam, *Plant Soil,* 62, 319, 1981.
978. **Reid, J. B. and Goss, M. J.,** Changes in the aggregate stability of a sandy loam effected by growing roots of perennial ryegrass (*Lolium perenne*), *J. Sci. Food Agric.,* 31, 325, 1980.
979. **Reid, M. S. and Pratt, H. K.,** Effects of ethylene on potato tuber respiration, *Plant Physiol.,* 49, 252, 1972.
980. **Reisenauer, H. M., Clement, C. R., and Jones, L. H. P.,** Comparative efficiency of ammonium and nitrate for grasses, in *Proc. 9th Plant Nutr. Coll.,* 1982, 539.
981. Report 1975 ICW, Wageningen, The Netherlands, Mededeling 153, 1975, 54.
982. **Reyes, D. M., Stolzy, L. H., and Labanauskas, C. K.,** Temperature and oxygen effects in soil on nutrient uptake in jojoba seedlings, *Agron., J.,* 69, 647, 1977.
983. **Richards, B. G. and Greacen, E. L.,** Mechanical stress on an expanding cylindrical root analogue in granular media, *Aust. J. Soil Res.,* 24, 393, 1986.
984. **Richards, B. G. and Greacen, E. L.,** An example of numerical modelling — expansion of a root cavity in soil, in impact of water and external forces on soil structure, *Catena Suppl.* 11, 101, 1988.

985. **Richards, D., Goubran, F. H., Garwoli, W. N., and Daly, M. W.,** A machine for determining root length, *Plant Soil,* 52, 69, 1979.
986. **Rickman, R. W., Letey, J., and Stolzy, L. H.,** Plant response to oxygen supply and physical resistance in the root environment, *Soil Sci. Soc. Am. Proc.,* 30, 304, 1966.
987. **Rikin, A., Blumenfield, A., and Richmond,** Chilling resistance as affected by stressing environment and AbA, *Bot. Gaz.,* 137, 307, 1976.
988. **Riley, D. and Barber, S. A.,** Effect of ammonium and nitrate fertilization on phosphorus uptake as related to root-induced pH changes of the root-soil interface, *Proc. Soc. Soil Sci. Am.,* 35, 301, 1971.
989. **Riley, H.,** A Review of Soil Compaction Studies in Norway, Rep. 60, Division of Soil Management, Swedish University of Agricultural Science, 1980, 1971.
990. **Riley, H.,** Cereal yields and soil physical properties in relation to the degree of compactness of some Norwegian soils, in *Proc. 11th Conf. ISTRO,* Edinburgh, Scotland, 1988, 109.
991. **Robards, A. W., Clarkson, D. T., and Sanderson, J.,** Structure and permeability of the epiderman/hypodermal layers of the sand sedge (*Carex arenaria* L.), *Protoplasma,* 101, 331, 1979.
992. **Robards, A. W., Jackson, S. M., Clarkson, D. T., and Sanderson, J.,** The structure of barley roots in relation to the transport of ions into the stele, *Protoplasma,* 77, 291, 1973.
993. **Robert, M. L., Taylor, H. F., and Wain, R. L.,** Ethylene production by cress roots and excised cress root segments and its inhibition by 3,5-diiodo-4-hydroxybenzoic acid, *Planta,* 126, 273, 1975.
994. **Rogers, H. H. and Bottomley, P. A.,** *In situ* nuclear magnetic resonance imaging of roots: influence of soil type, ferromagnetic particle content, and soil water, *Agron. J.,* 79, 957, 1987.
995. **Rogers, H. H., Bottomley, P. A., and Foster, T. H.,** Application of nuclear magnetic resonance imaging to plant root studies, in *Proc. Int. Conf. on Soil Dynamics,* Vol. 5, Auburn, AL, 1985, 1152.
996. **Roland-Fayardo, B. E. and Barea, J. M.,** VA-mycorrhizas in shrubs and trees, *An. Edafol. Agrobiol.,* 46, 229, 1987.
997. **Romo, R. and Diaz, D. H.,** Root system and nutritional status of peaches under drip or flood irrigation in warm climates, *Acta Hortic.,* 173, 167, 1985.
998. **Ron, M. M.,** Acid-base titration of barley roots, *J. Plant Nutr.,* 10, 403, 1987.
999. **Rose, C. W., Byrne, G. F., and Hansen, G. K.,** Water transport from soil through plant to atmosphere: a lumpee-parameter model, *Agric. Meteorol.,* 16, 171, 1976.
1000. **Rose, C. W. and Stern, W. R.,** Determination of withdrawal of water from soil by crop roots as a function of depth and time, *Aust. J. Soil Res.,* 5, 11, 1967.
1001. **Rosen, C. J. and Carlson, R. M.,** Influence of root zone oxygen stress on potassium and ammonium absorption by Myrobalan plum rootstock, *Plant Soil,* 80, 345, 1984.
1002. **Rosenberg, N. J. and Willets, N. A.,** Yield and physiological response of barley and beans grown in artificially compacted soil, *Soil Sci. Soc. Am. Proc.,* 26, 78, 1962.
1003. **Ross, C. W.,** The effect of subsoiling and irrigation on potato production, *Soil Till. Res.,* 7, 315, 1986.
1004. **Ross, D. J., Campbell, I. B., and Bridger, B. A.,** Biochemical activities of organic soils from subAntarctic tussock grasslands on Campbell Islands, *N. Z. J. Sci.,* 22, 161, 1979.
1005. **Roth, R. L. and Gardner, B. R.,** Root distribution of mature orange trees irrigated by pressurized systems, in *Proc. 3rd Int. Drip/Trickle Irrigation Congr.,* Vol. 2, American Society of Agriculture Engineering, St. Joseph, MI, 1985, 579.
1006. Report for 1987, Rothamsted Experimental Station, Harpenden, 125.
1007. **Roughly, R. J.,** The influence of root temperature, *Rhizobium* strain and host selection on the structure and nitrogen fixing efficiency of root nodules of *Trifolium subterraneum, Ann. Bot.,* 24, 631, 1970.
1008. **Rovira, A. D.,** Influence of crop rotation and tillage on *Rhizoctonia* patch of wheat, *Phytopathology,* 76, 669, 1986.
1009. **Rovira, A. D. and Campbell, R.,** A scanning electron microscope study of interactions between microorganisms and *Gaeumannomyces graminis* (syn. *Ophiobolus graminis*) on wheat roots, *Microb. Ecol.,* 2, 177, 1975.
1010. **Rovira, A. D. and Davey, C. B.,** Biology of the rhizosphere, in *The Plant Root and Its Environment,* Carson, E. W., Ed., University Press of Virginia, Charlottesville, 1974, 153.
1011. **Rovira, A. D., Foster, R. C., and Martin, J. K.,** Note on terminology: origin, nature and nomenclature of the organic materials in the rhizosphere, in *The Soil-Root Interface,* Harley, J. L. and Russell, R. S., Eds., 1970, 1.
1012. **Rovira, A. D., Newman, E. I., Bowen, H. J., and Campbell, R.,** Quantitative assessment of the rhizoplane microflora by direct microscopy, *Soil Biol. Biochem.,* 6, 211, 1974.
1013. **Rovira, A. D. and Venn, N. R.,** Effect of rotation and tillage on take-all and *Rhizoctonia* root rot in wheat, in *Ecology and Management of Soilborne Plant Pathogens,* Parker, C. A., Rovira, A. D., Mooree, K. J., Wong, P. T. W., and Kollmorgan, J. F., Eds., American Phytopathological Society, St. Paul, MN, 1985, 255.
1014. **Rowe, R. N. and Beardsell, D. V.,** Waterlogging of fruit trees, *Hortic. Abstr.,* 43, 533, 1973.
1015. **Rowe, R. N. and Catlin, J. B.,** Differential sensitivity to waterlogging and cyanogenesis by peach, apricot, and plum roots, *J. Am. Soc. Hortic. Sci.,* 96, 305, 1971.

1016. **Rowse, H. R.,** The effect of irrigation on the length, weight, and diameter of lettuce roots, *Plant Soil*, 40, 381, 1974.

1017. **Rowse, H. R. and Barnes, A.,** Weather, rooting depth and water relations of broad beans — a theoretical analysis, *Agric. Meteorol.*, 20, 381, 1979.

1018. **Rowse, H. R. and Goodmann, D.,** Axial resistance to water movement in broad bean (*Vicia faba*) roots, *J. Exp. Bot.*, 32, 591, 1981.

1019. **Rowse, H. R. and Phillips, D. A.,** An instrument for estimating the total length of root in a sample, *J. Appl. Ecol.*, 11, 309, 1974.

1020. **Ruark, G. A., Mader, D. L., and Tatter, T. A.,** The influence of soil compaction and aeration on the growth and vigor of trees. A literature review. I, *Arboricultural J.*, 6, 251, 1982.

1021. **Rufty, T. W., Jr., Raper, C. D., Jr., and Jackson, W. A.,** Nitrogen assimilation, root growth and whole plant responses of soybean to root temperature, and to carbon dioxide and light in the aerial environment, *New Phytol.*, 88, 607, 1981.

1022. **Russell, E. W.,** The effects of very deep ploughing and of subsoiling on crop yields, *J. Agric. Sci.*, 48, 129, 1956.

1023. **Wild, A., Ed.,** *Russell's Soil Conditions and Plant Growth*, 11th ed., Longman, London, 1988.

1024. **Russell, R. S.,** *Plant Root Systems: Their Function and Interaction with the Soil*, McGraw-Hill, London, 1977.

1025. **Russell, R. S.,** Cultivation, soil conditions and plant growth in temperate agriculture, in *Modification of Soil Structure*, Emerson, W. W., Bond, R. D., and Dexter, A. R., Eds., John Willey & Sons, New York, 1978, 353.

1026. **Russell, R. S. and Cooke, G. W.,** Contributions of chemistry to removing soil constraints to crop production, in *Chemistry and World Food Supplies, The New Frontiers*, Shemilt, L. W., Ed., Pergamon Press, Oxford, 1983, 1.

1027. **Russell, R. S. and Goss, M. J.,** Physical aspects of soil fertility. The response of roots to mechanical impedance, *Neth. J. Agric. Sci.*, 22, 305, 1974.

1028. **Rykbost, K. A., Boersma, L., Mack, H. J., and Schmisseur, W. E.,** Yield response to soil warming: vegetable crops, *Agron. J.*, 67, 738, 1975.

1029. **Rylski, I., Nothmann, J., and Spiegelman, M.,** Effects of soil temperature on the development of young eggplants, *Exp. Agric.* 12, 273, 1976.

1030. **Saglio, P. H. and Pradet, A.,** Soluble sugars, respiration and energy charge during aging of excised maize root tips, *Plant Physiol.*, 66, 516, 1980.

1031. **Saglio, P. H., Rancillac, M., Bruzen, F., and Pradet, A.,** Initial oxygen pressure for growth and respiration of excised and intact roots, *Plant Physiol.*, 76, 151, 1984.

1032. **Saglio, P. H., Raymond, P., and Pradet, A.,** Metabolic activity and energy charge of excised maize root tips under anoxia control by soluble sugars, *Plant Physiol.*, 66, 1053, 1980.

1033. **Saglio, P. H., Raymond, P., and Pradet, A.,** Oxygen transport and root respiration of maize seedlings, *Plant Physiol.*, 72, 1035, 1983.

1034. **Saini, G. R.,** Root elongation and plant growth in a basal till compact soil treated with 3,5-diiodo-4-hydroxybenzoic acid and gibberellic acid, *Agron. J.*, 71, 1067, 1979.

1035. **Sakkar, A. N., Jenkins, D. A., and Wyn Jones, R. G.,** Modifications to mechanical and mineralogical composition of soil within the rhizosphere, in *The Soil-Plant Interface*, Harley, J. L. and Russell, R. S., Eds., Academic Press, London, 1979, 125.

1036. **Samtsevitch, S. A.,** Gel-like excretions of plant roots and their influence upon soil and rhizosphere microflora, in *Methods of Productivity Studied in Root Systems and Rhizosphere Organisms*, Ghilarov, M. S., Ed., Nauka Publishing, Leningrad, 1968, 200.

1037. **Samui, R. P. and Kar, S.,** Simulation of root water uptake of groundnut under dry and wet soil water regimes, in *Trans. 13th Congr. Int. Soc. Soil Science*, Vol. 2, Hamburg, West Germany, 1986, 148.

1038. **Sanders, J. L. and Brown, D. A.,** A new fiber optic technique for measuring root growth of soybean under field conditions, *Agron. J.*, 70, 1073, 1978.

1039. **Sanderson, J.,** Water uptake by different region of the barley root. Pathways of radial flow in relation to development of the endodermis, *J. Exp. Bot.*, 34, 240, 1983.

1040. **Sands, R., Fiscus, E. L., and Reid, C. P. P.,** Hydraulic properties of pine and bean roots with varying degrees of suberization, vascular differentiation and mycorrhizal infection, *Aust. J. Plant Physiol.*, 9, 559, 1982.

1041. **Sands, R., Greacen, E. L., and Gerard, C. J.,** Compaction of sandy soils in *radiata* pine forests. I. A penetrometer study, *Aust. J. Soil Res.*, 17, 101, 1979.

1042. **Sandwell, I.,** Warm rooted tomatoes crop satisfactorily at 5 deg. C night, *Grower*, 88, 605, 1977.

1043. **Sattelmacher, B. and Gerendas, J.,** Root growth as affected by N-form and concentration, in *Abstr. of Symp. Plant Roots and Their Environment*, Uppsala, Sweden, 1988, 5.

1044. **Sauerbeck, D. R. and Helal, H. M.,** Plant root development and photosynthate consumption depending on soil compaction, in *Trans. 13th Congr. Int. Soc. Soil Science*, Vol. 3, Hamburg, West Germany, 1986, 948.

1045. **Sawada, Y.,** Rhizobial nodulation of alfalfa in soils conducive and suppressive to *fusarium* disease, JARQ, 16, 235, 1983.

1046. **Schaller, G.,** pH changes in the rhizosphere in relation to the pH-buffering of soils, *Plant Soil*, 97, 439, 1987.
1047. **Schiechtl, H. M.,** Plants as a mean to soil stabilization, in *Root Ecology and Its Practical Application*, Int. Symp. Gumpenstein, 1982, Bundesanstalt Gumpenstein, A-8952, 1983, 703 (in German).
1048. **Schleiff, U.,** Chloride content of onion roots and their adhering soil under irrigation with saline drainage water, *Z. Pflanzenernaehr. Bodenkd.*, 143, 638, 1980.
1049. **Schleiff, U.,** Water uptake by barley roots as affected by the osmotic and matric potential in the rhizosphere, *Plant Soil*, 94, 143, 1986.
1050. **Schleiff, U.,** A vegetation technique to study the water uptake by roots from salinized rhizospheric soils, *Z. Pflanzenernaehr. Bodenkd.*, 150, 139, 1987.
1051. **Schmidhalter, U., Evequoz, M., and Oertli, J. J.,** Adaptive mechanisms of root growth under drought stress, in *Abstr. Symp. Plant Roots and Their Environment*, Uppsala, Sweden, 1988, 1.
1052. **Scholefield, D. and Hall, D. M.,** Constricted growth of grass through rigid pores, *Plant Soil*, 85, 153, 1985.
1053. **Schroth, M. N. and Haricock, J. G.,** Disease-suppressive soil and root-colonizing bacteria, *Science*, 216, 1376, 1982.
1054. **Schubert, K. R. and Evans, H. J.,** Hydrogen evolution: a major factor affecting the efficiency of nitrogen fixation of nodulated symbionts, *Proc. Acad. Sci. U. S. A.*, 1973, 1207, 1976.
1055. **Schulte-Karring, H.,** Pneumatic deep loosening-function and effect, *Schweiz. Z. Obst Weinbau*, 123, 331, 1987 (in German).
1056. **Schumacher, T. E. and Smucker, A. J. M.,** Mechanical impedance effects on oxygen uptake and porosity of drybean roots, *Agron. J.*, 73, 51, 1981.
1057. **Schumacher, T. E. and Smucker, A. J. M.,** Effect of localized anoxia on *Phaseolus vulgaris* L. root growth, *J. Exp. Bot.*, 35, 156, 1984.
1058. **Schumacher, T. E., Smucker, A. J. M., Eshel, A., and Curry, R. B.,** Measurement of short-term root growth by prestaining with neutral red, *Crop Sci.*, 25, 1212, 1983.
1059. **Schuurman, J. J.,** Influence of soil density on root development and growth of oats, *Plant Soil*, 22, 352, 1965.
1060. **Schuurman, J. J. and Boer, J. J. H.,** The effect of soil compaction at various depths on root and shoot growth of oats, *Neth. J. Agric. Sci.*, 22, 133, 1974.
1061. **Schweitzer, L. E. and Harper, J. E.,** Effect of light, dark and temperature on root nodule activity (acetylene reduction) of soybeans, *Plant Physiol.*, 65, 51, 1980.
1062. **Scotter, D. R., Thurtell, G. W., and Tanner, C. B.,** Measuring oxygen uptake by the roots of intact plants under controlled conditions, *Soil Sci.*, 104, 374, 1967.
1063. **Sedgley, R. H. and Barley, K. P.,** Effects of root growth and decay on the capillary conductivity of a sandy loam at low soil-moisture tension, *Soil Sci.*, 86, 175, 1958.
1064. **Seggewiss, B. and Jungk, A.,** Influence of potassium dynamics at the soil-root interface on magnesium uptake of plants, *Z. Pflanzenernaehr. Bodenkd.*, 151, 91, 1988.
1065. **Sen, D. N.,** Environment and Root Behaviour, Geobios International, Jodhpur, India, 1980.
1066. **Sharp, R. E. and Davies, W. J.,** Solute regulation and growth by roots and shoots of water stressed maize plants, *Planta*, 147, 43, 1979.
1067. **Sheehy, J. E. and Peacock, J. M.,** Canopy photosynthesis and crop growth rate of eight temperate forage grasses, *J. Exp. Bot.*, 26, 679, 1975.
1068. **Sheppard, S. C. and Racz, G. J.,** Shoot and root response of wheat to band broadcast phosphorus at varying soil temperatures, *Can. J. Soil Sci.*, 65, 79, 1985.
1069. **Shierlaw, J. and Alston, A. M.,** Effect of soil compaction on root growth and uptake of phosphorus, *Plant Soil*, 77, 15, 1984.
1070. **Shoone, M. G. T. and Flood, A. V.,** Studies on uptake and loss of water by barley roots in relation to changes in root resistance, *J. Exp. Bot.*, 31, 1147, 1980.
1971. **Sikes, D. A. and Pattit, R. D.,** Soil temperature, oxygen, and water level effects on sand shinnery oak, *Soil Sci.*, 130, 344, 1980.
1072. **Silberbush, M., Gornat, B., and Goldberg, D.,** Effect of irrigation from a point source (trickling) on oxygen flux and on root extension in the soil, *Plant Soil*, 52, 507, 1979.
1073. **Sills, M. and Carrow, R. N.,** Turfgrass growth, N use, and water use under soil compaction and N fertilization, *Agron. J.*, 75, 488, 1983.
1074. **Simpson, J. R., Pinkerton, A., and Lazdovskis, J.,** Effects of subsoil calcium on the root growth of some lucerne genotypes (*Medicago sativa* L.) in acidic soil profiles, *Aust. J. Agric. Res.*, 28, 629, 1977.
1075. **Simpson, K.,** Factors influencing uptake of phosphorus by crops in southeast Scotland, *Soil Sci.*, 92, 1, 1961.
1076. **Singh, B. B. and Jones, J. P.,** Phosphorus sorption isotherm for evaluating phosphorus requirements of lettuce of five temperature regimes, *Plant Soil*, 46, 31, 1977.
1077. **Skidmore, E. L. and Stone, J. F.,** Physiological role in regulating transpiration rate of the cotton plant, *Agron. J.*, 56, 405, 1964.
1078. **Skinner, M. F. and Bowen, G. D.,** The penetration of soil by mycelial strands of ectomycorrhizal fungi, *Soil Biol. Biochem.*, 6, 57, 1974.

1079. **Slatyer, R. O.,** *Plant-Water Relationships,* Academic Press, New York, 1967, 198.
1080. **Slaviková, J.,** Compensation of root suction force within a single root system, *Biol. Plant.,* 9, 20, 1967a.
1081. **Slaviková, J.,** Dependence of the root suction force on the soil water content, *Biol. Plant.,* 9, 149, 1967b.
1082. **Slowik, K., Labanauskas, C. K., Stolzy, L. H., and Zentmyer, G. A.,** Influence of rootstocks, soil oxygen, and soil moisture on the uptake and translocation of nutrients in young avocado plants, *J. Am. Soc. Hortic. Sci.,* 194, 172, 1979.
1083. **Slusarczyk, E.,** An empirical model of the optimal physical properties of soil for the requirements of field crops, *Zesz. Probl. Postepow Nauk Roln.,* 220(I), 327, 1979 (in Polish, English summary).
1084. **Smika, D. E. and Klute, A.,** Surface area measurement of corn root systems, *Agron. J.,* 74, 1091, 1982.
1085. **Smiley, R. W.,** Rhizopshere pH as influenced by plants, soils, and nitrogen fertilizers, *Soil Sci. Soc. Am. Proc.,* 38, 795, 1974.
1086. **Smiley, R. W., Taylor, P. A., Clarke, R. G., Greenhalgh, F. G., and Trutmann, P.,** Simulated soil and plant management effects on root rots of subterranean clover, *Aust. J. Agric. Res.,* 37, 633, 1986.
1087. **Smith, D. L. O.,** Compaction by wheels: a numerical model for agricultural soils, *J. Soil Sci.,* 36, 621, 1985.
1088. **Smucker, A. J. M. and Atwell, B. J.,** Soil compaction modifications of root function, in *Abstr. Symp. Plant Roots and Their Environment,* Uppsala, Sweden, 1988, 1.
1089. **Smucker, A. J. M. and Erickson, A. E.,** An aseptic mist chamber system: a method for measuring root processes of peas, *Agron. J.,* 68, 59, 1976.
1090. **Smucker, A. J. M. and Ferguson, J. C.,** Production level image processing of video record plant root systems, in *Abstr. of Symp. Plant Roots and Their Environment,* Uppsala, Sweden, 1988, 4.
1091. **Smucker, A. J. M., McBurney, S. L., and Srivastava, A. K.,** Quantitative separation of roots from compacted soil profiles by the hydropneumatic elutration system, *Soil Sci. Soc. Am. J.,* 74, 500, 1982.
1092. **So, H. B., Aylmore, L. A. G., and Quirk, P.,** The resistance of intact maize roots to water flow, *Soil Sci. Soc. Am. J.,* 40, 222, 1976.
1093. **So, H. B. and Jayasekara, S.,** The effect of shoot/root ratios on the water relationship of sorghum (*Sorghum bicolor* L. Moench), in *Abstr. of Symp. Plant Roots and Their Environment,* Uppsala, Sweden, 1988, 3.
1094. **Soane, B. D., Blackwell, P. S., Dickson, J. W., and Painter, D. J.,** Compaction by agricultural vehicles: a review. II. Compaction under tires and other running gear, *Soil Till. Res.,* 1, 373, 1980/81.
1095. **Soane, B. D., Dickson, J. W., and Campbell, D. J.,** Compaction by agricultural vehicles: a review. III. Incidence and control of compaction in crop production, *Soil Till. Res.,* 2, 3, 1982.
1096. **Soane, B. D. and van Ouwerkerk, C.,** The role of field traffic studies in soil management research, *Soil Till. Res.,* 1, 205, 1980/81.
1097. **Soehne, W. H.,** Pressure distribution in the soil and soil deformation under tractor tires, *Grundlagen Landtech.,* 5, 49, 1953.
1098. **Soileau, J. M. and Englestad, O. P.,** Cotton growth in an acid fragipan subsoil. I. Effects of physical soil properties, liming, and fertilization on root penetration, *Soil Sci. Soc. Am. Proc.,* 33, 915, 1969a.
1099. **Soileau, J. M., Engelstad, O. P., and Martin, J. B., Jr.,** Cotton growth in an acid fragipan subsoil. II. Effects of soluble calcium, magnesium, and aluminum on roots and tops, *Soil Sci. Soc. Am. Proc.,* 33, 919, 1969b.
1100. **Sojka, R. E., Joseph, H. A., and Stolzy, L. H.,** Wheat response to short-term heat stress and to soil oxygen stress at three stages of growth, *Agron. J.,* 64, 450, 1972.
1101. **Sojka, R. E. and Stolzy, L. H.,** Soil-oxygen effects on stomatal response, *Soil Sci.,* 130, 350, 1980.
1102. **Sojka, R. E., Stolzy, L. H., and Kaufmann, M. R.,** Wheat growth related to rhizosphere temperature and oxygen levels, *Agron. J.,* 67, 591, 1975.
1103. **Sodatenkov, S. V. and Chao, H.s-T.,** The role of bean and corn leaves in respiration of oxygen-deprived roots, *Fiziol. Rast.,* 8, 385, 1961.
1104. **Solomos, T. and Laties, G. G.,** The mechanisms of ethylene and cyanide action in triggering the rise in respiration in potato tubers, *Plant Physiol.,* 55, 73, 1975.
1105. **Sommer, C.,** Soil compaction and water uptake of plants, in impact of water and external forces on soil structure, *Catena Suppl.* 11, 107, 1988.
1106. **Sommers, L. E., Harris, R. F., Dalton, F. N., and Gardner, W. R.,** Water potential relations of three root-infecting *Phytophtora* species, *Phytopathology,* 60, 932, 1970.
1107. **Souty, N.,** Roots behaviour and soils mechanical properties, in *Trans. 13th Congr. Int. Soc. Soil Science,* Vol. 2, Hamburg, West Germany, 1986, 165 (in French, English summary).
1108. **Souty, N.,** Mechanical behaviour of growing roots. I. Measurement of penetration force, *Agronomie,* 7, 623, 1987 (in French, English summary).
1109. **Souty, N. and Faure, A.,** Mechanical aspects of root growth, *Pedologie,* XXXV, 251, 1985 (in French, English summary).
1110. **Souty, N. and Stepniewski, W.,** The influence of external oxygen concentration on axial root growth force of maize radicles, *Agronomie,* 8, 295, 1988.
1111. **Spiers, J. M.,** Root distribution of "Trifblue" rabbiteye blueberry as influenced by irrigation, incorporated peatmoss and mulch, *J. Am. Soc. Hortic. Sci.,* 111, 877, 1986.

1112. **Spoor, G.,** Deep loosening and mole drainage on clay soils, *Trop. Agric.*, 59, 97, 1982.

1113. **Stamp, P.,** The root growth of maize cultivars at optimum and at extreme temperatures, in *Abstr. of Symp. Plant Roots and Their Environment*, Uppsala, Sweden, 1988, 3.

1114. **Stawinski, J.,** Interrelationships between the specific surface and some physico-chemical properties of soils, *Zesz. Probl. Postepow Nauk Roln.*, 197, 229, 1977.

1115. **Steen, E.,** The net stocking method for studying quantitative variation with time of grass roots, in *Proc. Int. Symp. Root Ecology and its Practical Application*, Böhm, W., Kutschera, L., and Lichenegger, E., Eds., Verlag Gumpenstein, Irdning, Austria, 1983, 63.

1116. **Steen, E.,** Variation of root growth in a grass ley studied with a mesh bag technique, *Swed. J. Agric. Res.*, 14, 93, 1984.

1117. **Steen, E.,** Root and rhizome dynamics in a perennial grass crop during an annual growth cycle, *Swed. J. Agric. Res.*, 15, 25, 1985.

1118. **Steen, E. and Håkansson, I.,** Use of ingrowth soil cores in mesh bags for studies of relations between soil compaction and root growth, *Soil Till. Res.*, 10, 363, 1987.

1119. **Steinert, P. and Körbs, P.,** Parameter for an estimation of rooting in field and rhizotron, *Arch. Acker-Pflanzenbau Bodenkd.*, 31, 327, 1987.

1120. **Steinman, F. and Brandla, R.,** Tolerance to flooding of the lake bulrush (*Schoenoplectus lacustris* (L.) Palla). III. Relations between acidity and oxygen concentration in the medium, *Flora*, 171, 307, 1081 (in German).

1121. **Stelzer, R. and Läuchli, A.,** Salt- and flooding tolerance of *Puccinellia peisonis*. IV. Root respiration and the role of aerenchyma in providing atmospheric oxygen to the roots, *Z. Pflanzenphysiol.*, 97, 171, 1980.

1122. **Stepniewski, W.,** The effect of irrigation and increased mineral fertilization on the aeration status of a cultivated soil, *Zesz. Probl. Postepow Nauk Roln.*, 197, 203, 1977.

1123. **Stepniewski, W.,** Oxygen diffusion and strength as related to soil compaction. IV. Choice of optimum parameters, *Pol. J. Soil Sci.*, XIX, 89, 1981.

1124. **Stepniewski, W. and Gliński J.,** Gas exchange and atmospheric properties of flooded soils, in *The Ecology and Management of Wetlands*, Hook, D. D., Ed., Croom Helm Ltd., London, 1987, 269.

1125. **Stepniewski, W. and Przywara, G.,** Influence of soil oxygen availability on dry matter and mineral composition of soybean and winter rye roots, *Zesz. Probl. Postepow Nauk Roln.*, 385, in press.

1126. **Steudle, E. and Jeschke, W. D.,** Water transport in barley roots, *Planta*, 158, 237, 1983.

1127. **Stevenson, C. A. and Laidlaw, A. S.,** The effect of moisture stress on stolon and adventitious root development in white clover (*Trifolium repens* L.), *Plant Soil*, 85, 249, 1985.

1128. **Stitt, R. E., Cassel, D. K., Weed, S. B., and Nelson, L. A.,** Mechanical impedance of tillage pans in Atlantic Coastal Plains soils and relationships with soil physical, chemical, and mineralogical properties, *Soil Sci. Soc. Am. J.*, 46, 100, 1982.

1129. **Sivakumar, M. V. K., Taylor, H. M., and Shaw, R. H.,** Top and root relations of field-grown soybeans, *Agron. J.*, 69, 470, 1977.

1130. **Stolzy, L. H. and Flühler, H.,** Measurement and prediction of anaerobiosis in soils, in *Nitrogen in the Environment*, Vol. 1, Nielsen, D. R. and McDonald, A. G., Eds., Academic Press, New York, 1978, 363.

1131. **Stolzy, L. H., Labanauskas, C. K., Klotz, L. J., and de Wolfe, T. A.,** Nutritional responses and root rot of *Citrus limon* and *Citrus sinensis* under high and low soil oxygen supplies in the presence and absence of *Phytophtora* spp., *Soil Sci.*, 119, 136, 1975.

1132. **Stolzy, L. H., Letey, J., Szuszkiewicz, T. E., and Lunt, O. R.,** Root growth and diffusion rates as functions of oxygen concentration, *Soil Sci. Soc. Am. Proc.*, 25, 463, 1961.

1133. **Stone, E. L., Feuer, R., and Wilson, H. M.,** *Approximate Adaptation of Plantation Species to Some Site Factors*, Suppl. Ext. Bull. No. 1075, Cornell University, Ithaca, N.Y., 1962.

1134. **Stone, J. A. and Taylor, H. M.,** Temperature and the development of the taproot and lateral roots of four indeterminate soybean cultivars, *Agron. J.*, 75, 613, 1983.

1135. **Stone, L. R., Teare, I. D., Nickell, C. D., and Mayaki, W. C.,** Soybean root development and soil water depletion, *Agron. J.*, 68, 677, 1976.

1136. **Stovold, G. E.,** Root rot caused by *Pythium irregulare* Buisman, an important factor in the decline of established subterranean clover pasture, *Aust. J. Agric. Res.*, 25, 537, 1974.

1137. **Strong, W. L. and La Roi, G. H.,** Root density — soil relationships in selected boreal forests of central Alberta, Canada, *For. Ecol.*, 12, 233, 1985.

1138. **Stucky, D. J. and Lindsay, T. C.,** Effect of soil compaction on growth and yield of soybeans grown in a greenhouse on several reconstructed soil profiles from prime farmland in Southern Illinois, *Reclam. Vegetation Res.*, 1, 297, 1982.

1139. **Stumpf, P. K.,** Biosynthesis of saturated and unsaturated fatty acids, in *The Biochemistry of Plants*, 4th ed., Stumpf, P. K., Ed., Academic Press, New York, 1980, 177.

1140. **Stypa, M., Nunez-Barrios, A., Barry, D. A., Miller, M. H., and Mitchell, W. A.,** Effects of subsoil bulk density, nutrient availability and soil moisture on corn root in the field, *Can. J. Soil Sci.*, 67, 293, 1987.

1141. **Sugiura, T. and Kobayashi, H.,** Relationship between soil physics and grass root elongation, in *Proc. XV Int. Grassland Congr.*, Science Council of Japan and Japanese Society of Grassland Science, Kyoto, 1985, 422.

1142. **Sumner, D. R.,** Influence of tillage practices on root diseases of southern pea (*Vigna unguiculata*), *Annu. Rep. Bean Imp. Coop.*, 20, 91, 1977.
1143. **Sumner, D. R., Doupnik, B., and Boosalis, M. G.,** Effects of reduced tillage and multiple cropping on plant diseases, *Annu. Rev. Phytopathol.*, 19, 167, 1981a.
1144. **Sumner, M. E. and Boswell, F. C.,** Alleviating nutrient stress, in *Modifying the Root Environment to Reduce Crop Stress*, Arkin, G. F. and Taylor, H. M., Eds., ASAE Monogr. No. 4, American Society of Agriculture Engineering, St. Joseph, MI, 1981b, 99.
1145. **Suslow, T. W. and Schroth, M. N.,** Role of deleterious rhizobacteria as minor pathogens in reducing crop growth, *Phytopathology*, 72, 111, 1982.
1146. **Sutton, C. D.,** Effect of flow soil temperature on phosphate nutrition of plants — a review, *J. Sci. Food Agric.*, 20, 1, 1969.
1147. **Sutton, J. C.,** Development of vesicular-arbuscular mycorrhizae in crop plants, *Can. J. Bot.*, 51, 2487, 1973.
1148. **Sutton, R. F.,** Form and Development of Conifer Root Systems, Tech. Commun. No. 7, Commonwealth Forestry Bureau, , 1969.
1149. **Sutton, R. F.,** Root system morphogenesis, *N. Z. J. For. Sci.*, 10, 264, 1980.
1150. **Svachula, V.,** Ecological factors affecting sugerbeet roots, in *Abstr. of Symp. Plant Roots and Their Environment*, Uppsala, Sweden, 1988, 3.
1151. **Swain, R. W.,** Subsoiling, in Soil Physical Conditions and Crop Production, Tech. Bull. No. 29, Ministry of Agriculture, Fisheries, and Food, Her Majesty's Stationery Office, London, 189, 1975.
1152. **Sytnik, K. M., Kniga, N. M., and Musatienko, L. I.,** Root Physiology, PWRiL, Warszawa, 1977 (in Polish).
1153. **Syversten, J. P., Zablotowicz, R. M., and Smith, M. L., Jr.,** Soil temperature and hydraulic conductivity, *Plant Soil*, 72, 3, 1983.
1154. **Szaniawski, R. K.,** Adaptation and functional balance between shoot and root activity of sunflower plants grown at different root temperatures, *Ann. Bot.*, 51, 453, 1983.
1155. **Szaniawski, R. K.,** Homeostasis in cucumber plants during low temperature stress, *Physiol. Plant.*, 64, 212, 1985.
1156. **Szaniawski, R. K.,** Plant, stress and homeostasis, *Plant. Physiol. Biochem.*, 25, 63, 1987.
1157. **Tachibana, Y. and Ohta, Y.,** Root surface area as a parameter in relation to water and nutrient uptake by cucumber plant, *Soil Sci. Plant Nutr.*, 29, 387, 1983.
1158. **Tackett, J. L. and Pearson, R. W.,** Oxygen requirements of cotton seedlings roots for penetration of compacted soil cores, *Soil Sci. Soc. Am. Proc.*, 28, 600, 1964.
1159. **Tackett, J. L. and Pearson, R. W.,** Effect of carbon dioxide on cotton seedling root penetration of compacted soil cores, *Soil Sci. Soc. Am. Proc.*, 28, 741, 1964.
1160. **Takeshima, H.,** Studies on the effects of soil temperature on rice plant growth. III. Effects of root temperature upon water and nutrient absorption at different stages and in alternating temperature, *Proc. Crop Sci. Soc.* Jpn., 32, 319, 1964.
1161. **Tardieu, F.,** Soil structure, root system and water uptake of maize. III. Availability of soil water reserves, *Agronomie*, 7, 279, 1987 (in French, English summary).
1162. **Tardieu, F.,** Analysis of the spatial variability of maize root density. I. Effect of wheel compaction on the spatial arrangement of roots, *Plant Soil*, 107, 259, 1988a.
1163. **Tardieu, F.,** Analysis of the spatial variability of maize root density. II. Distances between roots, *Plant Soil*, 107, 267, 1988b.
1164. **Tardieu, F.,** Effect of the structure of the ploughed layer on the spatial distribution of root density, in *Proc. 11th Conf. ISTRO*, Edinburgh, Scotland, 1988c, 153.
1165. **Tardieu, F.,** *In situ* study of the root system as a water sink, *Zesz. Probl. Postepow Nauk Roln.*, 385, in press (in French, English summary).
1166. **Tardieu, F. and Manichon, H.,** Characterization as a water sink of the maize root system in cultivated fields. II. A method for studying vertical and horizontal distribution of roots, *Agronomie*, 6, 415, 1986 (in French, English summary).
1167. **Tardieu, F. and Manichon, H.,** Soil structure, root system and water uptake of maize. II. Growth and spatial distribution of root systems, *Agronomie*, 7, 201, 1987 (in French, English summary).
1168. **Taylor, G. J., Crowder, A. A., and Rodden, R.,** Formation and morphology of an iron plaque on the roots of *Typha latifolia* L. grown in solution culture, *Am. J. Bot.*, 71, 666, 1984.
1169. **Taylor, H. M.,** Soil conditions as they affect plant establishment, root development, and yield. Effects of soil strength on seedling emergence, root growth and crop yield, in *Compaction of Agricultural Soils*, Barnes, K. K., Ed., American Society of Agricultural Engineering, St. Joseph, MI, 1971, 292.
1170. **Taylor, H. M.,** Root behavior as affected by soil structure and strength, in *The Plant Root and Its Environment*, Carson, E. W., Ed., University Press of Virginia, Charlottesville, 1974, 271.
1171. **Taylor, H. M.,** The effect of soil compaction on rooting patterns and water uptake of cotton, in *Proc. 7th Conf. ISTRO*, Vol. 43, Uppsala, Sweden, 1976, 1.

1172. **Taylor, H. M.,** Modifying root systems of cotton and soybean to increase water absorption, in *Adaptation of Plants to Water and High Temperatures Stress*, Turner, N. C. and Kramer, P. J., Eds., John Wiley & Sons, New York, 1980, 75.

1173. **Taylor, H. M. and Böhm, W.,** Use of acrylic plastic as rhizotron windows, *Agron. J.*, 68, 693, 1976.

1174. **Taylor, H. M. and Burnett, E.,** Influence of soil strength on root-growth habits of plants, *Soil Sci.*, 98, 174, 1964.

1175. **Taylor, H. M. and Gardner, H. R.,** Penetration of cotton seedling taproots as influenced by bulk density, moisture content and strength of soil, *Soil Sci.*, 96, 153, 1963.

1176. **Taylor, H. M., Huck, M. G., Klepper, B., and Lund, Z. F.,** Measurement of soil-grown roots in a rhizotron, *Agron. J.*, 62, 807, 1970.

1177. **Taylor, H. M. and Klepper, B.,** Water uptake by cotton roots during a drying cycle, *Aust. J. Biol. Sci.*, 24, 853, 1971.

1178. **Taylor, H. M. and Klepper, B.,** Rooting density and water extraction patterns for corn (*Zea mays* L.), *Agron. J.*, 65, 965, 1973.

1179. **Taylor, H. M. and Klepper, B.,** Water uptake by cotton root systems: an examination of assumptions in the single root model, *Soil Sci.*, 120, 57, 1975.

1180. **Taylor, H. M. and Klepper, B.,** The role of rooting characteristics in the supply of water to plants, *Adv. Agron.*, 30, 99, 1978.

1181. **Taylor, H. M. and Ratliff, L. F.,** Root elongation rates of cotton and peanuts as a function of soil strength and soil water content, *Soil Sci.*, 108, 113, 1969.

1182. **Taylor, H. M., Roberson, G. M., and Parker, J. F., Jr.,** Soil strength-root penetration relations for medium- to coarse-textured soil materials, *Soil Sci.*, 102, 18, 1966.

1183. **Taylor, H. M. and Terrell, E. E.,** Rooting pattern and plant productivity, in *Handbook of Agricultural Productivity*, Vol. 1. Releigh, M., Ed., CRC Press, Boca Raton, FL, 1982, 185.

1184. **Taylor, H. M. and Upchurch, D. R.,** New methods of root investigations, in A*bstr. of Symp. Plant Roots and Their Environment*, Uppsala, Sweden, 1988, 4.

1185. **Taylor, H. M. and Willatt, S. T.,** Shrinkage of soybean roots, *Agron. J.*, 75, 818, 1983.

1186. **Taylor, P. A., Clarke, R. G., Kelley, K., and Smiley, R.,** Root rot of irrigated subterranean clover in northern Victoria: significance and prospect for control, in *Ecology and Management of Soilborne Plant Pathogens*, Parker, C. A., Rovira, A. D., Moore, K. J., Wong, P. T. W., and Kollmorgan, J. F., Eds., American Phytopathological Society, St. Paul, MN, 1985, 271.

1187. **Teskey, R. O. and Hinckley, T. M.,** Influence of temperature and water potential on root growth of white oak, *Physiol. Plant.*, 52, 363, 1981.

1188. **Tester, M. and Morris, C.,** The penetration of light through soil, *Plant Cell Environ.*, 10, 281, 1987.

1189. **Carson, E. W., Ed.,** *The Plant Root System and Its Environment*, University Press of Virginia, Charlottesville, 1974.

1190. **Harley, J. L. and Russell, R. S., Eds.,** *The Soil-Root Interface*, Academic Press, London, 1979.

1191. **Tinker, P. B.,** Roots and water. Transport of water to plant roots in soil, *Philos. Trans. R. Soc. London Ser. B*, 273, 445, 1976.

1192. **Tisdall, J. M. and Oades, J. M.,** The management of ryegrass to stabilise aggregation of red-brown earth, *Aust. J. Soil Res.*, 18, 415, 1980.

1193. **Tjepkema, J.,** The role of oxygen diffusion from the shoots and nodule roots in nitrogen fixation by root nodules of *Myrica gale*, *Can. J. Bot.*, 56, 1365, 1978.

1194. **Tolley, M. D., DeLaune, R. D., and Patrick, W. H., Jr.,** The effect of sediment redox potential and soil acidity on nitrogen uptake root respiration, and growth of rice (*Oryza sativa*), *Plant Soil*, 93, 323, 1986.

1195. **Torbert, H. A., Edwards, J. H., and Pedersen, J. F.,** Growth and magnesium uptake of tall fescue clones with varying root diameters, *J. Plant Nutr.*, 8, 7321, 1985.

1196. **Torrey, J. G.,** Root hormones and plant growth, *Annu. Rev. Plant Physiol.*, 27, 435, 1976.

1197. **Marini-Bettolo, G. B., Ed.,** *Toward a Second Green Revolution. Development in Agricultural and Managed-Forest Ecology Series*, Vol. 19, Elsevier, Amsterdam, 1988.

1198. **Tripepi, R. R. and Mitchell, C. A.,** Metabolic response of river birch and European birch roots to hypoxia, *Plant Physiol.*, 76, 31, 1984.

1199. **Trolldenier, G.,** Influence of nitrogen and potassium fertilization of wheat and oxygen supply of roots on the number of bacteria, root respiration and denitrification in rhizosphere, *Zentralbl. Bakteriol. Parasitenkd. Infektionskr. Hyg. Abt. 2*, 126, 1971 (in German, English summary).

1200. **Trolldenier, G.,** Visualisation of oxidizing power of rice roots and of possible participation of bacteria in iron deposition, *Z. Pflanzenernaehr. Bodenkd.*, 151, 117, 1988.

1201. **Trolldenier, G. and Hecht-Buchholz, C.,** Effect of aeration status of nutrient solution on microorganisms, mucilage and ultrastructure of wheat roots, *Plant Soil*, 80, 381, 1984.

1202. **Trolldenier, G. and von Rheibaben, W.,** Root respiration and bacterial population of roots. I. Effect of nitrogen source, potassium nutrition and aeration of roots, *Z. Pflanzenaernehr. Bodenkd.*, 114, 366, 1981.

1203. **Trought, M. C. T. and Drew, M. C.,** The development of water-logging damage in wheat seedlings (*Triticum aestivum* L.). I. Shoot and root growth in relation to changes in the concentrations of dissolved gases and solutes in the solution, *Plant Soil,* 54, 55, 1980.

1204. **Troughton, A. and Whittington, W. J.,** The significance of genetic variation in root systems, in *Root Growth,* Whittington, W. J., Ed., Proc. 15th Easter School in Agricultural Science, University of Nottingham, 1968, 296.

1205. **Trouse, A. C., Jr.,** Soil conditions as they affect plant establishment, root development, and yield, in *Compaction of Agricultural Soils,* Barnes, K. K., Ed., American Society of Agriculture Engineering, St. Joseph, MI, 1971, .

1206. **Trouse, A. C., Jr.,** Soil physical characteristics and root growth, in *Proc. Int. Conf. on The Role of Soil Physical Properties in Maintaining Productivity of Tropical Soils,* Ibadan, Nigeria, 1977, 319.

1207. **Trouse, A. C., Jr.,** Effect of soil bulk density on root elongation, *Zesz. Probl. Postepow Nauk Roln.,* 220(I), 349, 1979.

1208. **Trouse, A. C., Jr.,** Observations on under-the-row subsoiling after conventional tillage, *Soil Till. Res.,* 3, 67, 1983.

1209. **Trouse, A. C., Jr.,** Crop root capabilities, in *Proc. 11th Conf. ISTRO,* Edinburgh, Scotland, 1988, 159.

1210. **Trubetskova, O. M. and Zhirnova, N. G.,** Diurnal rhythm of potassium transfer from the root system to the aerial organs of plants, *Fiziol. Rast.,* 6, 129, 1959 (in Russian, English summary).

1211. **Tu, J. C.,** Integrated control of the pea root rot disease complex in Ontario, *Plant Dis.,* 71, 9, 1987.

1212. **Turner, N. C. and Begg, J. E.,** Responses of pasture plants to water deficits, in *Plant Relations in Pastures,* Wilson, J. R., Ed., CSIRO, Melbourne, Australia, 1978, 50.

1213. **Turski, R., Domzal, H., Borowiec, J., Flis-Bujak, M., and Misztal, M.,** *Soil Science. Practicum for Students of Agronomy,* Agricultural University, Lublin, Poland, 1977, 103 (in Polish).

1214. **Unger, P. W.,** Effects of deep tillage and profile modification on soil properties, root growth and crop yields in the United States and Canada, *Geoderma,* 22, 275, 1979.

1215. **Unger, P. W. and Danielsen, R. E.,** Influence of oxygen and carbon dioxide on germination and seedling development of corn (*Zea mays* L.), *Agron. J.,* 57, 56, 1965.

1216. **Upchurch, D. R.,** Relationship Between Observations in Minirhizotrons and True Root Length Density, Ph.D. dissertation, Texas Tech University, Lubbock, 1985; *Diss. Abstr.,* 28594, 1985.

1217. **Upchurch, D. R., McMichael, B. L., and Taylor, H. M.,** Use of minirhizotrons to characterize root system orientation, *Soil Sci. Soc. Am. J.,* 52, 319, 1988.

1218. **Upchurch, D. R. and Ritchie, J. T.,** Root observations using a video recording system in minirhizotrons, *Agron. J.,* 75, 1009, 1983.

1219. **Upchurch, D. R. and Ritchie, J. T.,** Battery-operated color video camera for root observations in minirhizotrons, *Agron. J.,* 76, 1015, 1984.

1220. **Uppal, H. S., Cheema, S. S., and Prihar, S. S.,** Growth and yield of barley as affected by soil water storage and an early post-sowing irrigation, *J. Indian Soc. Soil Sci.,* 34, 1, 1986.

1221. **Vaadia, Y.,** Automatic diurnal fluctuations in rate of exudation and root pressure of decapitated sunflower plants, *Physiol. Plant.,* 13, 701, 1960.

1222. **Vaadia, Y. and Waisel, Y.,** Physiological processess as affected by water balance, in *Irrigation of Agricultural Lands,* Hagen, R. M., Haise, H. R., and Edminster, T. W., Eds., American Society of Agronomy, Madison, WI, 1967, 354.

1223. **van Bavel, C. H. M. and Ahmed, J.,** Dynamic simulation of water depletion in the root zone, *Ecol. Modelling,* 2, 189, 1976.

1224. **van Bavel, C. H. M., Brust, K. J., and Stirk, B. J.,** Hydraulic properties of a clay loam soil and the field measurement of water uptake by roots. II. The water balance of the root zone, *Soil Sci. Soc. Am. Proc.,* 32, 317, 1968b.

1225. **van Bavel, C. H. M., Stirk, G. B., and Brust, K. J.,** Hydraulic properties of a clay loam soil and the field measurement of water uptake by roots. I. Interpretation of water content and pressure profiles, *Soil Sci. Soc. Am. Proc.,* 32, 310, 1968a.

1226. **Van Custem, P. and Gillet, C.,** Proton-metal cation exchange in the cell wall of *Nitella flexilis, Plant Physiol.,* 73, 865, 1983.

1227. **Van den Honert, T. H.,** Water transport in plants as a caternary process, *Discuss. Faraday Soc.,* 3, 146, 1948.

1228. **Van der Heide, H., Boer-Bolt, B. M., and van Raalte, M. H.,** The effect of a low oxygen content of the medium on the roots of barley seedlings, *Acta Bot. Neerl.,* 12, 231, 1963.

1229. **Van der Valk, G. G. M.,** Effects of soil compaction, restricted aeration and frost injury on flowerbulb crops. II. Influence of short periods of restricted soil aeration on development of tulips, ICW, Wageningen, The Netherlands, Reprints 135, 333, 1970.

1230. **Van Diest, A.,** Effect of soil aeration and compaction upon yield, nutrient uptake and variability in a greenhouse fertility experiment, *Agron. J.,* 54, 515, 1962.

1231. **Van Huyssteen, L.,** Soil management for an optimal soil structure for grapevine roots, in *Proc. 11th Conf. ISTRO,* Edinburgh, Scotland, 1988, 177.

1232. **Van Loon, C. D. and Bouma, J.,** A case study on the effect of soil compaction on potato growth in a loamy sand soil. II. Potato plant responses, *Neth. J. Agric. Sci.*, 26, 421, 1978.

1233. **Van Noordwijk, M.,** Functional interpretation of root densities in the field for nutrient and water uptake, in *Proc. Int. Symp. on Root Ecology and Its Practical Application*, Böhm, W., Kutschera, L., and Lichenegger, E., Eds., Verlag Gumpenstein, Irdning, Austria, 1983, 207.

1234. **Van Noordwijk, M., de Jager, A., and Floris, J.,** A new dimension to observations in minirhizotrons. A stereoscopic view on root photographs, *Plant Soil*, 86, 447, 1985.

1235. **Van Noordwijk, M. and de Willigen, P.,** Mathematical models on diffusion of oxygen to and within plant roots, with special emphasis on effects of soil-root contact. II. Applications, *Plant Soil*, 77, 233, 1984.

1236. **Van Noordwijk, M. and de Willigen, P.,** Quantitative root ecology as element of soil fertility theory, *Neth. J. Agric. Sci.*, 34, 273, 1986.

1237. **Van Noordwijk, M., Floris, J., and de Jager, A.,** Sampling schemes for estimating root density distribution in cropped fields, *Neth. J. Agric. Sci.*, 33, 241, 1985.

1238. **Van Raalte, M. H.,** On the oxygen supply of rice roots, *Ann. Bot. Gard. Buitenzorg*, 51, 43, 1941.

1239. **Van Raalte, M. H.,** On the oxidation of the environment by the roots of rice (*Oriza sativa* L.), Hortus Botanicus Bogotiensis, Java, Syokubutu-Iho, 1, 15, 1943.

1240. **Varade, S. B., Letey, J., and Stolzy, L. H.,** Crop adaptation to high soil water conditions, *Calif. Agric.*, 24, 15, 1970.

1241. **Varade, S. B., Stolzy, L. H., and Letey, J.,** Influence of temperature, light intensity, and aeration on growth and root porosity of wheat, *Triticum aestivum*, *Agron. J.*, 62, 505, 1970.

1242. **Vartanian, N.,** Some aspects of structural and functional modifications induced by drought in root systems, *Plant Soil*, 63, 83, 1981.

1243. **Vartapetian, B. B., Andreeva, I. N., Maslova, I. P., and Davtian, N. G.,** The oxygen and ultrastructure of root cells, *Agrochimica*, XV, 1, 1970.

1244. **Vartapetian, B. B., Mazliak, P., and Lance, C.,** Lipid biosynthesis in rice coleoptiles grown in the presence or in the absence of oxygen, *Plant Sci. Lett.*, 13, 321, 1978.

1245. **Vazquez, A. and Gesto, D. V.,** Juvenility and endogenous rooting substances in *Castanea sativa* Mill., *Biol. Plant.* 24, 48, 1982.

1246. **Veen, B. W.,** The uptake of potassium, nitrate, water and oxygen by a maize root system in relation to its size, *J. Exp. Bot.*, 28, 1389, 1977.

1247. **Veen, B. W.,** Relation between root respiration and root activity, *Plant Soil*, 63, 73, 1981.

1248. **Veen, B. W. and Boone, F. R.,** The influence of mechanical resistance and phosphate supply on morphology and function of corn roots, *Plant Soil*, 63, 77, 1981.

1249. **Verpraskas, M. J. and Miner, G. S.,** Effects of subsoiling and mechanical impedance on tobacco root growth, *Soil Sci. Soc. Am. J.*, 50, 423, 1986.

1250. **Verbrugge, J. C. and Leclerq, J.,** Modelisation of stress-displacement interaction between root analogue and soil, in *Proc. 11th Conf. ISTRO*, Edinburgh, Scotland, 1988, 427.

1251. **Vesić, A. S.,** Expansion of cavities in infinite soil mass, *J. Soil Mech.* Found. Div., ASCE 98, (SM 3), 265, 1972.

1252. **Vine, P. N., Lal, R., and Payne, D.,** The influence of sands and gravels on root growth of maize seedlings, *Soil Sci.*, 131, 124, 1981.

1253. **Visser, J., Locher, J. Th., and Brouwer, R.,** Effects of aeration and mineral supply on growth and mineral content of shoots and roots of apple trees (var. "Golden Delicious" on M IX), *Neth. J. Agric. Sci.*, 19, 125, 1971.

1254. **Voorhees, W. B.,** Soil compaction, how it influences moisture, temperature, yield, root growth, *Crops Soils Mag.*, 29, 7, 1977.

1255. **Voorhees, W. B.,** Alleviating temperature stress, in *Modifying the Root Environment to Reduce Crop Stress*, ASAE Monogr. No. 4, Arkin, G. F. and Taylor, H. M., Eds., American Society of Agriculture Engineering, St. Joseph, MI, 1981, 217.

1256. **Voorhees, W. B., Amemiya, M., Allmaras, R. R., and Larson, W. E.,** Some effects of aggregate structure heterogeneity on root growth, *Agron. J.*, 35, 638, 1971.

1257. **Voorhees, W. B., Carlson, V. A., and Hallauer, E. A.,** Root length measurement with a computer-controlled digital scanning microdensitometer, *Agron. J.*, 72, 847, 1980.

1258. **Voorhees, W. B., Carlson, V. A., and Senst, C. G.,** Soybean nodulation as affected by wheel traffic, *Agron. J.*, 68, 976, 1976.

1259. **Voorhees, W. B., Farrell, D. A., and Larson, W., E.,** Soil strength and aeration effects on root elongation, *Soil Sci. Soc. Am. Proc.*, 3, 948, 1975.

1260. **Voorhees, W. B., Nelson, W. W., and Randall, G. W.,** Extent and persistence of subsoil compaction caused by heavy axle loads, *Soil Sci. Soc. Am. J.*, 50, 428, 1986.

1261. **Vos, J. and Groenwold, J.,** Estimation of root densities by observation tubes and endoscope, *Plant Soil*, 74, 295, 1983.

1262. **Vose, P. B. and Koontz, H. V.,** Uptake of strontium by pasture plants and its possible significance in relation to the fall-out of strontium-90, *Nature (London)*, 183, 1447, 1959.

1263. **Wacquant, J. P.,** Physicochemical selectivity for cations and CEC of grass roots, *Plant Soil*, 47, 257, 1977.
1264. **Waddington, D. V. and Bakker, J. H.,** Influence of soil aeration on the growth and chemical composition of three grass species, *Agron. J.*, 57, 253, 1965.
1265. **Waisel, Y.,** *Biology of Halophytes*, Academic Press, New York, 1972.
1266. **Waisel, Y.,** The stimulating effects of NaCl on root growth of Rhodes grass (*Chloris gayana*), *Physiol. Plant.*, 64, 519, 1985.
1267. **Walczak, R.,** Model investigations of water binding energy in soils of different compaction, *Zesz. Probl. Postepow Nauk Roln.*, 197, 11, 1977.
1268. **Walczak, R.,** Model investigations of the dependence of water retention on the parameters of solid phase of soil, *Probl. Agrofizyki*, 41, 1984 (in Polish).
1269. **Walczak, R.,** Basic problems of mass and energy transfer in the soil-plant-atmosphere system, *Zesz. Probl. Postepow Nauk Roln.*, 346, 11, 1987.
1270. **Waldron, L. J.,** The shear resistance of root-permeated homogeneous stratified *soil, Soil Sci. Soc. Am. J.*, 41, 893, 1977.
1271. **Waldron, L. J. and Dakessian, S.,** Effect of grass, legume and tree roots on soil shearing resistance, *Soil Sci. Soc. Am. J.*, 46, 894, 1982.
1272. **Walker, J. M.,** One-degree increments in soil temperature affect seedling behavior, *Soil Sci. Soc. Am. Proc.*, 33, 729, 1969.
1273. **Wallihan, E. F., Garber, N. J., Sharpless, R. G., and Printy, W. L.,** Effect of soil oxygen deficit on iron nutrition of orange seedlings, *Plant Physiol.*, 36, 425, 1961.
1274. **Walsh, K. B. and Layzell, D. B.,** Carbon and nitrogen assimilation and partitioning in soybeans exposed to low root temperatures, *Plant Physiol.*, 80, 249, 1986.
1275. **Wample, R. L. and Reid, D. M.,** Effect of aeration on the flood-induced formation of adventitious roots and other changes in sunflower (*Helianthus annuus* L.), *Planta*, 127, 236, 1975.
1276. **Wample, R. L. and Reid, D. M.,** The role of endogenous auxins and ethylene in the formation of adventitious roots and hypocotyl hypertophy in flooded sunflower plants (*Helianthus annuus*), *Physiol. Plant.*, 45, 219, 1979.
1277. **Wanner, H.,** Oxygen diffusion as a limiting factor of plant root respiration, *Vierteljahrsschr. Naturforsch. Ges. Zürich*, 90, 98, 1945.
1278. **Warkentin, B. P.,** Summary of the workshop, in impact of water and external forces on soil structure, *Catena Suppl.* 11, 167, 1988.
1279. **Wästerlund, I.,** Compaction of till soils and growth tests with Norway spruce and Scots pine, *For. Ecol. Manage.*, 11, 171, 1985.
1280. **Wästerlund, I.,** Machine forces that damage trees and ground in thinning operations, in *Proc.* IEA *Workshop*, Charleston, SC, 1986, 1.
1281. **Watson, E. R., Lapins, P., and Barron, R. J. W.,** Effect of waterlogging on the growth, grain and straw yield of wheat, barley and oats, *Aust. J. Exp. Agric. Anim. Husb.*, 16, 116, 1976.
1282. **Waughman, S. J.,** The effect of temperature on nitrogenase activity, *J. Exp. Bot.*, 28, 949, 1977.
1283. **Weber, D. F. and Miller, V. L.,** Effect of soil temperature on *Rhizobium japonicum* serogroup distribution in soybean nodules, *Agron. J.*, 64, 796, 1972.
1284. **Weimberg, R., Lerner, H. R., and Poljakoff-Mayber, A.,** Changes in growth and water soluble concentrations in sorghum bicolor stressed with sodium and potassium salts, *Physiol. Plant.*, 62, 427, 1984.
1285. **Weiming, S., Monxiong, H., and Zhiyu, L.,** The nutrient status of soil-root interface. IV. Study on sample preparation procedure for electron microprobe and its applications, *Acta Pedolog. Sin.*, 24, 290, 1987.
1286. **Weir, A. H. and Barraclough, P. B.,** The effect of drought on the root growth of winter wheat and on its water uptake from a deep loam, *Soil Use Manage.*, 2, 91, 1986.
1287. **Welbank, P. J.,** Soil conditions affecting root distribution, in *Soil Type and Land Capability*, Soil Survey Tech. Monogr. No. 4, Harpenden, 1974, 91.
1288. **Weller, D. M., Cook, R. J.,. MacNish, G. C., Bassett, E. N., Powelson, R. L., and Petersen, R. R.,** *Rhizoctonia* root rot of small grains favored by reduced tillage in the Pacific Northwest, *Plant Dis.*, 70, 70, 1986.
1289. **Werner, D. and Kuhlman, K. P.,** Calcium, iron and cobalt accumulation in root hairs of soybean (*Glycine max.*), *Z. Naturforsch., Teil C* 40, 912, 1985.
1290. **White, E. M.,** Note: soil compaction and contraction around plant roots, *Soil Sci.*, 119, 461, 1975.
1291. **White, J. M.,** Soil preparation effects on compaction, carrot yield, and root characteristics in organic soil, *J. Am. Soc. Hortic. Sci.*, 103, 433, 1978.
1292. **White, J. W. and Biernbaum, J. A.,** Effects of root-zone heating on growth and flowering of calceolaria, *HortScience*, 19, 289, 1984.
1293. **White, P. R.,** Seasonal fluctuations in growth rates of excised tomato root tips, *Plant Physiol.*, 12, 183, 1937.
1294. **Whitehead, D. and Jarvis, P. G.,** Coniferous forests and plantations, in *Water Deficits and Plant Growth*, Vol. 6, Kozlowski, T. T., Ed., Academic Press, New York, 1981, 49.

1295. **Whiteley, G. M. and Dexter, A. R.,** The dependence of soil penetrometer pressure on penetrometer size, *J. Agric. Eng. Res.*, 26, 467, 1981b.

1296. **Whiteley, G. M. and Dexter, A. R.,** Elastic response of the roots of field crops, *Physiol. Plant.*, 51, 407, 1981c.

1297. **Whiteley, G. M. and Dexter, A. R.,** Forces required to displace individual particles within beds of similar particles, *J. Agric. Eng. Res.*, 27, 215, 1982.

1298. **Whiteley, G. M. and Dexter, A. R.,** Behaviour of roots in cracks between soil peds, *Plant Soil*, 74, 153, 1983.

1299. **Whiteley, G. M. and Dexter, A. R.,** Displacement of soil aggregates by elongating roots and emerging shoots of crop plants, *Plant Soil*, 77, 131, 1984a.

1300. **Whiteley, G. M. and Dexter, A. R.,** The behaviour of roots encountering cracks in soil. I. Experimental methods and results, *Plant Soil*, 77, 141, 1984b.

1301. **Whiteley, G. M., Hewitt, J. S., and Dexter, A. R.,** The buckling of plant roots, *Physiol. Plant.*, 54, 333, 1982.

1302. **Whiteley, G. M., Utomo. W. H., and Dexter, A. R.,** A comparison of penetrometer pressures and the pressures exerted by roots, *Plant Soil*, 61, 351, 1981c.

1303. **Wiedenroth, E. M.,** Relations between photosynthesis and root metabolism of cereal seedlings influenced by root anaerobiosis, *Phytosynthetica*, 15, 575, 1981.

1304. **Wiegand, C. L. and Lemon, E. R.,** A field study of some plant-soil relations in aeration, *Soil Sci. Soc. Am. Proc.*, 22, 216, 1958.

1305. **Wiersma, D. and Mortland, M. M.,** Response of sugar beets to peroxide fertilization and its relationship to oxygen diffusion, *Soil Sci.*, 75, 355, 1953.

1306. **Wiersum, L. K.,** The relationship of the size and structural rigidity of pores to their penetration by roots, *Plant Soil*, 9, 75, 1957.

1307. **Wiersum, L. K.,** Uptake of nitrogen and phosphorus in relation to soil structure and nutrient mobility, *Plant Soil*, 16, 62, 1962.

1308. **Wiersum, L. K.,** Presumed aerotropic growth of roots of certain species, *Naturwissenschaften*, 54, 203, 1967a.

1309. **Wiersum, L. K.,** Potential subsoil utilization by roots, *Plant Soil*, 27, 383, 1967b.

1310. **Wiersum, L. K.,** Tulip root behaviour and aeration requirements, *Acta Hortic.*, 23, 318, 1971.

1311. **Wiersum, L. K.,** A comparison of the behaviour of some root systems under restricted aeration, *Neth. J. Agric. Sci.*, 27, 92, 1979.

1312. **Wiersum, L. K.,** The effect of soil physical conditions on roots and uptake, in *Mineral Nutrition of Fruit Trees*, Atkinson, D., Jackson, J., Sharples, M., and Waller, W., Eds., Butterworths, London, 1980, 111.

1313. **Wiesler, F. and Horst, W. J.,** Root growth and nitrogen utilization of maize cultivars under field conditions, in *Abstr. of Symp. Plant Roots and Their Environment*, Uppsala, Sweden, 1988, 3.

1314. **Wilhelm, W. W., Mielke, L. N., and Fentser, C. R.,** Root development of winter wheat as related to tillage practice in Western Nebraska, *Agron. J.*, 74, 85, 1982.

1315. **Wilkins, H., Wilkins, S. M., and Wain, R. L.,** Studies on plant growth regulating substances, XLVIII. The effect of 3,5-diiodo-4-hydroxybenzoic acid on seedling growth in compact and loose soil, *Ann. Appl. Biol.*, 87, 415, 1977.

1316. **Wilkins, S. M., Wilkins, H., and Wain, R. L.,** Chemical treatment of soil alleviates effects of soil compaction on pea seedling growth, *Nature (London)*, 259, 392, 1976.

1317. **Wilkins, S. M., Wilkins, H., and Wain, R. L.,** Studies on plant growth regulating substances. LI. The effect of 3,5-dibromo- and 3,5-dichloro-4-hydroxybenzoic acid on seedling growth in compact and loose soils, *Ann. Appl. Biol.*, 89, 265, 1978.

1318. **Willatt, S. T. and Taylor, H. M.,** Water uptake by soya-bean roots as affected by their depth and by soil water content, *J. Agric. Sci. Cambridge*, 90, 205, 1978.

1319. **Willatt, S. T.,** Root Growth of Winter Barley During a Wet Spring in a Controlled Traffic Experiment on a Sandy Loam Soil in Scotland, Dep. Note SIN/402, Scottish Institute of Agriculture Engineering, Penicuick, 1984, unpublished.

1320. **Willatt, S .T.,** Root growth of winter barley in a soil compacted by the passage of tractors, *Soil Till. Res.*, 7, 41, 1986.

1321. **Willatt, S. T. and Sulistyaningsih, N.,** Effect of plant roots on soil strength, in *Proc. 11th Conf. ISTRO*, Edinburgh, Scotland, 1988, 183.

1322. **Willey, C. R.,** Effect of short periods of anaerobic and near-aerobic conditions on water uptake by tobacco roots, *Agron. J.*, 62, 224, 1970.

1323. **Williams, J.,** Root density and water potential gradients near the plant roots, *J. Exp. Bot.*, 25, 669, 1974.

1324. **Williams, W. T. and Barber, D. A.,** The functional significance of aerenchyma in plant, in *Proc. 15th Symp. Society of Experimental Biology*, 1961, 132.

1325. **Williamson, R. E.,** Influence of gas mixtures on cell division and root elongation of broad bean, *Vicia faba* L., *Agron. J.*, 60, 317, 1968.

1326. **Williamson, R. E.,** Effect of soil gas composition and flooding on growth of *Nicotiana tabacum*, L., *Agron. J.*, 62, 80, 1970.

1327. **Williamson, R. E. and Splinter, W. E.,** Effects of light intensity, temperature, and root gaseous environment on growth of *Nicotiana tabacum* L., *Agron. J.*, 61, 285, 1969.
1328. **Willis, W. O.,** Soil temperature and tillage, in Conservation Tillage Research Progress and Needs, ARS-NC-57, Agriculture Research Service, U.S. Department of Agriculture, 1977, 19.
1329. **Willey, C. R.,** Effects of short periods of anaerobic and near-anaerobic conditions on water uptake by tobacco roots, *Agron. J.*, 62, 224, 1970.
1330. **Wind, G. P.,** Research and experience with mechanical soil improvement in The Netherlands, in *Proc. 9th Conf. ISTRO*, Osijek, Yugoslavia, 1982, 64.
1331. **Winters, E. and Simonson, W. R.,** The subsoil, *Adv. Agron.*, 3, 1, 1951.
1332. **Woltz, S. S.,** Nonparasitic plant pathogens, *Annu. Rev. Phytopathol.*, 16, 403, 1978.
1333. **Wong, D. H., Barbetti, M. J., and Sivasithampram, K.,** Effects of soil temperature and moisture on the pathogenicity of fungi associated with root rot of subterranean clover, *Aust. J. Agric. Res.*, 35, 675, 1984.
1334. **Woodward, R. A., Harper, K. T., and Tiedemann, A. R.,** An ecological consideration of the significance of cation-exchange capacity of roots of some Utah range plants, *Plant Soil*, 79, 169, 1984.
1335. **Woodroof, J. G.,** Pecan root growth and development, *J. Agric. Res.*, 49, 511, 1934.
1336. **Worker, G. F., Jr., Ayres, R. S., Karah, M. A., and Thomason, R. E.,** Effect of dense, stratified sandy soil on crop roots, *J. Agric. Sci. Cambridge*, 81, 513, 1973.
1337. **Yamasaki, T.,** Studies on the "excess moisture injury" of upland crops in overmoist from the viewpoint of soil chemistry and plant physiology, *Bull. Natl. Inst. Agric. Sci. Jpn. B*, 1, 1, 1952 (in Japanese, English summary).
1338. **Yamazaki, K. and Harada, J.,** The root system formation and its possible bearings on grain yield in rice plants, JARQ, 15, 153, 1982.
1339. **Yang, S. J. and de Jong, E.,** Effect of soil water potential and bulk density on water uptake patterns and resistance to flow of water in wheat plants, *Can. J. Sci.*, 51, 211, 1971.
1340. **Yanuka, M. and Elrick, D. E.,** Applications of minicomputer-based image digitization in soil and crop sciences, *Comput. Electron. Agric.*, 1, 59, 1985.
1341. **Yoshida, S., Bhattacharjee, D. P., and Cabuslay, G. S.,** Relationship between plant type and root growth in rice, *Soil Sci. Plant Nutr.*, 28, 437, 1982.
1342. **Yoshida, T. and Ancajas, R. R.,** Nitrogen-fixing activity in upland and flooded rice fields, *Soil Sci. Soc. Am. Proc.*, 37, 42, 1973.
1343. **Yu, P. T., Stolzy, L. H., and Letey, J.,** Survival of plants under prolonged flooded conditions, *Agron. J.*, 61, 844, 1969.
1344. **Yu, T.,** Oxidation-reduction properties of paddy soils, in *Proc. Symp. Paddy Soils*, China, Nanjing, China, 1980. Institute Soil Sci., Academia Science Press, Beijing, Springer-Verlag, Berlin, 1981, 95.
1345. Zach, M., Effects of different types of soil tillage on soil properties and yield parameters of wheat and sugar-beets, in *Proc. 8th Conf. ISTRO*, Stuttgart, West Germany, 1979, 249.
1346. **Zentmeyer, G. A.,** Effect of physical factors, host resistance and fungicide on root infection of the soil-root interface, in *The Soil-Root Interface*, Harley, J. L. and Russell, R. S., Eds., Academic Press, London, 1979, 315.
1347. **Zholkevich, V. N., Sinitsyna, Z. A., Peisakhzon, B. J., Abutalybov, V. F., and Dyachenko, I. V.,** On the nature of root pressure, *Fiziol. Rast.*, 26, 978, 1979.
1348. **Zimmerman, J. F. and Berry, L. J.,** Studies of onion root respiration. III. Amperometric titration as a method for the measurement of respiratory overshoot, *Biochem. Acta*, 3, 615, 1949.
1349. **Zisa, R. P., Halverson, H. G., and Stout, B. B.,** Establishment and Early Growth of Conifers on Compacted Soils in Urban Areas, USDA Res. Pap. Ne-451, Forestry Service, U.S. Department of Agriculture, 1980, 8.
1350. **Zyuz, N. S.,** Bulk density and hardness of the Hillocky Sands of the Middle Don., *Sov. Soil Sci.*, 13, 1769, 1968.

INDEX

A

Abscisic acid (ABA), 78, 151, 187
Absorptive capacity of roots, 41—46
Actinomycetes,72
Adenosine diphosphate (ADP), 138
Adenosine monophosphate (AMP), 138
Adenosine triphosphate (ATP), 136, 138
Adenylate energy charge (AEC), 138—139
ADH, see Alcohol dehydrogenase
Adhesive forces, 9
ADP, see Adenosine diphosphate
Adventitious roots, 33
AEC, see Adenylate energy charge
Aeration of soil, see Soil aeration
Age of roots, 41—42
Aggregates, see Soil aggregates
Agrophysical index, 31
Alcohol dehydrogenase (ADH), 129, 133, 137
Aminoethoxyvinylglycine (AVG), 137
Aminooxyacetic acid (AOA), 137
AMP, see Adenosine monophosphate
Anaerobiosis, 143, see also Atmospheric stress
Anatomical adaptation of roots to stress, 144—150
Anatomy of roots, 34—35, 84—85
Anoxia, see also Atmospheric stress
  energy charge of roots under, 133—139
  metabolic activity of roots under, 133—139
  root growth force pressure and, 141—142
  root growth rate and, 138—140
  root water uptake and, 141—142
AOA, see Aminooxyacetic acid
Argon, 17
Artificial media, 75—79
Atmospheric stress, 143—152, see also Anoxia; Waterlogging; specific types
  anatomical adaptation to, 144—150
  morphological adaptation to, 150
ATP, see Adenosine triphosphate
AVG, see Aminoethoxyvinylglycine

B

Bacteria, 61
Biological activity, see also specific types
  of roots, 90—91
  of soil, 180
Biological aspects of soil-root interface, 60—63
Breeding and root development, 189
Buckling of roots, 69, 78
Bulk soil, 57

C

Calcium, 157—158
Capillary forces, 5
Carbon dioxide
  root respiration and, 127
  in roots, 133
  in soil, 17, 19
Cation exchange capacity of roots, see Root cation exchange capacity (RCEC)
Chemical aspects of soil-root interface, 58—60
Chemical composition of roots, 87—89, 158
Chemical forces, 5, see also specific types
Chemical treatments, see also specific types
  root development and, 190—191
Classification of roots, 33
Clay, 1, 2, 90
COC, see Critical oxygen concentration
Cohesive forces, 8
Compaction of soil, see Soil compaction
Competition for nutrients, 50
Compression of soil, 21—25, 28—29
Cone index, 24—25,
Consistency of soil, 21—25, 28—29, 30
Conventional tillage, 181—183
Critical oxygen concentration (COC), 124—126
Cytochrome oxidase, 116

D

2,4-D, see 2,4-Dichlorophenoxyacetic acid
Darcy's law, 11, 12
Deep loosening of soil, 181—184
2,4-Dichlorophenoxyacetic acid (2,4-D), 189
Diffusion, 19
  coefficient of, 101
  of oxygen, see Oxygen diffusion
Diffusive properties of soil, 19
DIHB, see 3,5-Diiodo-4-hydroxy-benzoic acid
3,5-Diiodo-4-hydroxy-benzoic acid (DIHB), 187
Drainage, 108
Driving forces, 41, 48
Drought tolerance in roots, 100—107

E

Elasicity of roots, 39
Energy charge of roots, 133—138
Ethanol in roots, 133—136
Ethylene, 78
  root respiration and, 127
  in roots, 136—137
  in soil, 18
Excavation techniques, 195
Exudates, 40, see also specific types
Exudation of roots, 40—41

F

Fertilizer
  root development and, 190
  root respiration and, 127
Fick's equation, 118

Flooding conditions, 141, 142
Fungi, 66, 90—92, see also specific types

G

Gas exchange in soil, 19, see also specific types
Gibberellins, 152
Graviosmosis, 49
Gravitational forces, 9
Growth of roots, see Root growth

H

Half respiration concentration of oxygen (HRC), 126, 127
Hormones, 189—190, see also specific types
HRC, see Half respiration concentration of oxygen
Hydraulic conductance, 42—46, 62
Hydraulic conductivity, 11, 12, 66
Hydraulic potential, 11
Hydrogen bonds, 5

I

Impedance, see Mechanical impedance
Indol-3 ylacetic acid (IAA), 188
Intermolecular forces, 5
Ion movement, 178
Irrigation, 104–107
Isotherm BET equation, 3

L

LCML, see Lower critical mechanical limit
Light penetration through soil, 161
Lignification, 149
Longitudinal diffusion of oxygen, 119—122
Longitudinal water movement in roots, 49
Lower critical mechanical limit (LCML), 166
Lysates, 41

M

Macrodiffusion, 19
Magnesium, 157–158
Matric forces, 8, 9
Matric potential, 11
Mechanical composition of soil, 176–177
Mechanical impedance, 24, 68, 75—96, 161—171
  artificial media and, 75—78
  modeling of root growth and, 95—97
  oxygen and, 89
  penetrometer pressure and, 79, 91—95
  photosynthate requirements and, 89
  root anatomy and, 84—85
  root branching and, 83
  root chemical composition and, 87—89
  root diseases and, 90—91
  root distribution and, 84
  root growth and, 79—82, 91—95
  root hairs and, 83
  root morphology and, 82—83
  root nutrient uptake and, 87—88
  root respiration and, 128
  root water uptake and, 85—87
  soil aeration and, 164—166
  soil aggregate size distribution and, 167—170
  soil pore size distribution and, 167—170
  soil temperature and, 170—171
  soil water and, 164—167, 170—171
Mechanical properties, see also specific types
  of roots, 39—40
  of soil, 21
Metabolic activity of roots, 133—138
Metabolic adaptations of roots to stress, 151—152
Metabolism of roots, 40
Methane in soil, 17, 19
Microbes in roots, 138
Microdiffusion, 19, 20
Microorganisms, 61, 66, see also specific types
Mineralogical composition of soil, 176—177
Modeling, see also specific types
  of root growth, 39, 95—96
  of soil property effects on roots, 170
Moisture-retention curve, 9
Moisture in soil, see Soil water
Morphological adaptation of roots to stress, 150
Morphology of roots, 33, 82—84, 97
Mucillages, 40, 63
Mulching, 187
Mycorrhizal infection, 62

N

Nitrogen, 157
  in roots, 92, 159—60
  in soil, 17
Nitrogen fixation, 62, 158
Nitrous oxide in soil, 18
NLWR, see Nonlimiting water range
NMR, see Nuclear magnetic resonance
Nodulation of roots, 90—91, 158—160
Nonlimiting water range (NLWR), 166
Nuclear magnetic resonance (NMR), 195
Nutrients, see also specific types
  absorption of, 49—53
  availability of, 65
  competition for, 50
  in roots, 142
  soil temperature and, 156—158
  transport of, 49—53
  uptake of, 53, 62, 87—89, 142, 156—158

O

Organic matter, 19, see also specific types
Osmotic absorption of water, 46—47
Oxidizing capacity of roots, 150
Oxygen, see also Soil aeration
  deficiency in, 112
  diffusion of, see Oxygen diffusion
  half respiration concentration of, 126, 127
  lack of, see Anoxia
  root respiration and, 123—127

in roots, 90
in soil, 17, 19, 21, 30
Oxygenation of soil, 17
Oxygen diffusion, 112—122
longitudinal, 119—122
radial, 112—119
Oxygen diffusion rate (ODR), 20, 21, 24, 112, 118, 135, 141
Oxygen stress in soil, 142, 143

P

Particle size in soil, 1
Passive absorption of water, 47
Pasteur effect, 135, 151
Penetrometer pressure, 79, 91—94
Permeability of roots, 42—46
Phosphorus, 157
Photosynthate requirements, 89
Physical aspects of soil-root interface, 57—58
Physical characteristics of soil, 29—31
Physicochemical aspects of soil-root interface, 58—60
Physiological adaptations of roots to stress, 151—152
Plant breeding and root development, 189
Plant selection and root development, 189
Porosity
of roots, 97, 121
of soil, 7—18, 170, 178
Potassium, 157

R

Radial diffusion of oxygen, 112—119
Radial water movement in roots, 47—49
Radiography, 195
RCEC, see Root cation exchange capacity
Redox potential, 129
Reduced tillage, 179—181
Regeneration of soil, 28
Resistance coefficient, 44
Resistance to penetration, 24
Respiration
of root nodules, 133
of roots, see Root respiration
Respiratory quotient (RQ), 132
Rhizobacteria, 63
Rhizoctonia solani, 185
Rhizoplanes, 57, 63
Rhizospheres, 57, 62, 63, 66, 180
Root cation exchange capacity (RCEC), 50, 51, 52, 89, 92, 196
Root development, 179—190
chemical treatments and, 187—189
deep loosening and, 181—184
fertilizer and, 190
hormones and, 187—189
mulching and, 187
optimum root size and, 190
pattern of, 197
plant breeding and, 189
plant selection and, 189
root diseases and, 184—186
soil management and, 186
soil water and, 187
sowing date and, 186—187
tillage and, 184—186, 189—191
Root diseases, 90—91, see also specific types
soil temperature and, 160
soil water and, 109
tillage and, 184—w186
Root growth, 35—39
anoxia and, 138—140
force pressure and, 141
measurement of, 196
mechanical impedance of soils and,75—95
modeling of, 39, 95
rate of, 138—140
soil temperature and, 152—155
Root hairs, 33, 62, 68, 84, 98
Root nodules, 133
Root pressure relation, 93—95
Root respiration, 122—133
carbon dioxide and, 127
ethylene and, 127
external factors in, 123—130
fertilization and, 127
internal factors in, 130—133
mechanical impedance and, 129
oxygen and, 123—127
redox potential and, 129—130
soil salinity and, 129
temperature and, 128—129
Roots, 33—54
above-ground parts relationship to, 53—54
absorbing power of, 50
absorptive capacity of, 41—46
adaptations to stress in, 146—152, 200
adventitious, 33
age of, 41—42
anatomical adaptation to stress in, 144—149
anatomy of, 34—35, 84—85
artificial media and, 76—79
atmospheric stress and, see Atmospheric stress
biological activity of, 90—92
branching of, 83
buckling of, 69
carbon dioxide in, 133
chemical composition of, 87—88, 158
classification of, 33
counting techniques for, 195—196
development of, see Root development
diameter of, 103
diseases of, see Root diseases
distribution of, 84
drought tolerance in, 100—104
effects of on soil structure, 174
elasticity of, 40
energy charge of, 133—139
ethanol in, 134—136
ethylene in, 136—137
excavation of, 195

exudation of, 40—41
growth of, see Root growth
irrigation and, 107—108
light penetration and, 161
longitudinal water movement in, 49
measurement of processes of, 197
mechanical impedance and, see Mechanical impedance
mechanical properties of, 39—40
metabolic activity of, 133—139
metabolic adaptations to stress in, 151—152
metabolism of, 40
methods for studying, 193—197
microbes in, 138
modeling of soil property effects on, 172
modeling of water uptake by, 109—112
morphological adaptations to stress in, 151
morphology of, 33, 82—84, 98
nitrogen content of, 92
nitrogen fixation in, 159—160
nodulation of, 90—92, 159—160
nutrient absorption by, 49—53
nutrient content of, 143
nutrient transport in, 49—53
nutrient uptake by, 53, 62, 88—89, 143, 156—159
observation techniques for, 193—195
optimum size of, 190
osmotic absorption of water in, 46—47
oxidizing capacity of, 150
oxygen in, 90, 114
passive absorption of water in, 47
permeability of, 42—46
photosynthate requirements of, 90
physiological adaptations to stress in, 151—152
porosity of, 98, 122
prediction of growth of, 196
radial water movement in, 47—49
ratio of shoots to, 140
respiration of, see Root respiration
separation of from soil, 197
shallow, 151
shrinkage of, 98
soil aeration and, see Soil aeration
soil aggregate size and, 167—171
soil aggregates of, 176
soil biological activity and, 180
soil compaction and, 162
soil interface with, see Soil-root interface
soil mechanical composition and, 178—179
soil minerology and, 178—179
soil pore size and, 167—171
soil property effects on, 172
soil strength and, 176—178
soil structure effects on, 66, 68—70, 72—75
soil temperature and, 171—172
soil texture and, 65—66, 162—165
soil water and, 165—167, 171—172, 180
soil water and functions of, 98—101
soil water and growth of, 97
translocation of water in soil by, 49
water absorption by, 46—47, 47
water flow resistance in, 43
waterlogging of, 107
water movement in, 47—49
water potential of, 196
water uptake by, see under Water
water use by, 196—197
Root zone heating, 160
Root zone temperature (RZT), 160—161
RQ, see Respiratory quotient
RZT, see Root zone temperature

S

Sand, 1, 2
Secretions, 40
Shallow rooting, 151
Silt, 1, 2
Soil, see also specific types
acidity of, 129—130
argon in, 17
atmospheric stress in, see Atmospheric st
biological activity of, 180
bulk, 57
carbon dioxide in, 17, 19
compaction of, see Soil compaction
compression of, 21—25, 28—29
consistency of, 21—25, 28—29, 30
cracks in, 74, 77
deep loosening of, 181—184
defined, 1
diffusion in, 19
diffusive properties of, 19
ethylene in, 18
gas exchange in, 19
light penetration through, 162
macroaggregates of, 5
management of, 186
mechanical composition of, 178—179
mechanical impedance of, see Mechanica impedance
mechanical properties of, 21
methane in, 17, 19
microaggregates of, 5
mineralogical composition of, 178—179
moisture in, see Soil water
nitrogen in, 17
nitrous oxide in, 18
organic matter in, 19
oxygenation of, 17
oxygen in, 17, 19, 21, 30
oxygen stress in, 143
particle size in, 1
physical characteristics of, 29—31
physics of, 1
pore size distribution in, 167—171
porosity of, 7—18, 170, 178
regeneration of, 28
root interface with, see Soil-root interface
root separation from, 197
root uptake of nutrients from, 53
salinity of, 129

shrinkage of, 22
solid phase of, 1
strength of, 176—178, 182
structure of, see Soil structure
surface area of, 1, 2—4
temperature of, see Soil temperature
texture of, see Soil texture
as a three-phase system, 1
traction parameters of, 22
translocation of water in by roots, 49
water in, see Soil water
water flow resistance in, 43
Soil aeration, 15—21, 112—152, see also Oxygen
anoxia and, see Anoxia
atmospheric stress and, 144—152
root energy charge and, 133—139
root metabolic activity and, 133—139
root nutrients and, 143
root respiration and, see Root respiration
roots and, 165—167
soil acidity and, 129—130
Soil aggregates, 5, 6, see also specific types
root, 176
size distribution in, 167—170
Soil compaction, 21—25, 28—29, 90
degree of, 165
effects of, 199
intensity of, 199—200
reversal of effects of, 183
roots and, 162
Soil-root interface, 57—63
biological aspects of, 60—63
chemical aspects of, 58—60
physical aspects of, 57—58
physicochemical aspects of, 58—60
water relations and, 58
Soil structure, 5—17, 29, 31
effects on roots, 66, 68—70, 72—75
modification of, 181
stability of, 175—176
tillage and, 181
Soil temperature, 13—15, 30, 152—162
high, 161—162
low, 161—162
mechanical impedance and, 171—172
nitrogen fixation and, 159—160
nutrient uptake and, 156—159
root chemical composition and, 159
root diseases and, 160
root growth and, 152—155
root nodulation and, 159—160
root zones and, 160—161
soil texture and, 162—165
water uptake and, 155—156
Soil texture, 1—2, 29, 31, 65—66
roots and, 162—165
temperature and, 162—165
Soil water, 8—13, 30, 31, 79, 97—112
drainage and, 109
irrigation and, 107—108
mechanical impedance and, 171—172
regulatory practices for, 187
root development and, 187
root diseases and, 109
root drought tolerance and, 102—107
root functions and, 98—101
root growth and, 97
root morphology and, 98
root porosity and, 98
roots and, 165—167
root system effects on, 180
root water uptake and, 109—112
waterlogging and, 108—109
Solid phase of soil, 1
Sowing date, 186—187
Stress, 200, see also specific types
atmospheric, see Atmospheric stress
oxygen, 143
root adaptations to, 146—152, 200
Structure of soil, see Soil structure
Subsoil compaction, 22
Surface area of soil, 1, 2—4
Symbiotic fungi, 90—92
Symbiotic nitrogen fixation, 62

T

Temperature
root respiration and, 128—129
of soil, see Soil temperature
Texture of soil, see Soil texture
Tillage, 200, see also specific types
conventional, 181—183
reduced, 181—183
root diseases and, 186—188
Traction parameters of soil, 22

U

Upper critical mechanical limit (UCML), 166, 167

V

Van der Waals forces, 5
Void index (ratio), 7

W

Water
absorption of by roots, 46—47
movement of in roots, 47—49
resistance to flow of, 43
in soil, see Soil water
soil-root interface and, 58
translocation of in soil by roots, 49
uptake of, 42—46, 86—88, 142—143, 155—156, 180

modeling of, 109—112
soil temperature and, 155—156
use of by roots, 196—197
Waterlogging, 108—109, 112, 144, see also
Atmospheric stress
Wheat, 186

For Product Safety Concerns and Information please contact our EU representative GPSR@taylorandfrancis.com Taylor & Francis Verlag GmbH, Kaufingerstraße 24, 80331 München, Germany

Batch number: 10753415

Printed by Printforce, the Netherlands